Thermodynamics
and
Ecological Modelling

ENVIRONMENTAL *and* ECOLOGICAL MODELLING

Series Editor: **S.E. Jørgensen**

Handbook of Environmental and Ecological Modelling

S.E. Jørgensen, *The Royal Danish School of Pharmacy,*
Section of Environmental Chemistry, Copenhagen, Denmark

B. Halling-Sørensen, *Institute of Environmental Science and Engineering,*
The Technical University of Denmark, Copenhagen, Denmark

S.N. Nielsen, *The Royal Danish School of Pharmacy,*
Section of Environmental Chemistry, Copenhagen, Denmark

Handbook of Estimation Methods in Ecotoxicology and Environmental Chemistry

S.E. Jørgensen, *The Royal Danish School of Pharmacy,*
Section of Environmental Chemistry, Copenhagen, Denmark

B. Halling-Sørensen, *Institute of Environmental Science and Engineering,*
The Technical University of Denmark, Copenhagen, Denmark

H. Mahler, *Software Developer, Copenhagen, Denmark*

Handbook of Ecosystem Theories and Management

S.E. Jørgensen, *The Royal Danish School of Pharmacy,*
Section of Environmental Chemistry, Copenhagen, Denmark

F. Müller, *Kiel University, Kiel, Germany*

Thermodynamics and Ecological Modelling

S.E. Jørgensen, *The Royal Danish School of Pharmacy,*
Section of Environmental Chemistry, Copenhagen, Denmark

Thermodynamics and Ecological Modelling

Edited by
S.E. Jørgensen

LEWIS PUBLISHERS

Boca Raton London New York Washington, D.C.

Library of Congress Cataloging-in-Publication Data

Thermodynamics and ecological modelling / edited by Sven E. Jørgensen.
     p.   cm. (Environmental and ecological modeling)
     Includes bibliographical references.
     ISBN 1-56670-272-0
     1. Ecology. 2. Thermodynamics. I. Jørgensen, Sven Erik, 1934– II. Series.
QH541 .T493 2000
577′.14—dc21
                                         00-039127
                                                CIP

© 2001 by CRC Press LLC
Lewis Publishers is an imprint of CRC Press LLC

No claim to original U.S. Government works
International Standard Book Number 1-56670-272-0
Library of Congress Card Number 00-039127
Printed in the United States of America 1 2 3 4 5 6 7 8 9 0
Printed on acid-free paper

Preface

Thermodynamics is a holistic science, and it is therefore well suited to give descriptions of systems, including the most complex systems of all, namely, ecosystems. Thermodynamics can give a very useful description of, for instance, 10^{28} molecules in a room by use of a few equations — a much more useful description than if we measure (determine) the velocity and the place of all 10^{28} molecules. It is therefore not surprising that several system ecologists have applied thermodynamics to obtain a deeper understanding of how ecosystems are working as systems. During the last two decades of the 20th century, several thermodynamically based hypotheses in systems ecology were published and their interpretations have been compared with other theories in systems ecology, which are based on other scientific approaches, for instance, network theory, hierarchy theory, cybernetics, chaos theory, and so on.

It is the aim of this volume to present the entire (or rather almost the entire) spectrum of different thermodynamically based ecosystem theories and hypotheses to facilitate comparisons and interpretations of the various applications of thermodynamics in systems ecology. The readers, as I, will probably find that the various hypotheses to a certain extent are consistent. They uncover different properties of the very complex ecosystems as a result of the use of slightly different angles in the approaches. The various theories and hypotheses are therefore complementary and are all needed to be able to give a more complete description of the very complex ecosystems.

The chapters represent a wide spectrum of different thermodynamic approaches, although they all, of course, are based on the First through Third Laws of Thermodynamics. As it is not the aim to present a textbook on the application of thermodynamics in ecology, all authors have presented their ideas with their own scientific language. I have not, as editor, tried to homogenise the expressions and the languages, but have emphasised the authors' rights to present their ideas uncensored. The editorial work has been limited to ask the authors to clarify certain points and statements which the readers otherwise probably couldn't understand sufficiently well. In addition, I have written a summary of about one page for each chapter. I have emphasised what I see as the essential points in the chapters and/or how I see the content of the chapter is consistent with the other chapters. In each introduction I present an aphorism which attempts to summarise a part of the result in the chapter. All the aphorisms are translated from Danish aphorisms composed by F. Somogyi. It is hoped these short introductions to each chapter will demonstrate the scope of this book and make the reading of this theoretically difficult material easier to understand.

I would like to thank James Kay and Søren Nors Nielsen for the many fruitful discussions we have had during the preparation of the book. James Kay has been very open on a few occasions to discuss the content in detail. I am also grateful to Gerti Rosenfeld, for her professional transfer of the manuscript.

Contributors

Ichiro Aoki
Department of Systems Engineering
Faculty of Engineering
Shizuoka University
Hamamatsu, Japan

Hartmut Bossel
Zierenberg, Germany

Mark T. Brown
Environmental Engineering Sciences
    and Center for Wetlands
University of Florida
Gainesville, Florida, U.S.A.

Lionel Johnson
Sidney, British Columbia, Canada

Sven E. Jørgensen
Royal Danish School of Pharmacy
Department of Environmental Chemistry
Copenhagen, Denmark

S. A. L. M. Kooijman
Department of Theoretical Biology
Vrije Universiteit de Boelelaan
Amsterdam, The Netherlands

Ramon Margalef
University of Barcelona
Barcelona, Spain

R. M. Nisbet
Department of Ecology and Evolution
    and Marine Biology
University of California
Santa Barbara, California, U.S.A.

Yuri M. Svirezhev
PIK — Potsdam Institute for Climate
    Impact Research
Potsdam, Germany

Sergio Ulgiati
Department of Chemistry
University of Siena
Siena, Italy

Contents

Chapter 1
Introduction 1
Sven E. Jørgensen

Chapter 2
Exosomatic Structures and Captive Energies Relevant in Succession and Evolution 3
Ramon Margalef

Chapter 3
How Light and Nutrients Affect Life in a Closed Bottle 17
S. A. L. M. Kooijman and R. M. Nisbet

Chapter 4
Emergy Accounting of Human-Dominated, Large-Scale Ecosystems 61
Sergio Ulgiati and Mark T. Brown

Chapter 5
Thermodynamics and Theory of Stability...115
Yuri M. Svirezhev

Chapter 6
Application of Thermodynamic Concepts to Real Ecosystems:
Anthropogenic Impact and Agriculture 133
Yuri M. Svirezhev

Chapter 7
The Thermodynamic Concept: Exergy...153
Sven E. Jørgensen

Chapter 8
Entropy and the Exergy Principles in Living Systems 165
Ichiro Aoki

Chapter 9
Exergy and the Emergence of Multidimensional System Orientation 191
Hartmut Bossel

Chapter 10
Thermodynamics and Ecology: Far from Thermodynamic Equilibrium 211
Yuri M. Svirezhev

Chapter 11

Imperfect Symmetry: Action Principles in Ecology and Evolution 229
Lionel Johnson

Chapter 12

The Third Law of Thermodynamics Applied in Ecosystem Theory 287
Sven E. Jørgensen

Chapter 13

A Tentative Fourth Law of Thermodynamics 303
Sven E. Jørgensen

Chapter 14

Thermodynamics of the Biosphere 349
Yuri M. Svirezhev

Index 365

1 Introduction

Sven E. Jørgensen

CONTENTS

1.1 Energy and Ecology 1
1.2 A Short Overview of the Contents 2

1.1 ENERGY AND ECOLOGY

Energy became associated with ecology during the last years of the 1950s due to the brothers Odum, but not much interest for these approaches was shown by ecologists in the 1960s and 1970s with some few exceptions. During the last decade, an escalating interest for the use of thermodynamics in ecology has emerged with the result, that during the last 4 to 5 years, many papers on the application of thermodynamics to understand ecology have been published.

A number of theories based on thermodynamics of ecological systems have been proposed during the last decade. We felt it therefore timely to bring together these activities in one volume. We have asked the main proponents of each of these theories to present their work in this volume. The different theories are not completely consistent or even in some regard not necessarily compatible, but they represent different thermodynamic viewpoints on ecosystems, which on balance are complementary. We are of the opinion that a very complex system as an ecosystem requires plurality of perspectives to capture the richness of ecosystem dynamics. A simple physical phenomenon as light needs two descriptions: by waves and by particles. It is therefore not surprising that the very complex ecosystems need several complementary descriptions. We hope that this volume will contribute to the emergence of these complementary descriptions.

With this in mind we have chosen to let the authors speak for themselves, almost unfaded by the review process, which attempts to impose conformity. This way of presenting the various theories is well justified given the status of the selected authors. The readers will therefore have to judge for themselves on the applicability of the theories to achieve a better understanding of ecosystem behaviour. This implies, however, that the different authors may use different expressions to explain their ideas. It may even imply that different authors apply the same word to cover a different meaning. This has, however, often been the case with emergence of a new scientific field. To partly eliminate these ambiguities I decided to write a short introduction for each chapter to present the authors, the context for their work, the core ideas, and the terminology they use.

As already mentioned, we consider the different theoretical approaches as complementary. The final questions left for the readers to answer are: do we see consensus? Where do we see different perspectives, but not necessarily inconsistencies? Where do we see contradictions? Which theoretical approaches need further exploitation? Which theories need even further experimental testing?

We are also of the opinion that to a certain extent the different theories form a pattern of understanding, which we will attempt to reveal as far as it is possible in the introduction to each chapter.

1.2   A SHORT OVERVIEW OF THE CONTENTS

Each chapter has an introduction with a short overview of the content. It is therefore not the intent of this section to repeat the introductions, but try to draw a short overview of the topic "Thermodynamics and Ecology" at the edge of the 21st century.

We have made energy balances for ecosystems for many decades, and it has certainly given us new knowledge about the ecosystems and the role energy plays in ecosystems. This approach can still bring new knowledge depending on the considered system, as it is demonstrated in Chapter 3. It is, however, very characteristic that this approach, although very important, still gives surprising results. We would like, however, to include a quality measure or index of the energy — not only a record of the flow pattern. Emergy and exergy are energy expressions and use energy units, the first of which is joule, but the energy is multiplied by a weighting factor taking the energy quality into account. Emergy uses a weighting factor based on how much solar radiation measured in joules (the ultimate energy source on earth) is used to make 1 joule of energy embodied in a specific object. In other words emergy is based on a quality factor which accounts for the total energy cost expressed in solar energy. Exergy, on the other hand, considers the information or organisation carried by a specific organism, but the energy quality is already included in the definition of exergy: the work content of a system compared with a reference state. We distinguish between energy which can do work and energy which cannot do work. Both approaches are valuable, giving "two different sides of the same coin."

Another recent development in system ecology is the use of indicators or, as H. Bossel calls them, orientors. When we are using them in modelling context, we may call them goal functions. This development was initiated to assess the ecosystem health by use of indicators, or as it is called more frequently in Canada, ecosystem integrity. The environmental manager should consider himself a "doctor of the ecosystem" needing a diagnosis. As the doctor of medicine measures the blood pressure, makes biochemical analyses of the blood, listens to the heart and lungs, and analyses the urine, the doctor of ecosystem should have a list of ecosystem tests which could be used to assess the ecosystem health. Among the possible candidates as ecological health indicators, the thermodynamic-based orientors offer some advantages. Emergy, exergy, the energy flow pattern, entropy, and even ratios of these indicators (orientors) have been proposed and used to assess the ecosystem health. This application of thermodynamics in system ecology is mentioned and discussed in several chapters, particularly in Chapters 4, 8, 9, 12, and 13.

A third core topic in the application of thermodynamics in system ecology is the possibility to describe the ecosystem development by use of thermodynamic functions or, what would be even more beneficial, to set up rules or a theory on how an ecosystem will develop under given circumstances. Several propositions are presented in this volume and a more detailed discussion takes place in Chapters 6, 8, 9, 12, and 13. There seems to be general agreement that ecosystems use the available energy flow through the system to move away from thermodynamic equilibrium. The disagreement today in systems ecology is which of the possible thermodynamic functions are most appropriate to make this description. The discussion should not be repeated here; readers should make their own conclusions. There is, however, no doubt that this discussion will continue for several years, but also that a more complete theory for ecosystem development is around the corner and thermodynamics will play an important role in this theory.

Solutions close the possibilities
Problems disclose them

CHAPTER 2

It is natural to start this volume with a contribution by Ramon Margalef. He has used energy considerations on ecosystems for almost 50 years, and has as professor emeritus produced this chapter. He recently (1997) published a book named: *Our Biosphere* in the series Excellence in Ecology. This book and this chapter give a clear message to the readers: there is a long way to go before we have the integrated but urgently needed ecological theory, which would make it possible to understand the nature and the reactions of ecosystems. However, Margalef has many ideas which can be used to develop the present ecosystem theories or maybe rather the fragments of ecosystem theories in the right direction.

Margalef emphasises the importance of information: information multiplies its value when the unified support for it grows larger — a phrase which is completely in accordance with the thermodynamic approach based on exergy presented by Svirezhev in Chapter 14. Margalef is also in accordance with both Chapter 11 by Johnson and Chapter 13 by Jørgensen and with Chapter 14, when he presents a tentative driving principle: go as fast as possible to a condition of high information content. It is a formulation which is very close to the tentative Fourth Law of Thermodynamics proposed in Chapter 13 as a hypothesis.

Margalef distinguishes between external energy (exosomatic energy), for instance, up-welling and internal energy (endosomatic energy), for instance, photosynthesis. He refers to emergy as the energy (endosomatic and exosomatic) that has been involved in any process leading to a presently given structure or situation. Evolution is a question of mastering (being able to properly utilise) exosomatic energy. With this in mind, he can distinguish five peaks in the evolution: the emergence of stromalites, of corals, of macrophytes, of eusocial insects, and of humankind, where the cultural transmission also plays a major role. The exosomatic energy is used to organise space and to build gradients (the same idea as presented in Chapter 13 by Jørgensen, who is using exergy to interpret the ecological observations). The complexity associated with spatial organisation and the irreversibility of ecological processes and their relations to exosomatic energy would for Margalef be the concepts which could be the basis for further progress in ecosystem theory.

It is recommended that this chapter be read carefully, because it contains many ideas — also between the lines — which may give the readers inspiration to further progress in ecosystem theory.

2 Exosomatic Structures and Captive Energies Relevant in Succession and Evolution

Ramon Margalef

CONTENTS

2.1 Present Ecological Theory Needs to be Improved...5
2.2 Constraints of the Physical World 6
2.3 What is Life and How the Biosphere Becomes Organized 7
2.4 Ecological Succession..8
2.5 The Asymmetry of Change — Difficulties Concerning Prediction 9
2.6 Do We Need Just Plain Physics?...9
2.7 Back to Everyday Work..10
2.8 Properties of the Space Required for Obtaining Work from Exosomatic
    Energy Opens New Evolutionary Scenarios 12
2.9 Use of Exosomatic Energy Goes with the
    Capacity to Organize Space...14
References 14

2.1 PRESENT ECOLOGICAL THEORY NEEDS TO BE IMPROVED

The proper study of the biosphere requires a converging effort from many sciences. A balanced synthesis may be impossible to achieve, or at least has not been achieved, as, in function of intellectual fashion and of eventual breakthroughs being made in definite fields, separate aspects become emphasized in a nonsimultaneous way. Today, to speak of ecological theory is out of fashion, although there are many concepts and particular models. For example, theoretical ecology has continued the study of interaction among individuals of different species, perhaps in the continued hope to develop some kind of statistical mechanism, taking into account the eventual peculiarities of ecological relations. But creative insights into the working of large segments of the biosphere, in what concerns use of energy and capacity for organizations, are not being produced in the measure of the need for them, as it appears in relation with the pressing problems concerning global cycles and changes.

The bits and pieces of theory that have been proposed, even if some of them seem to be validated by experiments and observations, do not fit together easily and do not provide a general satisfying picture of how the biosphere has worked and evolved in the past and works and is evolving now. Attempts to find inspiration in more abstruse fields, like chaos, I consider as reactions taken in desperation when they come from ecologists. As ecology deals with physical systems, it seems reasonable that ecologists refrain from accepting theories that have not passed the proof of being of consequence in the field of physics.

I have often voiced these and other complaints (Margalef, 1980, 1991), being worried about the way the Lotka-Volterra approach has gone. Populations are treated as continuous variables and not enough attention was given to space and thermodynamics. The low reputation in which succession theory has fallen might show the lack of feeling for historical processes.

One of the subliminal reasons for the success of the word *ecosystem* may have been in the magic involved in the suffix "-system." Without going more formally into the concept, system implies a relatively close frame of reference for elements and events that play against each other and are bound together in a flexible way. Systems extend over space and time and can be mentally dissected into subsystems of different degrees of coherence, persistence, and flexibility. To set boundaries to any systems is a job for the observer, usually motivated by the convenience of the moment.

Concepts like matter and energy historical change need to be used in a consistent way. Others like ecosystem or succession are less precise or can be redefined by any "qualified" ecologist. It seems it would be easier to accept individuals as the smallest units to work with, but it is not so, because of the existence of such proteiform systems unified by common secretions as mucilages and sheats (stromatolithes), wood (vascular plants, in general, best exemplified by trees), materials like wax or the materials that build the constructions of termites, and even mortar and concrete in humans. This adds pungency to the task of defining boundaries.

2.2 CONSTRAINTS OF THE PHYSICAL WORLD

Passage of time is associated with the irreversible changes studied in thermodynamics. Entropy is a concept that has to be used with care: the increase of its value, measured in a conventional way, becomes an index of total physical change, but does not anticipate how the events are to influence the future of organisms, species, and ecosystems.

Interactions and changes in an ecosystem are irreversible. An example might concern the relation between phosphorus load and concentration of chlorophyll in the photic zone of eutrophic lakes. Relations between variables depend on the direction of change, and we should write $a = f(b, db/dt)$, or adopt complex numbers, $c = f(a + bi)$ (Volohonsky, 1985, 1986). As information may be related to the number of parameters (dimensions) necessary to describe a situation (Rosen, 1989), it is clear that the proper expression and the understanding of change becomes exceedingly complicated and practically impossible.

Spent energy leaves a trace in the record of history, usually in the form of an increase of complexity or its expression as information. Thermodynamics was born in science through consideration of the efficiency of steam machines as suppliers of power. In its beginnings it was much less influenced by historical changes in solid parts, for instance, in the moving pieces of the machine as a result of use, or, when replicable by external agents, like the constructors of the machines, in their repeated attempts to "improve" the artifacts. There is a thermodynamics of organisms, but also a thermodynamic ground for the results of natural selection and evolution. (Information is usually quantified by the length of a univocal message that completely describes the situation in an agreed language; success is not implied, although it is ordinarily bound to more extensive messages or requires larger blueprints. The proof is in survival, that means compatibility with other structures.)

The history of the solid Earth, even before life, shows the expected increase of complexity, as well as shifts in its style — earlier prevalence of folding, later of faulting — and asymmetries of change in the sedimentary cycles (from large boulders to fine-grained materials, then discontinuity, large materials again followed by the same regular change). Another good example, both of gradual accumulation of information and of asymmetrical and irreversible change, is found in the dynamics of meandering rivers.

An exceedingly suggestive example is offered by the calcareous concretions in caves that contain large amounts of information reflecting the energy involved in past percolation, circulation of water, dissolution of materials, and building of stalactitic structures. Their generation is in the accumulation of information associated with volcanic processes, which generate structures that at first sight appear less apt to record and preserve information.

Increase of entropy in the past is partially present now as information. This has been obvious in the development of technologies. The information which we now have access is not free; it has been paid for by the wholesale increase of entropy extended over time, just as biological evolution has been paid by the increase of entropy in the physical bodies of all the organisms that have lived.

Recovery of entropy as information is more efficient when exchanges happen close to the lower limit of the temperature scale, and this is a most important condition for life. There is also a relationship between the "machine" that can extract work, or better, utility from it. The submicroscopic machine of photosynthesis requires photons of relatively high quality energy.

2.3 WHAT IS LIFE AND HOW THE BIOSPHERE BECOMES ORGANIZED

Living systems are physical systems efficient in the task of recovering, as information (apt to be put into use sooner or later), one large fraction equivalent to the entropy accounted for by the physico-chemical reactions going on in the system and, we could probably add, in its immediate surroundings (trees, mankind, as notorious examples of such "overflow" over the close environment). Such entropy can be related to the total amount of energetic changes, as they become integrated in the notion of *emergy by H. T. Odum.*

It is hoped it might be possible to formulate some variational principle that should bear some relationship to what we call natural selection, and that could summarize biological history at every level (individual, ecosystems). Growth and development proceed relatively fast and wastefully at the beginning of each segment of history (individual life, or development, organization and succession in an ecosystem); later on, older structures, as information, are apt to deal more efficiently with the recent and new inputs. Such wishful expectation is behind the hope to attain the now popular future of "sustainable development," optimistically based in the possibility that requirements of energy might decrease, as information stored in the system continues to grow and more efficient use of it can be achieved. But is not easy to decide which direction of change merits the label "progress," a concept found by past generations of biologists, but now being turned against.

One tentative driving principle might be: go as fast as possible to a condition of high information content, maximizing the ratio (forwarded information)/(total accumulated – dissipated – increase of entropy). The entropy increase adds to nothing and is a consequence of mechanical and chemical work, under the Second Law of Thermodynamics, but information multiplies its value when the unified support for it grows larger. In consequence, the internal articulation of the system plays a major role, and hierarchical relations are a part of it. This means that larger organized blocks are made of smaller similarly organized blocks with the relevant information having shifted to become stored at the interfaces where they are best available. Information unfolds in successive layers with an increased participation of the capacity for handling information that most often could have hierarchical structure. Such organization in the information stores should pose an almost insoluble problem to the wish of measuring total information and to ascertain the best management that could be applied to such stores.

The first stages proceed fast with an apparently wasteful use of available energy; later, a higher efficiency along a defined direction can be suspected, because of competition, in the frame of the theory of natural selection. In a tentative way, the common basic principle applying to development, succession, and evolution may be summarized as follows: the system evolves, as fast as is allowed, toward a situation in which the capacity to carry over information is maximized, including information that actually is not more relevant in driving metabolic systems, and thus appears disconnected from its very origin.

Under relatively unchanging conditions, accumulated information may be large enough to anticipate the usual events, selection works no more and the system becomes almost closed and relatively fragile in front of some input that is really new. This is how (and the sad condition of) powerless indifference develops in formerly powerful systems, cleverly signaled, also in its risk for a thinking species, by Patten (1961).

Tendencies in evolution have been identified often, even named separately and dignified collectively as "laws of evolution." They relate to the general law of parsimony in the use of energy to forward an ever-increasing amount, that sometimes might become a burden of information along time and history. Relation between information content and the mass of its support is not straight-forward, since every unified system is able to carry information in an amount proportional to a power, larger than one, of its mass. Hence the advantage of larger individuals, and, in general, of larger unified systems as carriers and processors of information. This may provide a sufficient, if partial, explanation of the gradual increase of size that is often observed along phyletic lines — as well as of the tendency to utilize even nonorganic peripheral information carriers, as happens, at least in mankind, and probably also in trees.

The way to look at things and organisms proposed here implies that the same forces continue to operate in the successive scenarios that arise. Once evolution is accepted, we have to accept also that no total stasis is possible and life forms never persist without evolving or trying to evolve somehow. Evolution is not only necessary, but unavoidable as well. (The same, *mutatis mutandi*, applies to ecological succession.) In our species, genetical evolution is supplemented by cultural evolution, which implies a rapid and important transfer of information through imitative behavior, as well as its frequent storage in exosomatic contraptions. Such arrangement has been accepted by natural (super) selection and has been successful, as it is making total change faster and the potential stores of information much larger.

2.4 ECOLOGICAL SUCCESSION

Parallel reasoning and conclusions apply to ecological succession. Ecological successions lead also to the enrichment of information in living systems, here at the level of the organization of the ecosystems. If the system remains active, change cannot be avoided, although dominant organisms (trees, corals, mankind) may have the power and develop the tendency to shift the accretion of information to their own domain and depress biotic diversity — and with it, its information content — in the rest of the ecosystem.

Such approach, as proposed, provides a sensible rationale for ecological succession. In it the ratio of flow of energy (production) to biomass, biomass seen as the bearer of information and proportional to it is minimized as fast as possible. The complex form that takes the way of forwarding information toward the future might be advisable to widen the original concept of biomass. Plant succession on land is characterized by accumulation of wood, a large part of which consists of materials that are no longer alive. Instead of the primitive ratio production/biomass, we are led to consider the suitability of a new ratio of the form: (endosomatic energy + exosomatic energy)/(biomass proper + wood + necromass + exosomatic artifacts).

In ecological succession information is transferred from the present to the future and the shift is manifest in a historical way that has many aspects. One of them is that production and accumulation of biomass prevails at the beginning, and this results in what is often described as "bottom up control," a concept usually proposed with reference to plankton; later on, the high trophic levels take more control, "top down control" becomes more apparent. As a model I can imagine a coil, in which the region which is more compressed and tense shifts in the course of time, from bottom to top.

The deep functional parallelism between evolution and succession, both as historical sequences, leads to explore the possible development and expression of connections or interactions between them. I have always suspected (Margalef, 1959, 1991), that succession plays a role as providing recurrent and available or practicable paths for characteristic sequences of genetic change.

2.5 THE ASYMMETRY OF CHANGE — DIFFICULTIES CONCERNING PREDICTION

After all the past history, as reconstructed from the vestiges left, the recent biosphere does not look essentially different from any older one, although there have been ups and downs, times for sudden extinctions, and times apparently more appropriate for slow evolution and diversification. Undeniably there has been and continues to exist asymmetry in the speed to which both modalities of change proceed, that is, accelerated production of novelty, and late stasis with frequent extinctions here and there. Such asymmetry matters very much in what concerns prediction and thus becomes of direct interest, not only to paleontologists, but also to practicing ecologists.

Actually, the processes of self-organization and of accumulation of the kind of information that we can compare to "small talk" never go very far. From time to time, disturbances born in a larger space-time frame create havoc and reset the system, if it persists, to a lower degree of "organization," pushing back (partially) its historical development. Nevertheless, I suspect that most of the really new and especially meaningful properties acquired in evolution find in most situations some oportunity not to be lost. Indeed, many old interventions of life have found a way to persist until the present, but surely it is not possible to build a theory to reason their continued existence.

Asymmetries of change are present in all self-organizing systems. Gradual increase of organization is slow and, perhaps, relatively predictable, whereas changes in the opposite direction, implying simplification and loss of information, happen "suddenly" and appear to be unpredictable from inside, as being determined or filtered from outside. Anyways, loss of information can never be certified, since old structures, after a period of functional dormancy, often become significant again in the same or in different context.

Organisms grow and develop, but do not "de-grow"; they and we simply die. Succession is interrupted and reset in a catastrophic way, as is evidenced by the example closest at hand: A forest grows slowly for 50 or more years, but may be destroyed by fire in 1 day.

If properly censused and averaged over long stretches of time, strong disturbances appear to be less frequent than milder disturbances, their respective intensities being measured by the amount of external energy involved. That is all. The order of incidence of the disturbances is haphazard and defies prediction. I mean that useful precise prediction, in practical terms, may be rationally impossible.

2.6 DO WE NEED JUST PLAIN PHYSICS?

Any dynamic system, an organism or an ecosystem, cannot persist in a steady state. Sometimes one wonders why ecologists are so obdurate to keep a pre-Galilean mentality. A constant force produces acceleration, not constant movement. Forces operating in organic synthesis, during uninterrupted life, cannot result in steady states, and predator-prey systems do not persist in the way proposed by theory.

Development, evolution, and succession have common features, and the changes effective in each case, often expressed as an increase of information, appear in the place of the acceleration that physicists, since Galileo, associate with the persistence of acting forces. I would risk saying that plain growth and development, in the way they are usually considered in physiochemical terms, are not the most important consequences of the work done in metabolism. The most intimate recovering as information of a fraction of the implied entropy is what best defines such category of physical systems as apt to optimize such conversion. This might be their best characterization as "living beings." It is not a discontinuous and completely new property: the solid Earth, as mentioned before, also carries much information in its structures and bears witness to the past. The success of life has depended on working faster and at a scale that allows more detail (miniaturization).

The rules for selection remain the same. Mankind, now, after complementing and finally almost substituting cultural for genetic transfer of information, is taking wholesale information carrying contraptions in the exosomatic world, as well as trying to manipulate culturally the old genetic mechanisms.

We still need, after so many centuries of trying, a philosophical synthesis that might relate, in terms more satisfactory than the present, the events in living things and the biosphere with physical science, playing down the concern for "pattern" and "blueprints." Philosophers have discussed the implausibility of steady states, the trend to change, but most often in rhetorical terms (Bergson, 1907) although I am ready to accept that perhaps a deeper and more reasonable approach might be inspired by Whitehead (1929).

2.7 BACK TO EVERYDAY WORK

The quantitative approach to the understanding of populations of many species living together tries to combine processes going on in the separate populations. Decrease of the populations of one species (prey), by death or predation, is linked to the growth and multiplication of the population of the predator. The resulting set of equations that have to express such mutual relation are mathematically interesting and have proved perhaps too seductive.

It is really difficult properly to construct and quantify the influence of the environment and of so many other species on the population that is being considered. Space is the great organizer in nature, but rarely is adequately introduced in the models. Thermodynamics and irreversibility are forgotten. The degree of quantification and discontinuity at different levels, as a function of the average size of individuals of the different species is added, when considered at all, as an afterthought, and usually only as a reminder of the different possibilities of extinction, depending if the population is made by a larger number of small individuals or by a more limited number of large individuals.

In the pelagic world, solar energy is a key factor and appears at two scales and in two quite distinct roles: (1) as photosynthetically active radiation (PAR for short), and (2) as long wave environmental energy that defines mechanical properties, that extend from the level of the major patterns of oceanic circulation to the local mixing or vertical movements, and down to turbulence. Both forms of energy contribute to the support of life and this modelic situation can be characterized, respectively, as (1) endosomatic or set free and made available inside the organisms, through biological reactions, and (2) exosomatic, or generated outside the organism proper, through miscellaneous mechanisms, but effective as well in supporting its existence, as in placing and keeping the plankton organisms in a favorable level. Again, energy involved both in photosynthesis and in chemosynthesis and made available in respiration qualifies, of course, as endosomatic.

Phytoplankton "clouds" mark the areas where nutrients are advected, in a way that suggests at least two analogies: How tiny bubbles mark the path followed by charged particles inside a cloud chamber, or as the clouds in the sky reveal the patterns of the atmospheric disturbances. It is relatively easy to compute the energy involved in the supply of nutrients or their return to the photic zone, that amounts to between 20 to 50 times the energy that goes into photosynthesis. Ratios of the same order are found in terrestrial ecosystems, where rainfall and supply of nutrients through erosion and runoff are not free, and have to be computed also as external or exosomatic energy. One can wonder why only a small percentage of solar radiation (0.06%) is effective in photosynthesis worldwide. But if external, supporting or exosomatic energy is included, the percentage increases up to 3 and more.

It may be reasonable to split the relevant energy, making three categories:

1. The true far external energy, associated with the general climate.
2. The exosomatic energy directly involved with the processes of production, as may be the return of nutrients to the right place where they can be assimilated by phytoplankton, or the energy of evapotranspiration in terrestrial plants.

3. Endosomatic energy, used in photosynthesis or transferred inside or between organisms in the form of the chemical bond.

The concept of exosomatic energy leads directly to the notion of exosomatic contraptions (tools, buildings by eusocial insects, cars, houses, highways). They become especially significant when they are helpful in managing exosomatic energy, like in powered machine tools, cars, and other vehicles, etc.

There is a less straightforward situation when there is a gradient between the more biologically active parts of the organisms and some hard or skeleton-like differentiation of them, selected for durability (wood in trees, calcium carbonate in corals), and continuing to allow the use of such parts even after actual life has ceased in them (dead wood in trunks). Actually, the system of support and transport of a tree is more comparable to a human civilization than to an individual.

H. T. Odum has been a pioneer in his appreciation of the relevance of energy and I have been inspired by his writings. The notion of external energy that I prefer to call *exosomatic energy*, and its role in human civilization, supported also by contraptions created by our own culture, is well presented in many of his publications.

His notion of *emergy*, as I have understood it, refers to the energy (endosomatic and exosomatic, or endosomatic at least) that has been involved in any process (evolutive, successional) leading to a presently given structure or situation, that evidently has been paid for. Emergy could be associated with the entropy increased somewhere during the acquisition of the relevant information and of the information bearers on which our interest focus. This may be a minimum evaluation, however, if only endosomatic energy is being considered. Including exosomatic energy as well might be more just, but also more uncertain.

The following sketch might help to clarify my interpretation of these related concepts.

Climate energy

> Far exosomatic energy
> (example, upwelling)

>> Close, "quasi-internalized"
>> Exosomatic energy
>> (example, evapotranspiration)

>>> Endosomatic energy
>>> (example, photosynthesis)

Odum's emergy, integrated also over time

Some suggestions to improve models concerning pelagic populations follow.

Oceanographers and marine biologists have attempted to ameliorate the original models of the Lotka-Volterra type, introducing them in space. This is done, so to speak, through the back door, accepting the result of sedimentation of organisms in water, plus turbulence of water (Riley, Stommel, and Bumpus, 1949). Turbulence, at least, begs to consider exosomatic energy. I tried to generalize the same points of view and in doing so I found unexpected inspiration in a study about chemical reactions going on in colloidal systems, when subjected to an increasing compartmentalization of the space where reactions happen (Smoluchowsky, 1918). It is a matter of course to analogize such subdivision in more restricted domains to what happens in association with water stratification and thermocline formation, and, in some situations, also with the development of fronts.

This approach leads to the following tentative expression:

	DB/dt	$=$	A	$\bullet$	C
	Biological production		Exosomatic Energy flow		Covariance in the distribution of factors of production
dimensions	$m(t^{-1})$ (speed)		$ml^2\,t^{-1}$ (kinematic viscosity)		l^{-2} (covariance)

Of course, other forms of expression might be proposed and could be more appropriate to express actual physical phenomena (and dimensions). Biological production (P) is power (dimensions $ml^2\,t^{-3}$, watts); dividing power by biomass (B, gives the work done, $ml^2\,t^{-2}$, in joules), and P/B appears obviously as an inverse of time, t^{-1}. Available exosomatic energy must enter as power, and emergy as work done. Covariance (C) in the distribution of the factors of production appears properly as an inverse of length, squared. This length, in the particular condition of plankton, may be imagined as a (weighted) distance, along the vertical, between the respective baricenters of the distributions of light, nutrients, and cells (chlorophyll). Light cannot be mechanically pushed and appears independent from the exosomatic energy that relates — in our example — to mixing and turbulence, that is, to mechanics of the environment.

The precedent expression continues to be consistent and even more satisfying through derivation with respect to time. Assuming in this presentation the usual trend to deceleration, it provides an abridged description of the normal ecological succession:

$$d^2B/dt^2 \qquad = \qquad dP/dt \;\; = \;\; C(dA/dt) \quad + \quad A(dC/dt)$$

slowing down of production and of turnover speed	decrease in available exosomatic energy	segregation of reactants

The alternative possibility is the increase of available exosomatic energy, leading to more overlap in the distribution of reactants and acceleration of turnover.

In the application to the pelagic ecosystem, slowing down of change is associated with stratification (decrease of energy invested in mixing) and with the segregation of reactants (where there is light, nutrients have been reduced to a minimum).

2.8 PROPERTIES OF THE SPACE REQUIRED FOR OBTAINING WORK FROM EXOSOMATIC ENERGY OPENS NEW EVOLUTIONARY SCENARIOS

Models based on pure biological interactions between populations lack realism, as they suggest a world in which there is no space in which exosomatic energy is effective. As shown, there are ways to integrate the notion of exosomatic energy and space use (introduced as the covariance in distributions) as a matter of course.

In continental ecosystems, the importance of exosomatic energy is obvious in the supply of nutrients. They come from erosion and are liberated, dissolved, and made available by rainfall, runoff, and the fluvial systems. Evapotranspiration is dependent on nonphotosynthesis energy and makes nutrients available to the leaves of higher plants; water comes in the second place of interest and often is available in surplus, although it is absolutely necessary as a vehicle for nutrients in solution. Energy involved in evapotranspiration may be considered as a quality lower than light and its amount is several times higher than the amount of high quality energy involved in the process of photosynthesis, at the atomic and molecular scale.

Evapotranspiration energy in vascular plants may be compared, up to a certain point, to exosomatic energy, as this concept is applied to humans, although it has a particular position. Its amount, related to the energy involved in photosynthesis, may approach the same relation between the energy that brings back nutrients to the marine or the lacustrine photic zone and the energy involved in actual photosynthesis of plankton.

Primary production per unit surface on the continents is, on the average, three times higher than primary production in the oceans. The same important difference between the respective ecosystems is the existence of wood and trunks in terrestrial plants, as structures of support and transport, made by materials that have been subjected to a long selection for durability and efficiency, under rather hard specifications, as resistance to moisture, presence of molds and bacteria, etc. In fact, very durable wood, found in moist and bacteria-rich environments, is characterized by a slow turnover and, perhaps, by a relative decrease in the speed of evolution of the trees that were a source of such "precious" woods, that are indeed becoming nonrenewable in an accelerated world.

Comparable strategies in the management of quasi-exosomatic and exosomatic energies have developed independently in separate branches of the evolutionary "tree." They consist in the apparition and development of particular morphologies, special contrivances or aptitudes that facilitate the manipulation of external energies available in the respective environments. Along such lines of reasoning might be explained perhaps several qualitative leaps in evolution.

They appear to have arrived in unrelated lines of evolution. This is a noteworthy aspect of evolution to which scarce attention has been given. The success to which it leads, and the specialization it requires, may have allowed a slowing down of the speed of change in the respectives lines. Trees are long lived and rather conservative. Corals also. Mankind has stopped genetic evolution almost altogether, substituting it by cultural or imitative transmission, apt to work much faster, but — for the moment — does not touch the old genetic store very much.

At least five disconnected peaks in evolution may be singled out by their outstanding mastering of exosomatic energy.

1. Stromatolites. From flimsy oscillating films made of thread-like bacteria and cyanophytes immersed in their mucilaginous secretions to hard structures that maintain inside steep chemical gradients.
2. Corals. The production of masses of calcium carbonate places the photosynthetically active parts at levels of illumination suitable for the symbionts. Interaction of the mineral mass with water improves the access of nutrients. Recently the possibility of internal upward transportation of nutrients is being considered (endo-upwelling).
3. Macrophytes or superior plants specifically in trees. The relevant structures, effective in support and transportation, are made of materials that have evolved for durability. Combinations of cellulose and lignine have been adopted. Part of the structures are not living at all, although they continue to be functional. This has to be taken into account when comparing ratios like P/B or the Redfield proportions C:N:P, both in pelagic and in continental vegetation. On land, the ratio P/B is shown to decline dramatically along succession, as wood is accounted as B. No two trees of the same species are alike, and each tree can be compared, in its function and in its diversification, to a whole human culture, with its town, highways, etc.; botanists face serious difficulties when they wish to apply a brand of demography, born in the field of animal science, to plant populations.
4. Eusocial insects. Construction of auxiliary structures has had a definite effect on evolution, as is especially well documented in the case of termites and also in what concerns development of characteristic landscapes.
5. Mankind. *Homo sapiens* have outdistanced themselves from primate ancestors in ways that combine the strategies of corals, trees, and termites. Natural selection is a general principle

and not restricted to the area of genetic transmission. It applies to information transfer and we know that "units" of information are more valuable if integrated or interacting in the context of larger entities. Small genetic change may have wide consequences if integrated in large genomes.

2.9 USE OF EXOSOMATIC ENERGY GOES WITH THE CAPACITY TO ORGANIZE SPACE

Energy does work moving materials over space and its effects result in changes in natural gradients. The role of shoots and roots in macrophytes is typical. Stromatolites, or at least some forms of them, build lasting structures that maintain inside them persistant and rather sharp redox gradients.

It could only be expected that the control of low quality energy, that is, energies outside those of photons and of the chemical bond, have to be mastered by *machines many order of dimension larger than cells* and, consequently, the job is carried over through some form of restructuring of the space. Think only of trees and of all the contraptions of human civilization. In the back of the mind of everyone remains the images of windmills and hydroelectric plants (artifacts geared to low quality energy), the expectations of nuclear energy (high quality energy), and the success of fossil fuels (chemical bond energy, also of high quality).

I think it was Lotka who introduced the adjective exosomatic, applied to "tools." The hand is an endosomatic tool; pencils, pliers, etc. are exosomatic tools. It is only natural to extend the use of the concept and speak, in general, of exosomatic artifacts that are often channels for circulating exosomatic energy, as in transportation systems and controlled flows in human civilization. Other organisms, as corals, trees, etc. come very close in displaying a comparable capacity for using materials that have lived or not, but that are actually effective in deflecting flows and organize the world around them. So, the concept of exosomatic energy needs the complementary notion of exosomatic artifacts. Further, adoption of exosomatic energy and exosomatic artifacts is subjected to natural selection, and can introduce an independent evolution of their own, when considered as "secondary organisms" (cars, instruments, etc., Cavalli-Sforza and Feldman, 1981) that may seem disconnected or not more related with the eventual evolution of the genuine organism behind all.

REFERENCES

Bergson, H. L., *L'Evolution creatice*. Paris, Alcan, 1907.

Cavalli-Sforza, L. L. and Feldman, M. W., *Culural transmission and evolution, A quantitative approach*. Princeton University Press, Princeton, NJ, 1981.

Jørgensen, S. E., Patten, B. C., and Straskraba, M., Ecosystems emerging: toward an ecology of complex systems in a complex future. *Ecol. Model.*, 62:1–27, 1992.

Margalef, R., Mode of evolution of species in relation to their places in ecological succession. *XV Intern. Congr. Zool. London, Linn. Soc. London,* pp. 787–789, 1959.

Margalef, R., *La biosfera entre la termodinámica y el juego.* Omega, Barcelona, 1980.

Margalef, R., *Teoria de los sistemas ecológicos.* Publ. Univ. Barcelona, 1991.

Margalef, R., *Our Biosphere.* Ecology Institute, Oldendorf/Luhe, 1997.

Odum, H. T., Energy in ecosystems, 337–369 in N. Polunin (Ed.) *Ecosystems theory and application.* John Wiley & Sons, New York, 1986.

Patten, B. C., Competitive exclusion, *Science,* 134:1599–1601, 1961.

Riley, G. A., Stommel, H., and Bumpus, D. F., Quantitative ecology of the plankton of the Western North-Atlantic, *Bull. Bingham Ocean. Coll.,* 12:1–69, 1949.

Rosen, R., Similitude, similarity, and scaling. *Landscape Ecol.,* 3:207–216, 1989.

Rosen, R., *Life Itself.* Columbia University Press, New York, 1991.

Smoluchowsky, M. von., Versuch einer mathematischen Theorie von Koagulationskinetik kolloider Lösungen, *Z. Phys. Chem.,* 92:130–168, 1918.

Volohonsky, H., Free energy flow in the process of collecting material for body build up of aquatic organisms, *Ecol. Model.,* 15:313–329, 1982.

Volohonsky, H., 1985. Form-building potencies of photons and the structural dynamics of ecosystems. *Ecol. Model.,* 29:139–154, 1985.

Volohonsky, H., Ecosystems memory in the context of structural dynamics, *Ecol. Model.,* 33:59–75, 1986.

Whitehead, A. N., *Process and reality. An essay in cosmology.* The Free Press, a division of Macmillan, New York, 1929.

CHAPTER 3

This chapter will present how dynamical mass and energy balances can be used on a canonical community to build a model, offering more information than the usual population dynamic models. The so-called DAB community (Daphnia, Algae, and Bacteria) is used to exemplify the ideas. The dynamics of the simplified community is constrained by setting up a dynamic mass-and-energy balance for all state variables in the model. This strategy leads to a disclosure of several processes and compound properties that other models, particularly black box models, cannot capture. It is, for instance, clear that assimilation, maintenance, and growth contribute to nitrogen waste. Respiration is usually converted into energy by use of a fixed coefficient, and interpreted as maintenance costs. In the presented dynamic energy budget model all three basic energy fluxes contribute to respiration while the ratio between oxygen consumption and carbon dioxide production is not necessarily constant.

The model is presented in all details which makes it possible for readers to build the model on their own computer and thereby verify the many results obtained by use of these detailed dynamical mass and energy balances. The message from this chapter is clearly, "use detailed and dynamical thermodynamic constraints on biological components and their interactions" and you will inevitably get some new interesting results. As we do know that the thermodynamic constraints are in accordance with the laws of nature, this strategy will lead to observations that are in accordance with the laws of nature.

The two authors of this chapter have both previously published their basic theoretical considerations in books. A detailed treatment of the idea presented here can be found in *Dynamic Energy and Mass Budgets in Biological Systems* by S. A. L. M. Kooijman (Cambridge University Press, 2000). *Ecological Dynamics* by W. S. C. Gurney and R. M. Nisbet (Oxford University Press, 1998) focuses on population dynamic models. The book is unique by its treatment of physiologically and spatially structured populations.

3 How Light and Nutrients Affect Life in a Closed Bottle

S. A. L. M. Kooijman and R. M. Nisbet

CONTENTS

3.1 Introduction 19
3.2 A DEB Representation of the DAB Community 22
3.3 The DAB System 27
    3.3.1 Notation and Symbols..27
3.4 The SU-Extended Monod Model 29
    3.4.1 Degenerated Simplifications 29
3.5 The DEB Model..34
    3.5.1 Feeding and Time Budgets 35
        3.5.1.1 Daphnia 38
        3.5.1.2 Alga 40
        3.5.1.3 Bacterium 40
    3.5.2 Dissipating Power 40
    3.5.3 Growth and Reserve Kinetics 41
        3.5.3.1 Daphnia 41
        3.5.3.2 Daphnia Individuals 42
        3.5.3.3 Daphnia Reproduction and Aging 43
    3.5.4 Daphnia Population Structure..44
    3.5.5 Alga 45
    3.5.6 Bacterium 46
    3.5.7 Mineral Fluxes and Mass Balance...46
3.6 Mass Turnover..47
3.7 Energy Turnover and Dissipation Heat 51
3.8 Top-Down vs. Bottom-Up Control 53
3.9 From Canonical to Natural Communities 55
3.10 Discussion 56
Acknowledgments 57
References 57

3.1 INTRODUCTION

Figure 3.1 illustrates the structure of an idealized, simple, three-component ecosystem. Producers use light and nutrients to produce organic matter, which is transformed by consumers, while decomposers release nutrients from the organic matrix.[73] The system is "open" to energy flow, but closed to inputs or removal of elemental matter. The structure, which might represent a closed bottle containing daphnids, algae, and bacteria, is found in many ecosystems; for example, it is very similar to the one used for material turnover in microbial flocs in seawater plankton systems.[23,50] It can,

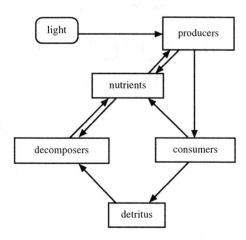

FIGURE 3.1 The canonical community consists of three components: producers that gain energy from light and take up nutrients to produce biomass, consumers that feed on producers, and decomposers that recycle nutrients from producers and consumers. The community is nearly closed for nutrients, but requires a constant supply of energy. Influx and efflux of nutrients largely determine the long-term behavior of the community.

therefore, be regarded as a canonical community. In this chapter, we describe an approach to developing parsimonious models of the flow of energy and nutrients in the canonical community.

Nutrient dynamics affects not only the functioning of an ecosystem, but also governs structural aspects, such as the succession of algal species in seawater plankton.[76] The main nutrients that usually affect standing crops most drastically are nitrogen compounds, phosphate, iron, and silicon for diatoms,[19,82] while light can also be limiting.[14,30,72] The production of organic matter by phytoplankton has been suggested to play an important role in the euphotic zone,[5] and implies a coupling between carbon and mineral fluxes. Model studies and mesocosm data, however, could not confirm the importance of this "microbial loop."[18,77] Excretion of inorganic nutrients, such as carbon dioxide and ammonia, by consumers can also be important "nonconsumptive top-down effects" in aquatic ecosystems.[74,85] Andersen[4] raised the question whether zooplankton acts as a source or a sink of nutrients, and concluded that this depends on the time scale and the fate of the individuals; a set of models suggested that the fraction of nutrients locked into zooplankton initially increases with increasing nutrient loading until a maximum is reached at some intermediate loading level. The roles of bacteria as consumers of minerals (i.e., competitors of algae) vs. remineralizers (symbionts of algae) are still under discussion.[35,49] It largely depends on the composition of detritus relative to that of bacteria; bacteria must take up more ammonia from the environment if the nitrogen content of detritus is poor. Thingstad and Pengerud[80] found that coexistence of algae and bacteria is only possible if the influx of nitrogen-poor organic matter from outside the system does not exceed a threshold value.

Ecosystem modeling commonly addresses questions that relate to a much larger scale in space and time than is relevant for individuals.[90,91] Processes on totally different space/time scales combine poorly into a single model, implying that the fate of individuals will seldom be important at the ecosystem level. Our study explores strategies of discarding detailed information without affecting the overall dynamics of the system too much.

There are many existing model studies for DAB-type communities. Some "strategic" models[32,33,52,61,63,84] focus on the relative stability of closed ecosystems vs. those that exchange elemental matter with their environment. Many more practical population, community, and ecosystem models incorporate nutrient flow, because the effect of nutrient availability on algal growth is well recognized.[6,7,10,16,17,20,36,62,78] The models that we will study in this contribution differ from their predecessors in being fully closed for mass flow while still recognizing more than one element, and being fully dynamic. Moreover, we will study *all* chemical transformations and the detailed energetic aspects

TABLE 3.1
Table of Frequently Used Symbols for Variables

Symbol	Dimension	Interpretation
X_j	mol^{-3}	Concentration of compound (substrate) j
x_{ji}	—	Scaled substrate concentration j for reserve i: $X_j/X_{K,ji}$
V	l^3	Structural body volume: $M_V/[M_V]$
M_{Ei}, M_V	mol	Mass of reserve i, structural biomass in C-moles
m_{Ei}	$\dfrac{mol\ Ei}{mol\ struc}$	Structure-specific reserves i: M_{Ei}/M_V
$\dot{J}_i, \dot{J}_{i,J}$	$mol\ t^{-1}$	Flux of compound i (involved in transformation j)
$\mathscr{J}_i$	$\dfrac{mol\ i}{mol\ struc}t^{-1}$	Structure-specific flux of compound i: $\dot{J}_i/M_V$
$\dot{p}_i$	et^{-1}	Power i
$\dot{p}_T$	et^{-1}	Dissipating heat
f_i	—	Scaled functional response for reserve i
$\dot{r}_{V*,G*}$	t^{-1}	Sp. growth rate for structural mass *: $\mathscr{J}_{V*,G*}/M_{V*}$
$\dot{h}$	t^{-1}	Hazard rate

Note: Dots refer to rates (dimension: time^{-1}), not to derivatives with respect to time, which are indicated by d/dt. Index m refers to the maximum value. In the dimension column, l means length, t time. Table 3.5 gives labels of compounds and transformations, which are used as indices, Table 3.3 gives parameters. Since volumes will only be used here for *Daphnia*, the index D will be suppressed.

TABLE 3.2
Chosen Values for the Mass-Mass Couplers $y_{i,j}$ = mole of compound i/mole of compound j

$y_{ED,VA}$	0.3	y_{ED,E_1A}	0.8	y_{ED,E_2A}	0.8	$y_{ED,VB}$	0.3	$y_{ED,EB}$	0.8
$y_{PA,VA}$	0.7	$y_{PB,VB}$	0.7	$y_{EB,PA}$	0.7	$y_{EB,PB}$	0.7	$y_{EB,PV}$	0.7
$y_{EB,PE}$	0.9	$y_{ED,VD}$	2.5	$y_{E_1A,VA}$	1.5	$y_{E_2A,VA}$	1.7	$y_{EB,VB}$	1.25

of a (very much) simplified system. In contrast, many of the previous studies of closed systems only considered one element (e.g., carbon or nitrogen), while others assume fixed grazing rates by zooplankton, or fixed production rates by phytoplankton, implying fixed population sizes, while we consider these population sizes as part of the system.

The models in this chapter are based on Kooijman's Dynamic Energy Budget (DEB) theory,[41] which is described in more detail in Section 3.2. This theory specifies all mass transformations (feeding, growth, reproduction, etc.) at the individual level, and how they change during the life cycle. The transformations at the population level directly follow via addition of fluxes. The flux of organic matter through food webs is mainly set by body-size scaling relationships, which may be derived directly from the DEB theory.[41] A promising DEB-based approach to an extension to the ecosystem level is via mass conversion at the population level,[42,44] in combination with energy and mass conservative laws at the whole system level. The gist of this approach is that the link between the metabolic properties of individuals and community and ecosystem performance is still preserved. A particularly simple situation, discussed later in this chapter, arises if organisms change shape during growth in such a way that surface area is proportional to volume. For these organisms, called 1D-isomorphs, the DEB theory becomes really simple, and can be considered as a generalization of the models by Monod (no maintenance or reserves), Marr-Pirt (no reserves), and Droop (no maintenance).

To separate the community and ecosystem level questions from population dynamics, and to study the effects of different model components, we will discuss a (short) series of related models of the DEB system within the DEB framework that span a range of levels of complexity. The extended Monod model for the three biota represents the simplest model. A DEB model for 1D-isomorphs with one or two types of reserves for the algae takes an intermediate position, and, finally, we consider the daphinds to be 3D-isomorphs (defined in the next section), which is much more realistic, but also more complex.

3.2   A DEB REPRESENTATION OF THE DAB COMMUNITY

A closed bottle, completely filled with water containing daphnids, algae, and bacteria constitutes the DAB community, which is completely closed from its surroundings in terms of mass exchange, while dissipating heat that is generated from metabolic processes and from absorbed light can leave the bottle with such ease that the temperature in the bottle equals that of its surroundings, which is assumed to be constant.

The foundation of our representations of DAB *ecosystem* dynamics is a dynamic energy budget (DEB) model incorporating information on the physiology of *individuals* (Reference 41, and references therein). DEB models use differential equations to describe the rates at which individual organisms assimilate and utilize energy from food for maintenance, growth, reproduction, and development. These rates depend on the state of the organism (age, size, sex, nutritional status, etc.) and the state of its environment (food density, temperature, toxicant levels, etc.). Solutions of the model equations represent the life history of individual organisms in a potentially variable environment.

In Kooijman's model, input of energy to an organism is assumed to involve transfer of material across surfaces (gut wall, membranes of cells and organelles, etc.), before it is spent on volume-dependent processes, such as growth and maintenance. The geometry of the organism determines how the areas of the critical surfaces scale with size; thus organisms are characterized as "1D-isomorphs" (where organisms change shape during growth in such a way that surface area is proportional to volume) or "3D-isomorphs" (where surface area is proportional to volume$^{2/3}$). Somatic and reproductive tissues compete for available energy, and the organism aims at a stable internal environment (homeostasis), in which the relative proportions of structural tissue and energy reserves are related to the food environment. Structural material and reserves are conceived as different generalized compounds, i.e., rich mixtures of chemical compounds, mainly consisting of proteins, lipids, and carbohydrates, which allow relatively simple relationships between different size measures (volumes, weights, C-moles).

Here, we shall assume that all living components of the DAB system daphnids, algae, and bacteria) have one type of structural body mass, and one type of reserve, but that the algae have two types of reserves (one with and one without nitrogen, so a protein-rich reserve and a carbo-hydrate/lipid-rich one). The daphnids feed on bacteria and algae; algae feed on light, carbon dioxide, and ammonia; bacteria feed on feces and dead corpses of daphnids.

For simplicity's sake, we make a series of not always extremely realistic assumptions about the DAB system. Our main purpose is not to maximize on realism, but to show which processes contribute to mass and energy turnover, and how. The less realistic assumptions can be avoided or replaced by more realistic ones, but this needs extra parameters. The assumptions are:

- The volume change involved in any of the chemical transformations is negligibly small, which implies that the pressure in the bottle is constant.
- Although the energy content of each photon depends on its wavelength, we neglect this diversity and only count photons that can be used by algae to drive their metabolism. Biochemical evidence indicates that the photosynthetic system extracts a fixed amount

TABLE 3.3
Parameters and Their Chosen Values

Symbol	Unit	Value	Interpretation
$[M_{VD}]$	$\dfrac{\text{mol}}{\text{cm}^3}$	0.04	Structural mass density of *Daphnia*: M_{VD}/V
$X_{K,VA}$	$\dfrac{\text{mol}}{\text{dm}^3}$	2	Half-saturation constant of algae for *Daphnia*
$X_{K,VB}$	$\dfrac{\text{mol}}{\text{dm}^3}$	5	Half-saturation constant of bacteria for *Daphnia*
$X_{K,L1}$	$\dfrac{\text{mol}}{\text{dm}^3}$	5	Half-saturation constant of light for algal reserves 1
$X_{K,C1}$	$\dfrac{\text{mol}}{\text{dm}^3}$	1	Half-saturation constant of CO_2 for algal reserves 1
$X_{K,L2}$	$\dfrac{\text{mol}}{\text{dm}^3}$	2	Half-saturation constant of light for algal reserves 2
$X_{K,C2}$	$\dfrac{\text{mol}}{\text{dm}^3}$	2	Half-saturation constant of CO_2 for algal reserves 2
$X_{K,N2}$	$\dfrac{\text{mol}}{\text{dm}^3}$	2	Half-saturation constant of ammonia for algal reserves 2
$X_{K,PA}$	$\dfrac{\text{mol}}{\text{dm}^3}$	2	Half-saturation constant of alga-feces for bacteria
$X_{K,PB}$	$\dfrac{\text{mol}}{\text{dm}^3}$	2	Half-saturation constant of bacterium-feces for bacteria
$X_{K,PV}$	$\dfrac{\text{mol}}{\text{dm}^3}$	2	Half-saturation constant of *Daphnia*-struc. mass for bacteria
$X_{K,PE}$	$\dfrac{\text{mol}}{\text{dm}^3}$	1	Half-saturation constant of *Daphnia*-reserves for bacteria
$\{\dot{J}_{VA,AD,m}\}$	$\dfrac{\text{mol}}{\text{h mm}^2}$	1.5	Maximum specific ingestion rate of algae by *Daphnia*
$\{\dot{J}_{VB,AD,m}\}$	$\dfrac{\text{mol}}{\text{h mm}^2}$	1.25	Maximum specific ingestion rate of bacteria by *Daphnia*
$j_{E_1A,A_1A,m}$	$\dfrac{\text{mol}}{\text{h mol}}$	6.0	Maximum specific synthesis rate of algal reserves 1
$j_{E_2A,A_2A,m}$	$\dfrac{\text{mol}}{\text{h mol}}$	4.0	Maximum specific synthesis rate of algal reserves 2

(*Continued*)

TABLE 3.3
Parameters and Their Chosen Values (Continued)

Symbol	Unit	Value	Interpretation
$\dot{j}_{PA.A_1B.m}$	$\dfrac{mol}{h\,mol}$	3.0	Maximum specific uptake rate of alga-feces by bacteria
$\dot{j}_{PB.A_2B.m}$	$\dfrac{mol}{h\,mol}$	3.0	Maximum specific uptake rate of bacterium-feces by bacteria
$\dot{j}_{PV.A_3B.m}$	$\dfrac{mol}{h\,mol}$	2.0	Maximum specific uptake rate of *Daphnia*-struc. mass by bacteria
$\dot{j}_{PE.A_4B.m}$	$\dfrac{mol}{h\,mol}$	3.0	Maximum specific uptake rate of *Daphnia*-reserves by bacteria
$\dot{j}_{ED.MD}$	$\dfrac{mol}{h\,mol}$	0.8	Specific maintenance costs for *Daphnia* reserves
$\dot{j}_{E_1A.MA}$	$\dfrac{mol}{h\,mol}$	0.5	Specific maintenance costs for algal reserves 1
$\dot{j}_{E_2A.MA}$	$\dfrac{mol}{h\,mol}$	0.1	Specific maintenance costs for algal reserves 2
$\dot{j}_{EB.MB}$	$\dfrac{mol}{h\,mol}$	0.5	Specific maintenance costs for bacterial reserves
$\dot{k}_{E_1A}$	h^{-1}	6.0	Turnover rate of algal reserves 1
$\dot{k}_{E_2A}$	h^{-1}	6.0	Turnover rate of algal reserves 2
$\dot{k}_{EB}$	h^{-1}	3.0	Turnover rate of bacterial reserves
$V_b^{1/3}$	mm	0.42	Volumetric length at birth (*Daphnia*)
$V_p^{1/3}$	mm	1.32	Volumetric length at puberty (*Daphnia*)
g	—	1.0	Energy investment ratio (*Daphnia*)
κ	—	0.3	Fraction to growth + somatic maint. (*Daphnia*)
κ_R	—	0.8	Fraction of rep. flux to embryonic reserves (*Daphnia*)
κ_{R1}	—	0.8	Return fraction of rejected reserves 1 (alga)
κ_{R2}	—	0.7	Return fraction of rejected reserves 2 (alga)

Note: The mass-mass couplers are given in Table 3.2, chemical indices in Table 3.5. Length measures refer to volumetric lengths, which equal 0.526 times body lengths for *Daphnia*.

of energy from each usable photon; the excess energy is lost as heat. The DAB community is optically thin enough to neglect self-shading. We will not include the processes of photo-adaptation and photo-inhibition.

- Apart from light and carbon dioxide, we only consider ammonia as a possibly limiting mineral nutrient. Although nitrate would be the most frequent nitrogen nutrient in natural systems, (aquatic) animals excrete ammonia, which can efficiently block nitrate uptake by algae, even at extremely low concentrations. This interaction would complicate our analysis. Although phosphate is frequently limiting in freshwater systems,[25] and its inclusion gives no theoretical complications, we again exclude it for simplicity's sake. For the same reason, we also exclude production of organic excretions by algae.

- No chemical transformations occur other than via organisms; the pH in the environment is assumed to be constant. The transformations occur within cells, which keep all chemicals that participate directly in chemical transformations at constant and low levels. Carbonates and bicarbonates are included in the variable that stands for carbon dioxide, despite the fact that these compounds are not taken up by algae. This translates into the assumption that the association/dissociations in the $CO_2/HCO_3^-/H_2CO_3$-complex is fast with respect to algal uptake, so CO_2 is a constant fraction of the complex. Oxygen and water are assumed to be nonlimiting. *Daphnia* can survive moderate oxygen stress, by making more hemoglobin, and is probably able to ferment. Fermentation, however, comes with a considerable reduction of the energy yield of food, which would complicate our analysis.

- Aging in bacteria and algae is so sufficiently slow in relation to predation by *Daphnia*, that these processes can be neglected, so only *Daphnia* ages. We need this aging process to make sure that the life span of all organisms is limited.

- Digestion of the reserves of bacteria and algae by *Daphnia* is complete, which means that these compounds do not contribute to feces; feces is considered to be a leftover of the digestion of structural body mass of algae and bacteria. This assumption does not imply a high reserve yield. A substantial fraction of structural algal biomass consists of cellulose, which cannot be digested by *Daphnia*. Digestion of feces and *Daphnia* corpses by bacteria is complete, and only water, carbon dioxide, and ammonia is formed. The result is that the molecules of all compounds have a limited life span. This is essential to make sure that the DAB system has a finite memory. We also assume that digestion is instantaneous, which implies that only structural body mass and reserves result from *Daphnia* that die from aging, and we do not have to consider their partially digested gut contents.

- Apart from the feeding relationships, we excluded all other forms of interactions between organisms, such as social interactions, which might reduce feeding rate at high population density.

- No spatial structure exists. The minerals and organic products (*Daphnia* corpses and feces) are homogeneously distributed at the molecular level, daphnids, algae, and bacteria at the individual level. Although *Daphnia* corpses and fecal pellets do not dissolve instantaneously in reality, the resulting reduction in biodegradation is partially taken into account via the transfer parameters.

- We assume that the total amounts of all compounds in the DAB system have a point attractor in the parameter region of interest, and that the number of daphnids is sufficiently large that the size distribution is close to the stable size distribution. We expect, however, that the compounds have a stable limit cycle in the case of a structured consumer population, but the mean biomass levels will probably be close to the ones that we will derive for the point attractor. This value will be helpful anyway as a reference for the dynamical results.

- Algae and bacteria behave as 1D-isomorphs, daphnids as 3D ones (see the Section 3.5 on the DEB model), while reproduction is continuous in time. In reality, *Daphnia* reproduction is in clutches, which are produced in synchronization with the molting cycle. All continuous-time models for allocation require a buffer for allocation to reproduction and rules specifying how this buffer converts to new offspring. We refrain from considering these details, however, which means that reproduction in the model occurs slightly earlier than it should, and that *Daphnia* corpses have less reserves for bacteria to digest. *Daphnia* usually reproduces parthenogenetically (diploid females give birth to new ones, without interference by males). Although they can reproduce sexually, we assume that they don't do this in our bottle.

TABLE 3.4
Conversions between Volume-Based, Molar-Based, and Energy-Based Quantities of the DEB Model

struc volume	$V = \dfrac{M_V}{[M_V]}$	l^3	max struc v	$V_m = \left(\dfrac{\dot{v}}{\dot{k}_M g}\right)^3 = \dfrac{M_{Vm}}{[M_V]} = \left(\dfrac{\kappa y_{VX}\{\dot{J}_{XAm}\}}{\dot{k}_M\ [M_V]}\right)^3$	l^3
reserve ener	$E = \mu_E M_E$	e	max reserve	$E_m = [E_m]V_m = \mu_E M_{Em}$	e
max res dens	$[E_m] = \mu_E[M_e]$	$\dfrac{e}{l^3}$	struc vol ener	$[E_G] = \mu_{GV}[M_V]$	$\dfrac{e}{l^3}$
scaled length	$l = \left(\dfrac{M_V}{M_{Vm}}\right)^{1/3} = \left(\dfrac{V}{V_m}\right)^{1/3}$	—	res density	$e = m_E\dfrac{[M_V]}{[M_e]} = m_E \kappa g y_{VE}$	—
max spec assim	$\{\dot{p}_{Am}\} = \mu_{AX}\{\dot{J}_{XAm}\}$	$\dfrac{e}{l^2 t}$	spec maint	$[\dot{p}_M] = \dot{k}_M \mu_{GV}[M_V]$	$\dfrac{e}{l^2 t}$
ener conduct	$\dot{v} = \dfrac{\{\dot{p}_{Am}\}}{[E_m]} = y_{EX}\dfrac{\{\dot{J}_{XAm}\}}{[M_e]}$	$\dfrac{l}{t}$	Invest ratio	$g = \dfrac{[E_G]}{\kappa[E_m]} = \dfrac{y_{EV}[M_V]}{\kappa\ [M_e]}$	—
maint rate	$\dot{k}_M = \dfrac{[\dot{p}_M]}{[E_G]} = j_{EM}y_{VE}$	$\dfrac{1}{t}$	relative res	$m_E = \dfrac{M_E}{M_V} = \dfrac{e}{\mu_E}\left[\dfrac{E_m}{M_V}\right]$	—
struc mass	$M_V = V[M_V]$	m	max struc m	$M_{Vm} = V_m[M_V] = \left(\dfrac{\kappa y_{VX}\{\dot{J}_{XAm}\}}{\dot{k}_M\ [M_V]}\right)^3[M_V]$	m
reserve mass	$M_E = \dfrac{E}{\mu_E} = M_V\dfrac{e\,y_{EV}}{\kappa g}$	m	max reserve	$M_{Em} = \dfrac{E_m}{\mu_E} = M_{Vm}\dfrac{[M_e]}{[M_V]}$	m
assim/food coupl	$\mu_{AX} = \dfrac{\{\dot{p}_{Am}\}}{\{\dot{J}_{Xm}\}} = \dfrac{\mu_E}{y_{XE}}$	$\dfrac{e}{m}$	assim/prod coupl	$\mu_{AP} = \dfrac{\{\dot{p}_{Am}\}}{\{\dot{J}_{Pm}\}} = \dfrac{\mu_E}{y_{PE}}$	$\dfrac{e}{m}$
res chem pot	$\mu_E = \dfrac{[E_m]}{[M_e]}$	$\dfrac{e}{m}$	growth/struc coupl	$\mu_{GV} = \dfrac{[E_G]}{[M_V]} = \dfrac{\mu_E}{y_{VE}}$	$\dfrac{e}{m}$
prod/res coupl	$y_{PE} = \dfrac{\mu_E}{\mu_{AP}} = \dfrac{\dot{J}_{PA}}{\dot{J}_{EA}}$	$\dfrac{m}{m}$	prod/food coupl	$y_{PX} = \dfrac{\mu_{AX}}{\mu_{AX}} = \dfrac{\dot{J}_{PA}}{\dot{J}_{XA}}$	$\dfrac{m}{m}$
struc/res coupl	$y_{VE} = \dfrac{\mu_E}{\mu_{VG}} = \dfrac{\dot{J}_{VE}}{\dot{J}_{EG}}$	$\dfrac{m}{m}$	food/res coupl	$y_{XE} = \dfrac{\mu_E}{\mu_{AX}} = \dfrac{\dot{J}_{XA}}{\dot{J}_{EA}}$	$\dfrac{m}{m}$
spec assim fl	$j_{EA} = \dot{J}_{XA}y_{EX}$	$\dfrac{m}{mt}$	spec maint fl	$j_{EM} = \dot{k}_M y_{EV}$	$\dfrac{m}{mt}$
assim flux	$\dot{J}_{EA} = j_{EA}M_V = \dot{J}_{XA}y_{EX}$	$\dfrac{m}{t}$	food flux	$\dot{J}_{XA} = j_{XA}M_V$	$\dfrac{m}{t}$

Note: Coefficients $[M_*]$ convert volume to C-moles (dimension: mole volume^{-1}; the brackets [ ] refer to volume^{-1}, while the braces { } refer to surface area^{-1}). The energy-mass coupler $\mu_{*_1*_2}$ couples energy flux $*_1$ to mass flux $*_2$ (dimension energy per mole). The chemical potential μ_* also has dimension energy per mass, but cannot be interpreted as ratio of fluxes. The mass-mass coupler $y_{*_1*_2}$, also known as yield or stoichiometric coefficient, is a ratio of mass fluxes and taken to be constant, just like other couplers. We have $y_{*_1*_2} = y_{*_2*_1}^{-1}$, $y_{*_1*_2}\,y_{*_2*_3} = y_{*_1*_3}$ and $\eta_{*_1*_2} = \mu_{*_2*_1}^{-1}$ is a mass-energy coupler. Volumes are indicated with V, masses in C-moles with M, structure-specific masses with $m_* = M_*/M_V$. Mass fluxes in C-moles per time are indicated with $\dot{J}_*$ (the dot refers to time^{-1}), structure-specific mass fluxes with $j_* = \dot{J}_*/M_V$. Energy fluxes (i.e., powers) are indicated with $\dot{p}$. Index X refers to food, P to product (feces). The following conversions between volume-based and mole-based quantities hold, where the dimensions are indicated with l (length), m (mass), e (energy), t (time).

In spite of this long list of simplifying assumptions, what is left is still rather complex and involves 16 compounds and 16 chemical transformations. Four of the compounds represent reserves, so the system might seem to allow further simplification, in view of the fact that internal reserves are commonly ignored in ecosystem models. Indeed, it is one of our purposes to uncover the most relevant processes involved in mass transformations in the system. Several arguments call for an inclusion of reserves in dynamic models, the most fundamental one being for consistency reasons. Models which include maintenance, but no reserves, may have internal inconsistencies, which become apparent during transient states. Since reserve densities (amounts of reserves per unit of structural biomass) are related to food densities, reserves decrease the substrate/structural biomass conversion, which means that they are also relevant for steady-state phenomena at steady state. Furthermore, reserves of algae play an essential role in *Daphnia* nutrition, as explained in the assumption about digestion; algae can accumulate large amounts of carbohydrate reserves under nitrogen-limiting conditions. Total algal carbon is not necessarily an adequate quantifier of the nutritional value for daphnids and reserves of algae substantially affect the nutritional value of algae for *Daphnia*.[87] Finally, comparison of the dynamics of models with and without explicit representation of reserves shows that reserves may profoundly influence the dynamics of the system.[65,80]

3.3 THE DAB SYSTEM

The chemical compounds and their transformations in the DAB community are presented in Table 3.5. When we replace the signs by model-dependent quantitative expressions, as discussed below, this turns Table 3.5 into a matrix that is known as scheme matrix,[69] which will be indicated by matrix $\boldsymbol{\dot{J}}$; element i, j of matrix $\boldsymbol{\dot{J}}$ called $\dot{J}_{i,j}$, gives the flux of compound i involved in transformation j. We quantify the compounds in terms of moles (for minerals) or C-moles (for organic compounds and biomass), and indicate the vector of moles of all compounds by $\boldsymbol{M}$. The concept of C-mole is a theoretical construct, where each "molecule" has one C-atom and fractional amounts of the elements H, O, and N. These fractions do not change in time, and other elements are neglected.

When the transformations can be written as functions of the total amount of moles of the various compounds, $\boldsymbol{M}$, the dynamics of $\boldsymbol{M}$ can be written $(d/dt)\boldsymbol{M} = \boldsymbol{\dot{J}}\mathbf{1}$, which just states that the change in masses equals the sum of the columns of the scheme matrix.

This type of model is called an unstructured population model, because it does not explicitly allow for differences between individuals. If individuals differ too much, and the frequency distribution of the various types of individuals changes rapidly, it becomes essential to turn to more advanced methods to quantify the processes,[58,83] but this is beyond the scope of this chapter. It is possible, however, to evaluate some properties of structured populations at steady state without resort to mathematically sophisticated techniques, because the frequency distribution of individuals among physiological states can be identified if it does not change in time.

Although we will formulate the full transient dynamics of the DAB community, the steady-state dynamics is of special interest, because it is much simpler, due to the fact that a lot of information about the initial conditions becomes lost. The Jacobian $(d/d\boldsymbol{M}^T)\boldsymbol{\dot{J}}\mathbf{1}$ at steady state contains interesting information about the possible behavior of the system close to the steady state.

3.3.1 NOTATION AND SYMBOLS

Table 3.1 presents the notation and frequently used symbols, while Table 3.5 labels compounds and transformations. The substantial variety of compounds and processes forced us to use a rather elaborate notation that might seem clumsy at first glance, but minimizes stress on memory. The present notation allows an easy distinction between, e.g., the transformation of bacterial structural mass into feces (by daphnids feeding on bacteria) and the backward transformation (by bacteria using feces as substrate).

The leading symbol refers to the nature of the variable (e.g., M stands for a molar mass, J for a molar flux). Dots refer to the dimension "per time," and *not* to a time-derivative, which is indicated by d/dt. The first index identifies the compound, the second identifies the process (for fluxes). Organic compounds are coded in two characters (type and species): V for structure, E for reserves, P for product (detritus), and D for daphnids, A for algae, B for bacteria. Processes are also coded in two characters (type and species): A for assimilation, D for dissipation, G for growth, H for hazard, and again D, A, B for daphnids, algae, bacteria. So $\dot{J}_{VD,GD}$ represents the molar flux ($\boldsymbol{\dot{J}}$)

TABLE 3.5
The Chemical Compounds of the DAB Community and Their Transformations and Indices

Transformations		Minerals					Detritus				Daphnia		Alga			Bacteria	
		L	C	H	O	N	PA	PB	PV	PE	VD	ED	VA	E_1A	E_2A	VB	EB
		Light	Carbon Dioxide	Water	Oxygen	Ammonia	Alga-Feces	Bact-Feces	Dead Daph-Struc	Dead Daph-res	Structure	Reserves	Structure	Reserves 1	Reserves 2	Structure	Reserves
Daphnia																	
Assim 1	A_1+		+	+	−	+	+					+	−	−	−		
Assim 2	A_2+		+	+	−	+		+				+				−	−
Growth	$G+$		+	+	−	+					+	−					
Dissip	$D+$		+	+	−	+						−					
Death	$H+$								+	+	−	−					
Alga																	
Assim 1	A_1A	−	−	−	+									+			
Assim 2	A_2A	−	−	−	+	−									+		
Growth	GA		+	+	−	+							+	−	−		
Dissip 1	D_1A		+	+	−	+								−			
Dissip 2	D_2A		+	+	−	+									−		
Bacterium																	
Assim 1	A_1B		+	+	−	+	−										+
Assim 2	A_2B		+	+	−	+		−									+
Assim 3	A_3B		+	+	−	+			−								+
Assim 4	A_4B		+	+	−	+				−							+
Growth	GB		+	+	−	+										+	−
Dissip	DB		+	+	−	+											−
Carbon	C		1				1	1	1	1	1	1	1	1	1	1	1
Hydrogen	H			2		3	1.6	1.6	1.8	1.8	1.8	1.8	1.6	2	1.6	1.6	1.6
Oxygen	O		2	1	2		0.4	0.4	0.5	0.5	0.5	0.5	0.4	1	0.4	0.4	0.4
Nitrogen	N					1	0.1	1	0.2	0.2	0.2	0.2	0.2		0.4	0.2	0.4

Note: The + signs mean appearance; the − signs, disappearance. The signs of the mineral fluxes depend on the chemical indices and parameter values. The labels on rows and columns serve as indices to denote mass fluxes and powers. So J_C denotes the vector of C-fluxes, while $J_{C,GB}$ denotes the C-flux associated with the growth of bacteria. (Note that the table shows J^+, rather than J, if the signs are replaced by quantitative expressions.) The flux J_C^+ adds all positive contributions in J_C, and J_C^- all negative ones, so $J_C^+ + J_C^- = 0$. This applies to all compounds, but not for light. Index $\mathcal{M}$ collects the 4 minerals, $\mathcal{O}$ the 11 organic compounds; $J_{\mathcal{M},GB}$ denotes the 4 minerals fluxes that are associated with bacterial growth, $J_{\mathcal{O},GB}$ does the same for the 11 organic fluxes; $n_{\mathcal{M}}$ collects the 4 × 4 chemical indices for minerals, $n_{\mathcal{O}}$ is the 4 × 11 matrix of chemical indices for the organic compounds. Index D refers to individual daphnids; indices A and B refer to individual algae and bacteria, as well as to their populations.

of structural mass (V) of daphnids (D) that is involved in growth (G) of daphnids (D). If there is more than one compound or flux of the same type, they are simply numbered; E_1A is the first reserve of algae, A_2D is the second assimilation process for *Daphnia*. The sign of the flux indicates appearance ($+$) or disappearance ($-$), and can change in time. Yield coefficients y have two two-character indices that refer to compounds; $y_{PA,VA}$ stands for the molar mass of algal product (feces), PA, that is produced per mole of algal structural mass, VA, by daphnids. These yield coefficients are constants. Chemical indices n have two indices as well; the first one specifies the chemical element, the second two-character index the compounds; $n_{N,VA}$ stand for the number of nitrogen atoms, per carbon atom, in algal structural mass. These chemical indices are constants.

Table 3.4 gives some useful volume-mass conversions. Volumes (or lengths) are basic to the DEB model because of the assumption that uptake is proportional to surface area; molar masses are essential for mass balance equations.

3.4 THE SU-EXTENDED MONOD MODEL

Table 3.6 gives the simplest possible characterization of the compounds and transformations in the DAB system. These are quantified in Table 3.7, as an introduction to the more extended discussion in the next section. The simple model results as a special case of the specification by the DEB theory, if the maintenance costs are small, the reserve turnover rates high, the reserve capacities low, while all living components can be conceived as 1D-isomorphs. The latter assumption is a realistic approximation for unicellular consumers that propagate by division, such as ciliates. The resulting model, the Monod model,[38] boils down to the simple assumptions that food uptake is a hyperbolic function of food density and proportional to body mass, and growth is proportional to food uptake. The chemical composition of body mass is implicitly assumed to be constant, and the conversion from food into body mass and feces is assumed to be fixed. Since this model basically handles just one type of food, we need to extend it to allow feeding on more than one type of food.

To apply the Monod model, we omit the four reserves, which leaves us with an eight-dimensional system: two types of minerals, three types of detritus, and three populations. We also have to include growth into the process of assimilation, and assume a constant hazard rate for consumers. Oxygen and water are not included, because they are assumed to be nonlimiting. They can be easily included, however, on the basis of the conservation law for elements. The prime in $y'_{VB,PB}$ in Table 3.7 just indicates that, generally, $y'_{VB,PB} \neq y'_{PB,VB}$, a notational rule that otherwise would apply; one conversion relates to the digestion process of daphnids, the other to the assimilation by bacteria. Figure 3.2 illustrates how the steady states of masses depend on the environmental parameters total carbon, total nitrogen, and light.

The Monod model cannot have nontrivial steady states for $y^{-1}_{PB,VB} \leq y'_{VB,PB}$. This can be seen from the system in Table 3.7, when we set $(d/dt)M_{PB} = (d/dt)M_{VB} = 0$, and try to solve the equilibrium explicitly. Such an exercise reveals that no positive solution can exist for the three detritus species, which implies cyclic nontrivial attractors (if any). The same holds for the equivalent Lotka-Volterra model, where the hyperbolic functional responses are replaced by linear ones. It is not really likely, however, that $y^{-1}_{PB,VB} \leq y'_{VB,PB}$, because it would imply that daphnids can extract more energy from the conversion of bacterial structural mass to fecal product, than bacteria invest to reverse the reaction.

The details of the model, such as the concept of the Synthesizing Unit, are discussed in the next section, but we shall first mention degenerated models that allow mathematical analysis.

3.4.1 DEGENERATED SIMPLIFICATIONS

A number of authors[61,67] have studied a degenerate closed system with a single nutrient, say nitrogen, one producer and one consumer, where decomposition is instantaneous, the amount of decomposers negligibly small, and functional responses linear, as in the Lotka-Volterra model. It can be obtained from the hyperbolic functional response by increasing the maximum intake, as well as the saturation

TABLE 3.6

A Simplified Table of Compounds and Transformations Results from Table 3.5, If We Apply the Monod Model, and So Omit Maintenance and Reserves and Merge Assimilation and Growth

		Light L	Carbon Dioxide C	Water H	Oxygen O	Ammonia N	Alga-Feces PA	Bact-Feces PB	Dead Daphnids PV	Daphnids VD	Algae VA	Bacteria VB
Daph assim 1	AD		+	+	−	+	+			+	−	−
Daph assim 2	AD		+	+	−	+		+		+		−
Daph death	HD								+	−		
Alga assim	AA	−	−	−	+	−	−				+	
Bact assim 1	AB		+	+	−	+	−					+
Bact assim 2	AB		+	+	−	+		−				+
Bact assim 3	AB		+	+	−	+			−			+
Carbon	C		1				1	1	1	1	1	1
Hydrogen	H			2		3	1.6	1.6	1.8	1.8	1.8	1.6
Oxygen	O		2	1	2		0.4	0.4	0.5	0.5	0.5	0.4
Nitrogen	N					1	0.1	0.1	0.2	0.2	0.2	0.2

TABLE 3.7
The SU-Extended Monod Model for Daphnids (VD), Algae (VA), and Bacteria (VB) in a Confined Environment (no maintenance or reserves, all three components conceived as 1D-isomorphs)

$$\dot{J}_{VA.A_1D} = -M_{VD}\dot{j}_{VA.AD.m}\frac{x_A}{1 + x_A + x_B}$$

$$\dot{J}_{PA.A_1D} = -y_{PA.VA}\dot{J}_{VA.A_1D}$$

$$\dot{J}_{VB.A_2D} = -M_{VD}\dot{j}_{VB.AD.m}\frac{x_B}{1 + x_A + x_B}$$

$$\dot{J}_{PB.A_2D} = -y_{PB.VB}\dot{J}_{VB.A_2D}$$

$$\dot{J}_{VD.GD} = -\dot{J}_{VA.A_1D}y_{VD.VA} - \dot{J}_{VB.A_2D}y_{VD.VB}$$

$$\dot{J}_{PV.HD} = M_{VD}\dot{h}$$

$$\dot{J}_{VD.HD} = -M_{VD}\dot{h}$$

$$\dot{J}_{VA.GA} = M_{VD}\dot{j}_{VA.GA.m}f_A \quad \text{with for } * \in \{L, N, C\}$$

$$f_A = \left[1 + \sum_* x_*^{-1} - \left(\sum_{*\neq L} x_*\right)^{-1} - \left(\sum_{*\neq N} x_*\right)^{-1} - \left(\sum_{*\neq C} x_*\right)^{-1} + \left(\sum_* x_*\right)^{-1}\right]^{-1}$$

$$\dot{J}_{PA.A_1B} = -M_{VB}\dot{j}_{PA.AB.m}\frac{x_{PA}}{1 + x_{PA} + x_{PB} + x_{PV}}$$

$$\dot{J}_{PB.A_2B} = -M_{VB}\dot{j}_{PB.AB.m}\frac{x_{PB}}{1 + x_{PA} + x_{PB} + x_{PV}}$$

$$\dot{J}_{PV.A_3B} = -M_{VB}\dot{j}_{PV.AB.m}\frac{x_{PV}}{1 + x_{PA} + x_{PB} + x_{PV}}$$

$$\dot{J}_{VB.GB} = -\dot{J}_{PA.A_1B}y_{VB.PA} - \dot{J}_{PB.A_2B}y_{VB.PB} - \dot{J}_{PV.A_3B}y_{VB.PV}$$

$$\frac{d}{dt}M_{PA} = \dot{J}_{PA} = \dot{J}_{PA.A_1D} + \dot{J}_{PA.A_1B}$$

$$\frac{d}{dt}M_{PB} = \dot{J}_{PB} = \dot{J}_{PB.A_2D} + \dot{J}_{PB.A_2B}$$

$$\frac{d}{dt}M_{PV} = \dot{J}_{PV} = \dot{J}_{PV.HD} + \dot{J}_{PV.A_3B}$$

$$\frac{d}{dt}M_{VD} = \dot{J}_{VD} = \dot{J}_{VD.GD} + \dot{J}_{VD.HD}$$

$$\frac{d}{dt}M_{VA} = \dot{J}_{VA} = \dot{J}_{VA.GA} + \dot{J}_{VA.A_1D}$$

$$\frac{d}{dt}M_{VB} = \dot{J}_{VB} = \dot{J}_{VB.A_2D} + \dot{J}_{VB.GB}$$

$$\frac{d}{dt}M_C = \dot{J}_C = -\dot{J}_{PA} - \dot{J}_{PB} - \dot{J}_{PV} - \dot{J}_{VD} - \dot{J}_{VA} - \dot{J}_{VB}$$

$$\frac{d}{dt}M_N = \dot{J}_N = -0.1\dot{J}_{PA} - 0.1\dot{J}_{PB} - 0.2\dot{J}_{PV} - 0.2\dot{J}_{VD} - 0.2\dot{J}_{VA} - 0.2\dot{J}_{VB}$$

Note: Detritus includes alga-feces (PA), bacterium-feces (PB), and dead daphnids (PV). Carbon dioxide (C) and ammonia (N) are obtained from the balance equation for carbon and nitrogen. The variables x refer to the scaled mass densities: $x_{PA} = M_{PA}/X_{K.PA}$, $x_{PB} = M_{PB}/X_{K.PB}$, $x_{PV} = M_{PV}/X_{K.PV}$, $x_A = M_{VA}/X_{K.VA}$, $x_B = M_{VB}/X_{K.VB}$, $x_L = \dot{J}_L/\dot{J}_{K.L}$, $x_N = M_N/X_{K.N}$, $x_C = M_C/X_{K.C}$, where $\dot{J}_L$ is the light flux that is supplied to the system to keep it going.

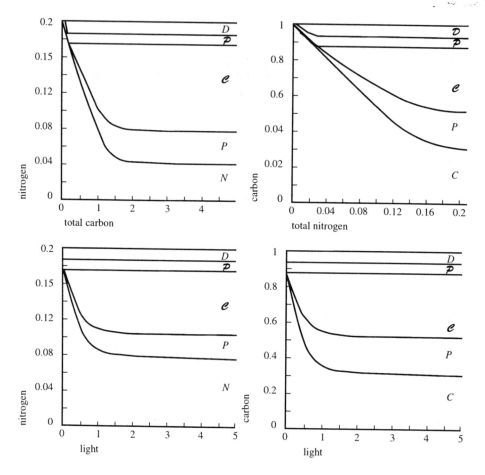

FIGURE 3.2 The steady-state distribution of carbon and nitrogen in the DAB community while increasing the total amount of carbon (upper left), nitrogen (upper right), or light (lower panels), using the extended Monod model. The nonchanged amounts are 1 unit for carbon, 0.2 units for nitrogen, and 5 for light. The amounts of carbon and nitrogen are plotted cumulatively, from bottom to top, across the minerals (carbon dioxide, C, or ammonia, N), detritus (P, three types), daphnids (i.e., consumers $\mathcal{C}$), algae (i.e., producers $\mathcal{P}$), and bacteria (i.e., decomposers $\mathcal{D}$). Parameters: $j_{VA,AD,m} = 1.25$, $j_{VB,AD,m} = 1.25$, $X_{K,VA} = 2$, $X_{K,VB} = 5$, $j_{VA,GA,m} = 1.8$, $X_{K,L2} = 5$, $X_{K,C2} = 2$, $X_{K,N2} = 0.2$, $j_{PA,AB,m} = 3$, $j_{PB,AB,m} = 3$, $j_{PV,AB,m} = 2$, $X_{K,PA} = 2$, $X_{K,PB} = 2$, $X_{K,PV} = 2$, $h = 0.01$, $y_{VD,VA} = 0.2$, $y_{VD,VB} = 0.2$, $y_{PA,VA} = 0.8$, $y_{PB,VB} = 0.8$, $y_{VB,PA} = 0.35$, $y'_{VB,PB} = 0.3$, $y_{VB,PV} = 0.35$.

constant, such that their ratio remains constant. Rewritten in the present notation, that model reduces to

$$\frac{d}{dt} M_{VA} = (j_{VA,GAm}M_N / X_{K,N} - j_{VA,AD,m}M_{VD} / X_{K,VA})M_{VA}$$

$$\frac{d}{dt} M_{VD} = (y_{VD,VA}j_{VA,AD,m}M_{VA} / X_{K,VA} - h)M_{VD}$$

$$\frac{d}{dt} M_N = -n_{N,VA} \frac{d}{dt} M_{VA} - n_{N,VD} \frac{d}{dt} M_{VD}$$

The model is degenerated because it neglects bacteria in the mass balance, while it assumes that their action, decomposition, is infinitely fast. The equilibrium of this system can easily be calculated, for any given (constant) amount of total nitrogen in the system. The Jacobian of the system always

TABLE 3.8
The DEB Model for the Daphnids (VD), Algae (VA), and Bacteria (VB)
System in a Confined Environment, When All Three Components Are
Conceived as 1D-Isomorphs

$$\dot{J}_{ED.A_iD} = -\sum_* y_{ED.*}\dot{J}_{*.A_iD} \quad \text{for} \quad (i,*) \in \{(1,VA),(1,E_1A),(1,E_2A),(2,VB),(2,EB)\}$$

$$\dot{J}_{E_iA.A_iD} = m_{E_iA}\dot{J}_{VA.A_iD} \quad \text{for} \quad i \in \{1,2\}; \quad \dot{J}_{EB.A2D} = m_{EB}\dot{J}_{VB.A_2D}$$

$$\dot{J}_{VD.GD} = M_{VD}\frac{m_{ED}\dot{k}_{ED} - \dot{J}_{ED.MD}}{m_{ED} + y_{ED.VD}}; \quad \dot{J}_{ED.DD} = -\dot{J}_{ED.MD}M_{VD}$$

$$\dot{J}_{PV.HD} = \dot{h}_aM_{VD}\frac{y_{VD.ED}m_{ED}}{1 + y_{VD.ED}m_{ED}}; \quad \dot{j}_{VD.HD} = -\dot{J}_{PV.HD}$$

$$\dot{J}_{PE.HD} = m_{ED}\dot{J}_{PV.HD}; \quad \dot{j}_{ED.HD} = -\dot{J}_{PE.HD}$$

$$\dot{J}_{E_1A.A_1A} = M_{VA}\dot{j}_{E_1A.AA.m}f_{A_1} \quad \text{with} \quad f_{A_1} = \left[1 + \sum_* x_*^{-1} - \left(\sum_* x_*\right)^{-1}\right]^{-1} \quad \text{for } * \in \{L,C\}$$

$$\dot{J}_{E_2A.A_2A} = M_{VA}\dot{j}_{E_2A.AA.m}f_{A_2} \quad \text{with for } * \in \{L,N,C\}$$

$$f_{A_2} = \left[1 + \sum_* x_*^{-1} - \left(\sum_{*\neq L} x_*\right)^{-1} - \left(\sum_{*\neq N} x_*\right)^{-1} - \left(\sum_{*\neq C} x_*\right)^{-1} + \left(\sum_* x_*\right)^{-1}\right]^{-1}$$

$$\dot{J}_{VA.GA} = \dot{r}_{VA.GA}M_{VA} \quad \text{with} \quad \dot{r}_{VA.GA} = \left[\sum_i \dot{r}_{E_i}^{-1} - \left(\sum_i \dot{r}_{E_i}\right)^{-1}\right]^{-1} \quad \text{and}$$

$$\dot{r}_{E_i} = \frac{m_{E_iA}(\dot{k}_{E_iA} - \dot{r}_{VA.GA}) - \dot{j}_{E_iA.MA}}{y_{E_iA.VA}}$$

$$\dot{J}_{E_iA.D_iA} = -\dot{j}_{E_iA.MA}M_{VA} - (1 - \kappa_{R_i})((\dot{k}_{E_iA} - \dot{j}_{VA.GA})M_{E_iA} - (\dot{j}_{E_iA.MA} + \dot{j}_{VA.GA}y_{E_iA.VA})M_{VA})$$

$$\dot{J}_{E_iA.GA} = -y_{E_iA.VA}\dot{j}_{VA.GA} \quad \text{for } i \in \{1,2\}$$

$$\dot{J}_{*.A_iB} = -M_{VB}\dot{j}_{*.AB.m}\frac{x_*}{1 + x_{PA} + x_{PB} + x_{PV} + x_{PE}}$$

$$\dot{J}_{EB.A_iB} = -\dot{J}_{*.A_iB}y_{EB.*} \quad \text{for } (i,*) \in \{(1,PA),(2,PB),(3,PV),(4,PE)\}$$

$$\dot{J}_{VB.GB} = M_{VB}\frac{m_{EB}\dot{k}_{EB} - \dot{J}_{EB.MB}}{m_{EB} + y_{EB.VB}}; \quad \dot{j}_{EB.DB} = -\dot{J}_{EB.MB}M_{VB}$$

$$\frac{d}{dt}M_{PE} = \dot{J}_{PE} = \dot{J}_{PE.HD} + \dot{J}_{PE.A_4B}$$

$$\frac{d}{dt}M_{ED} = \dot{J}_{ED} = \dot{J}_{ED.GD} + \dot{J}_{ED.HD} + \dot{J}_{ED.DC} + \dot{J}_{ED.A_1D} + \dot{J}_{ED.A_2D}$$

$$\frac{d}{dt}M_{E_iA} = \dot{J}_{E_iA} = \dot{J}_{E_iA.A_1D} + \dot{J}_{E_iA.A_iA} + \dot{J}_{E_iA.GA} + \dot{J}_{E_iA.D_iA} \quad \text{for } i \in \{1,2\}$$

$$\frac{d}{dt}M_{EB} = \dot{J}_{EB} = \dot{J}_{EB.A_2C} + \dot{J}_{EB.A_1B} + \dot{J}_{EB.A_2B} + \dot{J}_{EB.A_3B} + \dot{J}_{EB.A_4B} + \dot{J}_{EB.GB} + \dot{J}_{EB.DB}$$

$$\frac{d}{dt}M_C = \dot{J}_C = -\dot{J}_{PA} - \dot{J}_{PB} - \dot{J}_{PV} - \dot{J}_{PE} - \dot{J}_{VD} - \dot{J}_{ED} - \dot{J}_{VA} - \dot{J}_{E_1A} - \dot{J}_{E_2A} - \dot{J}_{VB} - \dot{J}_{EB}$$

$$\frac{d}{dt}M_N = \dot{J}_N = -0.1\dot{J}_{PA} - 0.1\dot{J}_{PB} - 0.2\dot{J}_{PV} - 0.2\dot{J}_{PE} - 0.2\dot{J}_{VD} - 0.2\dot{J}_{ED} +$$

$$-0.2\dot{J}_{VA} - 0.4\dot{J}_{E_2A} - 0.2\dot{J}_{VB} - 0.4\dot{J}_{EB}$$

Note: Only the equations that replace or supplement the ones in Table 3.7 are given. The additional scaled mass densities: $x_{PE} = M_{PE}/X_{K.PE}$, $x_{Li} = \dot{J}_L/\dot{J}_{K.Li}$, $x_N = M_N/X_{K.N2}$, $x_{Ci} = M_C/X_{K.Ci}$.

has one eigenvalue 0 implying that the equilibrium is locally stable. As a general rule, simple prey–predator systems tend to oscillate and systems with additional degrees of freedom may exhibit a wide repertoire of exotic dynamics. The analysis of this simple model suggests that the explicit inclusion of nutrient recycling might have a stabilizing effect in more complex models. This point is explored in more detail by Gurney and Nisbet (Reference 24, pages 195–200).

3.5 THE DEB MODEL

The assumptions on which the DEB theory is based are listed in Table 3.11. This theory aspires to apply to all organisms, but the selection of assumptions listed here does not cover *inter alia* endotherms, fetal development (see Reference 41 for them), and plants (which require roots as well as shoots as types of structural biomass V, see Reference 41).

Tables 3.9 and 3.10 show that the simplest version of the DEB model can be summarized in compact form, on the basis of relationships between energy fluxes and volumes (or lengths). We here need to include mineral fluxes, and introduce the DEB model stepwise on the basis of molar fluxes. To this end, we will specify all organic (i.e., nonmineral) fluxes and the mineral fluxes involved in algal assimilation as consequences of these assumptions, while the remaining mineral fluxes and dissipating heat will follow from the mass and energy conservation laws. We discuss the fluxes as if the densities of all compounds in the bottle are known, and later show how they can be obtained from mass balance considerations.

TABLE 3.9
The Energy Fluxes as Specified by the DEB Model for the Consumer of Scaled Length l and Scaled Reserve Density e at Scaled Functional Response f, Where X Denotes the Food Density

power $\overline{\mu_E M_{Em} k_{Mg}}$	Embryo $0 < l \le l_b$	Juvenile $l_b < l \le l_p$	Adult $l_p < l < 1$
Assimilation, $\dot{p}_A$	0	fl^2	fl^2
Catabolic, $\dot{p}_C$	$el^2 \dfrac{g+l}{g+e}$	$el^2 \dfrac{g+l}{g+e}$	$el^2 \dfrac{g+l}{g+e}$
Somatic maintenance $\dot{p}M_s$	κl^3	κl^3	κl^3
Maturity maintenance $\dot{p}M_m$	$(1-\kappa)l^3$	$(1-\kappa)l^3$	$(1-\kappa)l_p^3$
Somatic growth, $\dot{p}_G$	$\kappa l^2 \dfrac{e-l}{1+e/g}$	$\kappa l^2 \dfrac{e-l}{1+e/g}$	$\kappa l^2 \dfrac{e-l}{1+e/g}$
Maturity growth, $\dot{p}G_m$	$(1-\kappa)l^2 \dfrac{e-l}{e/g+1}$	$(1-\kappa)l^2 \dfrac{e-l}{1+e/g}$	0
Reproduction, $\dot{p}_R$	0	0	$(1-\kappa)\left(l^2 \dfrac{e-l}{1+e/g} + l^3 - l_p^3\right)$

Note: The powers $\dot{p}X = \dot{J}_X \mu_X$ and $\dot{p}p = \dot{J}_p \mu_p$ for ingestion and defecation occur in the environment, not in the individual. The DEB model assumes that $\dot{J}_X \propto \dot{J}_P \propto \dot{p}_A$. The table gives scaled powers of an ectotherm, where μ_E denotes the chemical potential of the reserves. Parameters: g investment ratio, $\dot{k}_M$ maintenance rate coefficient, κ partitioning parameter for catabolic power. Dissipating power amounts to $\dot{p}_D = \dot{p}_{M_m} + \dot{p}_{M_s} + \dot{p}_{G_m} + (1-\kappa_R)\dot{p}_R$. The implied dynamics for $e > l > l_b$: $\dfrac{d}{dt} e = \dfrac{f-e}{l} \dot{k}_M g$ and $\dfrac{d}{dt} l = \dfrac{e-l}{e/g+1} \dfrac{\dot{k}_M}{3}$.

TABLE 3.10

The Energy Fluxes as Specified by the DEB Model for a Decomposer of Scaled Length l and Scaled Reserve Density e at Scaled Functional Response f

power $\dfrac{}{\mu_E M_{Em} k_M g}$	Juvenile $2^{-1/3} l_d < l \leq l_d$
Assimilation, $\dot{p}_A$	$l^3 f / l_d$
Dissipating, $\dot{p}_D$	l^3
Somatic growth, $\dot{p}_G$	$l^3 \dfrac{e/l_d - 1}{e/g + 1}$

Note: We take $\dot{p}_R = 0$, so that $\dot{p}_D = \dot{p}M_s + \dot{p}M_m + \dot{p}G_m$. An individual of structural volume $V \equiv M_V / [M_V]$ takes up substrate at rate $[J_{Xm}] f V$. The implied dynamics for e and l: $\dfrac{d}{dt} e = \dfrac{f - e}{l_d} \dot{k} Mg$ and $\dfrac{d}{dt} l = \dfrac{\dot{k}_M l}{3} \dfrac{e/l_d - 1}{e/g + 1}$.

The DEB model is built on two state variables:

- Structural biomass, quantified as volume V (maximum volume V_m), mass M_V (maximum mass M_{Vm}), or scaled length $l \equiv (V/V_m)^{1/3}$ (maximum scaled length 1),
- Reserves, quantified as energy density $[E]$ (maximum energy density $[E_m]$), mass M_E (maximum mass M_{Em}), or scaled reserve density $e \equiv [E]/[E_m]$.

Diagrams for mass fluxes through the three types of organisms in the DAB community are presented in Figure 3.3; Table 3.9 gives the basic powers for each individual, which can be classified into three groups for each reserve that organize all mass fluxes:

$$\dot{p} \equiv \begin{pmatrix} \dot{p}_A \\ \dot{p}_D \\ \dot{p}_G \end{pmatrix} = \begin{matrix} \text{assimilation power (coupled to food intake)} \\ \text{dissipating powers (no net synthesis)} \\ \text{anabolic power (somatic growth)} \end{matrix}$$

Most of the maintenance power, and part of the growth and assimilation power will end up as dissipating heat, because of the overhead costs; growth and assimilation do not occur with 100% efficiency. Dissipating heat can be written as a weighted sum of the three organizing energy fluxes.

3.5.1 FEEDING AND TIME BUDGETS

Uptake of algae and bacteria by *Daphnia* and of the four types of organic compounds by bacteria are sequential processes for substitutable compounds, while the uptake of light, carbon dioxide, and ammonia by algae are parallel processes for complementary compounds. Uptake rates for sequential and parallel processing can be derived with reference to time budgets.

The uptake system (a receptor in the outer membrane of bacteria, or the whole food grabbing/mouth/gut-system of animals) can be in a substrate (food) "binding" stage, or in a substrate processing stage. When it is in the latter stage, it cannot bind other substrate items, which sets an upper limit to the substrate uptake capacity. The system cannot do two things at the same time,

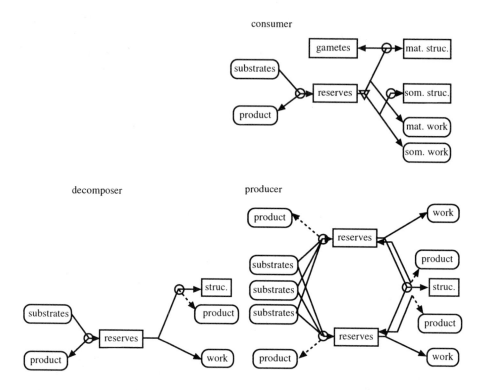

FIGURE 3.3 A diagram of the structure of the DEB model for a consumer, producer, and a decomposer. The boxes represent state variables and the rounded ones sources or sinks. The circles indicate SUs and the triangle indicates allocation. The substrates for the decomposer and the consumer are mixtures of substrates.

which means that we use the conservation law for time at this place. This mechanism directly leads to a hyperbolic functional response for a single substrate, if the substrate "particles" arrive randomly with a certain intensity, and if the length of the processing stage does not depend on the intensity of the substrate arrival process (e.g., this length is constant, or exponentially distributed, or proportional to the size of the substrate "particle" that arrived). Substrate rejection is, therefore, an essential feature of biological uptake systems.

We assume that the handling time is proportional to cell size, which implies that the food densities of algae and bacteria, x_A and x_B, for *Daphnia*, can be taken proportional to their structural biomasses, M_{VA} and M_{VB}, because the volume of the bottle remains constant. All other choices would involve the cell size structure of algae and bacteria populations in the feeding rates of daphnids.

If uptake of different types of substrate is by the same uptake systems uptake is sequential and directly interferes because this system has some maximum capacity. Each substrate has its own substrate product (reserves) conversion efficiency. (We think of reserves as generalized compounds, i.e., rich mixtures of carbohydrates, lipids, proteins, ribosomal RNA, etc. that do not change in chemical composition.) If each type of substrate has its own uptake system, and uptake is parallel rather than sequential, such as nutrients in algae, the interaction of the uptake rates is more complex if a single generalized compound has to be produced, due to stoichiometric constraints.

The Synthesizing Unit (SU) is a simple concept[43] that can be conceived as an enzyme or set of enzymes (or more abstract generalizations thereof), that follows classical association/dissociation kinetics, except that the dissociation rates of the substrate-SU complex are taken to be small in comparison with the dissociation of the product-SU complex. The maximum production rate is then set by the product-SU dissociation rate. This leads to very simple quantitative expressions for the production rate of the SU. While the full description of the association/dissociation kinetics of an enzyme that binds one molecule of A and one molecule of B to produce one molecule of C requires nine parameters, this SU only requires three: two saturation constants and one maximum production rate. For a single substrate, the SU follows the Michaelis Menten kinetics and uptake is a hyperbolic function of substrate concentration (Holling type II functional response).

The saturation constant (for nutrients) combines the process of transportation and acceptance (i.e., binding probability, if the SU is in the binding stage). Transportation is required to bring molecules in the environment to the SU, and converts a concentration to an arrival flux, which is frequently taken to be proportional to concentration. The conceptual separation of the transportation from the acceptance process is essential to combine nutrient and light inputs, which must be in comparable units (arrival frequency of "particles": molecules and photons). The saturation constant for light reflects acceptance only, since the process of transportation is already included in the nature of light. Light can be quantified in different ways, all relating to the number of photons that pass a unit of surface area that is perpendicular to the direction of the photons (assuming parallel light). Differences in light measures relate to differences in the way the energy contents of the photons are weighted. Solar radiation peaks between wavelength of 400 and 700 nm. Photons with wavelength between 550 and 720 nm can be used for photosynthesis. When these photons are coming in, part of them is reflected, and most of them is absorbed by compounds that cannot extract useful work from light. Ignoring photochemical reactions other than those involving photo pigments, this light energy only affects temperature. A (small) part of the photons that can be used by the photo pigments actually reaches the pigments. We here assume that this part represents a fixed fraction of incoming light per unit of structural algal mass, when changing the light intensity; the relative frequency distribution of wavelengths among photons is thus taken to be independent of the intensity. We here quantify light as number of incoming photons times this fraction.

Note the fundamental difference between substitutable (as for bacteria and daphnids) and complementary (as for algae) substrates. If substrates are substitutable, $f \to 1$ if one of the substrates becomes abundant, which does not hold for complementary substrates, due to stoichiometric constraints on product (reserve) formation. At high light, $f_{A_1} \to (1 + x_{C_1}^{-1})^{-1}$, for instance; f_{A_1} only approaches 1 if both light and carbon dioxide are abundant. Tilman[81] gives a classification of substrates.

Substrate (food) uptake is proportional to surface area, according to the DEB model, which means that it is proportional to squared length for organisms that do not change in shape during growth. Such organisms are called 3D-isomorphs. If organisms do change in shape during growth, surface area can be a complex function of volume. If it is proportional to volume, the organism is called a 1D-isomorph. Since maintenance is also assumed to be proportional to volume, size structure is irrelevant to much of the dynamics, and we arrive at an attractive simplicity, where the population behavior of a few large individuals is identical to that of many small ones, if the total biovolumes are equal.[60] It can be argued that size structure is of minor importance for organisms that reset their volume when it increased by a factor two since birth, which is the reason that we assume 1D-isomorphism for bacteria and algae. Detailed change in shape of growing bacteria resembles a mixture of a 1D- and a 0D-isomorph (surface area does not change during growth of a 0D-isomorph, where increase in maintenance-requiring structural biomass is at the expense of internal vacuoles, see Reference 41), while algae are approximately 3D (green algae) or 0D (diatoms[51]) isomorphs. We assume 3D-isomorphism for *Daphnia* only, because the ratio of the maximum length and the length at birth is about a factor 6, so the ratio of the corresponding volumes is about $6^3 = 216$, which would deviate too much from 1D-isomorphism.

The feeding process has, in an ecosystem setting, two aspects: the disappearance of substrate, and the increase of reserves. These processes are quantified in Tables 3.7, 3.8, and 3.12, and will be briefly discussed in the following sections.

TABLE 3.11
The Assumptions, in Their Simplest Form, That Lead to the DEB Model

General

1. Structural body mass and one or more reserves are the state variables of the individual and they do not change in compositon (strong homeostasis).
2. Substrate (food) is converted into feces and these compounds do not change in composition at steady state (just convenience, not essential).
3. Assimilates derived from food are added to reserves, which fuel all other metabolic processes: synthesis of structural body mass, of (embryonic) reserves (i.e., reproduction), and processes that are not associated with synthesis of biomass.
4. If the individual propogates via reproduction (rather than via division), it starts in embryonic stage that initially has a negligibly small structural body mass (but a substantial amount of reserves).

Specific

4a. The reserve density of the hatchling equals that of the mother at egg formation.
5. The transition from embryo to juvenile initiates feeding, and from juvenile to adult initiates reproduction, which is coupled to ceasing maturation. The transitions occur when the cumulated energy invested in maturation exceeds a threshold value. Unicellulars divide a fixed time after initiation of DNA duplication, which occurs when the cumulated energy invested in maturation exceeds a threshold value.
6. Somatic and maturity maintenance are proportional to structural volume, but maturity maintenance does not increase after a given cumulated investment in maturation.
7. The feeding rate is proportional to surface area and depends on food density according to the rules for synthesizing units.
8. The reserves must be partitionable, such that the dynamics is not affected and the energy density at steady state does not depend on structural body mass (weak homeostasis).
9. A fixed fraction of energy, utilized from the reserves, is spent on somatic maintenance plus growth, the rest on maturity maintenance plus maturation or reproduction (the κ-rule).
9a. If more than one reserve contributes to growth, the growth rate follows the rule of a fast synthesizing Unit, while fixed fractions of the rejected reserves fluxed are returned to the reserves from which the fluxed are mobilized, the remainder being excreted.
10. Under starvation conditions, individuals always give priority to somatic maintenance and follow one of two possible strategies:
    They do not change the reserve dynamics (so continue to reproduce).
    They cease energy investment in reproduction and maturity maintenance (thus changing reserve dynamics).
    Most unicellulars and some animals shrink during starvation.
11. Death by aging follows a hazard rate that is proportional to the cumulated concentration of modified proteins that accumulate at a rate proportional to the amount of changed DNA; the increase in changed DNA is proportional to the catabolic rate.

3.5.1.1 Daphnia

Embryonic *Daphnia* does not feed; it maintains, grows, and develops at the expense of reserves. The scaled functional response for juvenile and adult *Daphnia* amounts to

$$f_D = (1 + (x_A(1 + \varepsilon_{E_1A}m_{E_1} + \varepsilon_{E_2A}m_{E_2A}) + x_B(1 + \varepsilon_{EB}m_{EB}))^{-1})^{-1} \qquad (1)$$

where $x^* = X^*/X_{K,*}$, for $* \in \{A, B\}$, are the scaled densities of the structural body mass of algae and bacteria, and $X_{K,*}$ the corresponding saturation constants, so $X_* = M_*/V_{env}$, for a bottle of volume V_{env}. The parameters ε_* quantify the extension of the prey-handling time by the reserves of the prey,

TABLE 3.12
The Specification of Fluxes When the Consumer Is Considered as a 3D-Isomorph, Rather than a 1D One

$$j_{v*}A_iD = \{j_{v*}A_iD\}V^{2/3} \quad \text{for}*\varepsilon \in \{A,B\}$$

$$\frac{d}{dt}M_{ED} = j_{ED.AD} - j_{ED.CD} = j_{ED.AD} - (\dot{v}V^{-1/3} - j_{VD.GD})M_{ED}$$

$$\frac{d}{dt}M_{VD} = J_{VD.GD} = \frac{j_{ED.GD}}{y_{ED.VD}} = \frac{\kappa j_{ED.CD} - j_{ED.MD}}{y_{ED.VD}}$$

$$j_{VD.GD} = \frac{\kappa(\dot{v}V^{-1/3} - j_{VD.GD})m_{ED} - j_{ED.MD}}{y_{ED.VD}} = \frac{\kappa m_{ED}\dot{v}^{-1/3} - j_{ED.MD}}{\kappa m_{ED} + y_{ED.VD}}$$

$$j_{ED.CD} = m_{ED}\frac{y_{ED.VD}\dot{v}V^{-1/3} + j_{ED.MD}}{\kappa m_{ED} + y_{ED.VD}}$$

$$J_{ED.DD} = J_{ED.MD} + (1-\kappa)J_{ED.CD} \quad \text{for embryos and juveniles}$$

$$J_{ED.DD} = J_{ED.MD} + (1-\kappa_R)J_{ED.RD} + \frac{1-\kappa}{\kappa}J_{ED.MD}V_p[M_{VD}] \quad \text{for adults}$$

$$J_{ED.RD} = (1-\kappa)J_{ED.CD} - \frac{1-\kappa}{\kappa}J_{ED.MD}V_p[M_{VD}]$$

$$\dot{h}(a) = \ddot{h}_a V(a)^{-1}\int_{a_b}^a \left(V(a_1) - V_b + \dot{k}_{MD}\int_{a_b}^{a_1}V(a_2)da_2\right)da_1$$

$$\overset{a > a_x}{\simeq} \dot{h}(a_x) + \ddot{h}_a(1 - V_b/V_x + \dot{k}_{MD}(a_x V_a/V_x + 0.5(a - a_x)))(a - a_x)$$

$$j*_{,A+} = N\varepsilon j*_{,AD} \quad \text{for}* \in \{VA,E_1A,E_2A,VB,EB,PA,PB\}$$
$$j_{PV.H+} = N\varepsilon \dot{h}M_{VD} \quad \text{and} \quad J_{PE.H+} = m_{ED}J_{PE.H+}$$
$$J_{ED.D+} = N_e\varepsilon_e J_{ED.DD} + N\varepsilon J_{ED.DD}$$
$$j_{*,G+} = N_e\varepsilon_e j_{*.GD} + N\varepsilon j_{*.GD} \quad \text{for} * \in \{VD,ED\}$$
$$M_{*+} = N_e\varepsilon_e M_{*D} + N\varepsilon M_{*D} \quad \text{for} * \in \{V,E\}$$

Index D now relates to individuals and index $+$ to the population of daphnids. These equations replace the corresponding ones in Table 3.8.

e.g., by elongating the time required for digestion. The quantities m_{E_1A}, m_{E_2A}, and m_{EB} denote the molar densities of the reserves of algae and bacteria.

Since feeding is assumed to be proportional to surface area, and *Daphnia* is a 3D-isomorph, we have that $J_{*.AD.m} = \{J_{*.AD.m}\}V^{2/3}$ for $* \in \{VA, VB\}$, where the surface area-specific maximum ingestion rate $\{J_{*.AD.m}\}$ is treated as a parameter (i.e., a constant), and $V = M_{VD}/[M_{VD}]$ is the structural biovolume, where $[M_{VD}]$ is treated as a parameter. Since the nutritional value of algae and bacteria for *Daphnia* does not need to be the same, *Daphnia* might have a preference for one of them. Such preferences turn up in the values for the saturation constants, which are hidden in the scaled food densities x_A and x_B.

Let $J_{VA.AD.m}$ and $J_{VB.AD.m}$ denote the maximum ingestion rates of algae and bacteria by *Daphnia*, $y_{ED.*}$ the food-reserve coupling, and $y_{P*.V*}$ the food-feces coupling on the basis of C-moles. We give the maximum ingestion rates positive values, which means that the feeding fluxes have negative signs, because food disappears. The feeding rates for *Daphnia* amount to

$$J_{VA,A_iD} = -f_D\frac{x_*J_{V*,AD,m}}{x_A(1 + \varepsilon_{E_1A}m_{E_1A} + \varepsilon_{E_2A}m_{E_2A}) + x_B(1 + \varepsilon_{EB}m_{EB})} \tag{2}$$

for $(i,*) \in \{(1, A), (2, B)\}$. As a first approximation, we might neglect the effect of reserves on the prey-handling time and set $\varepsilon_* = 0$. Note that the ingestion of prey reserves is coupled to the prey's structural body mass via the reserve densities. Although daphnids are feeding on two types of prey, the chemical composition of the prey can change in time and depends on the nutritional condition of the prey. The fecal production is taken to be proportional to the ingestion of structural mass.

3.5.1.2 Alga

As explained in References 43, 51, and 45, the scaled functional responses that are presented in Table 3.8 directly follow from the dynamics of the Synthesizing Unit. It has the property of being numerically close to a minimum model: if a nutrient exceeds others in availability, relative to the needs, it hardly affects assimilation. The concentration ranges where several nutrients limit assimilation simultaneously is rather small.

The maximum assimilation rates $\dot{J}_{E_iA. A_iA. m}$ are proportional to the total structural body mass of the algae, as a consequence of the assumption of 1D-isomorphism, so $\dot{J}_{E_iA. A_iA. m} = \dot{j}_{E_iA. A_iA. m}M_{VA}$, where the mass-specific maximum assimilation rates $\dot{j}_{E_iA. A_iA. m}$ are treated as parameters. Nutrients and light are complementary, rather than substitutable, which urges the specificiation of synthesis of reserves in terms of nutrient densities (and light availability), and evaluates uptake rates of nutrients from synthesis of reserves via stoichiometric couplings. We assume that the assimilation process of nutrients does not involve overhead costs, other than paid from light, which implies that the mass-mass couplers equal the chemical coefficients, so $y_{C. E_1A} = y_{C. E_2A} = 1$, and $y_{N. E_2A} = n_{N. E_2A}$. We conceive $y_{L. E_iA}$ as a parameter, which indirectly specifies the heat loss that is associated with assimilation via the energy difference between light plus nutrients and reserves.

3.5.1.3 Bacterium

The scaled functional response for the bacterium follows from the assumption that substrates are fully substitutable and are converted into reserves. Since mineral fluxes are obtained from mass balance equations, they can become negative, meaning that minerals are taken up from the environment. Depending on the C/N ratio of substrate relative to reserves, ammonia is taken up from or excreted into the environment, for instance. Minerals do not occur in the expression for the assimilation flux, which implies that they must be available *ad libitum*. If this is actually not the case, mineral masses can become negative in the expressions for the change of the system. This unrealistic situation can only occur with unusual combinations of parameter values and mass distributions within the DAB system. If, in any particular system, the assumption of *ad lib* minerals for bacteria is not realistic, model modification is required.

The maximum assimilation rates $\dot{j}_{*. A. B. m}$ are proportional to the total structural body mass of the bacteria, as a consequence of the assumption of 1D-isomorphism, so $\dot{J}_{*. A_iB. m} = \dot{J}_{*. A_iB. m}M_{VB}$, where the mass-specific maximum assimilation rates $\dot{J}_{*. A_iB. m}$ are treated as parameters.

3.5.2 DISSIPATING POWER

Somatic maintenance is used to fuel protein turnover, to maintain concentration gradients of ions and compounds in cells and in the body, and to fuel behavior, movements, and other forms of activity. It is conceived as an energy drain, or a drain of reserves, that eventually is fully mineralized and leaves the body in the form of dissipating heat, carbon dioxide, ammonia, and water, without fueling any net synthesis. Overhead costs in growth and/or reproduction account for processes that directly relate to synthesis. Maintenance has priority over growth, which ceases as soon as the allocation to growth plus maintenance matches the maintenance costs.

Somatic maintenance costs are taken proportional to structural mass, and the specific maintenance costs j_{E*M*} are considered as parameters. The dissipating power for bacteria equals the maintenance power. Algae excrete a fraction $(1 - \kappa_{Ri})$ of the reserves that are rejected by the growth SU, which contribute to dissipating power as well. Daphnids invest in development and have overhead costs for reproduction, which also contribute to dissipating power (see later).

Dissipating power should not be confused with dissipating heat; it stands for the power involved in metabolic processes that are not directly related to processes of synthesis. Dissipating power, assimilation power, and growth power all contribute to dissipating heat. The overhead costs of assimilation and growth contribute to dissipating heat, but not to dissipating power. This is just a matter of bookkeeping and largely related to convenience, rather than being fundamentally different.

3.5.3 GROWTH AND RESERVE KINETICS

The weak homeostasis assumption (8 in Table 3.11) and the partitionability requirement for reserves imply that the reserve density, i.e., the ratio of the reserves and the structural mass follows first-order kinetics, at a rate that is inversely proportional to length for a 3D-isomorph, or constant for a 1D one. The DEB model assumes that a constant fraction of the catabolic flux, i.e., the flux that is released from the reserves, is spent on somatic maintenance plus growth, the rest on maturity maintenance plus the increase in the state of maturity (development) or reproduction (if the state of full maturity is reached). Since both somatic and maturity maintenance are taken proportional to structural mass, the two maintenance fluxes and the two growth fluxes can be taken together in the juvenile stage, where feeding occurs, but not reproduction. This means that for unicellulars (algae, bacteria), allocation can be simplified to the simple rule: maintenance is subtracted from the catabolic flux; the rest is invested into growth.

Since uptake and maintenance are both proportional to structural mass in 1D-isomorphs, the distinction between the individual and the population vanishes. For algae and bacteria M_{V*} can stand for the structural mass of a single cell, as well as for the sum of these structural masses of all cells in the bottle; the specific growth rate equals the population growth rate: $J_{V*, G*} = \dot{r}_{V*, G*}, * \in \{A, B\}$. This does not hold for 3D-isomorphs, where M_{VD} stands for the structural mass on an individual *Daphnia*. We need the frequency distribution of these masses in the populations to evaluate the population performance. This will be done in the section on mass turnover.

3.5.3.1 Daphnia

When daphnids are considered to be 1D-isomorphs, and we do not have to distinguish the individual and population levels, growth is proportional to the difference between the catabolic and maintenance rates, and the catabolic rate follows from the first-order dynamics of the reserve density. The parameter $\dot{k}_{ED}$ stands for the reserve turnover, and $\dot{J}_{ED.EM}$ for the specific maintenance costs (see Table 3.8). 1D-isomorphic growth is commonly an acceptable idealization for dividing organisms, such as ciliates; if substrate density is constant, mass (or volume) increases exponentially in time, but it is reset to the volume at "birth" as soon as it increased by a factor of two. Realism is lost at the level of the individual if it does not divide. The dynamics for the consumers in Table 3.8 is assumed to be realistic at the individual level, e.g., ciliates, rather than daphnids. The main reason for its presentation is to reveal the link of physiologically structured populations with the better-known unstructured ones[60] for conceptual reasons, as well as to study the trade-off between realism and model complexity. It is commonly true that the structured populations hardly differ from the much simpler unstructured ones.[59]

Aging in unicellulars differs from that of multicellulars, according to the DEB theory, because it is a binary process at the cellular level; multicellulars consist of a mixture of cells that are and are not hit by the aging process. Because unaffected cells grow and divide, if energetics allows, the fraction of affected cells can vary in time. For unicellulars, the aging rate works out to be proportional to the oxygen consumption that is associated with the catabolic rate. If the elemental composition of reserves and structural mass are identical, oxygen consumption is proportional to the catabolic rate.

3.5.3.2   Daphnia Individuals

The daphnids are considered to be reproducing multicellular 3D-isomorphs; we also have to distinguish the life embryo stages, juvenile, and adult. The equations in Table 3.12 now apply, but these flux specifications cannot be used for time-integrations; they only specify the steady state on the assumption that the age distribution among individuals in the population is that at steady state.

During the embryonic stage, we set $\dot{J}_{AD,AD} = 0$ as the only difference between the juvenile and adult stages. Juveniles do not differ from adults in reserve kinetics and growth.

The balance equation for reserves is the difference between assimilation and catabolism. The reserve density follows a first-order process with rate $\dot{\nu}V^{-1/3}$, which decreases for increasing size as a consequence of the assumption of weak homeostasis. The parameter $\dot{\nu}$ is an energy conductance, which can be interpreted as the ratio of the maximum surface area-specific assimilation rate and the (maximum) volume-specific reserve density. The expression for maximum assimilation can easily be obtained in the case of feeding on one chemical compound, but we now have to consider five compounds, a complication that will be discussed below.

The investment in growth is a fixed fraction of the catabolic flux minus the somatic maintenance costs, which gives the growth rate and the catabolic flux. The role of $\dot{k}_{ED}$ for 1D-isomorphs is now taken over by $\dot{\nu}V^{-1/3}$, which changes in time during growth.

At steady state we have $d/dt\, M_{ED} = \dot{J}_{VD,GD}\, M_{ED}$, or $\dot{j}_{ED,AD} = \dot{\nu}V^{-1/3}\, M_{ED}$, or

$$m_{ED} = \frac{\{\dot{J}_{ED,AD}\}}{[M_{VD}]\dot{\nu}} \tag{3}$$

In other words, assimilation balances catabolism, and the reserve density does not change during growth at steady state.

The structural mass settles ultimately at $M_{VD\infty} = \kappa\dot{J}_{ED,AD}/\dot{J}_{ED,MD}$, where

$$\dot{J}_{ED,AD} = \{\dot{J}_{ED,AD}\}(M_{V,D\infty}/[M_{VD}])^{2/3}$$

which gives

$$V_\infty^{1/3} = \frac{e\dot{\nu}}{g\dot{k}_M} = \left(\frac{M_{V,D\infty}}{[M_{VD}]}\right)^{1/3} = \frac{m_{ED}\kappa\dot{\nu}}{\dot{J}_{ED,MD}} = \frac{\kappa\{\dot{J}_{ED,AD}\}}{[M_{VD}]\dot{J}_{ED,AD}} = \frac{\kappa\{\dot{J}_{ED,AD}\}}{[\dot{J}_{ED,MD}]} \tag{4}$$

The energy conductance is given by $\dot{\nu} = \kappa g y_{VD,ED}\{\dot{J}_{ED,AD,m}\}[M_{VD}]$, with $\{\dot{J}_{ED,AD,m}\}$ being the maximum value of $\{\dot{J}_{ED,AD}\}$ for all possible combinations of food intake, leading to $\{\dot{J}_{ED,AD,m}\} = \max\{\{\dot{J}_{ED,A_1D,m}\}, \{\dot{J}_{ED,A_2D,m}\}\}$. The maximum assimilation from bacteria occurs when bacteria have maximum reserves, which leads to

$$\{\dot{J}_{ED,A_2D,m}\} = \{\dot{J}_{VB,AD,m}\}\left(y_{ED,VB} + \frac{\dot{J}_{BB,AB,m}}{\dot{k}_{EB}}y_{ED,EB}\right) \tag{5}$$

with $\dot{J}_{EB,\,AB,\,m} = \max_* \{y_{EB},\,\dot{J}_{ED,\,A_iD,\,m}\}$, $(*, i) \in \{(PA, 1), (PB, 2), (PV, 3), (PE, 4)\}$. The maximum assimilation from algae is more complex, because the reserves of the algae are coupled; depending on parameter values, one reserve can be maximal if the other is limiting growth to the extent that growth is ceased. It can be written as

$$\{\dot{J}_{ED,\,A_iD,\,m}\} = \max_{m_{E_i}} = \{\dot{J}_{VA,\,AD,\,m}\}(y_{ED,\,VA} + m_{E_1A}y_{ED,\,E_1A} + m_{E_2A}y_{ED,\,E_2A}) \tag{6}$$

The maximum is likely to be reached in one of three possible combinations of nutrient inputs for the algae: minimal for CO_2 and maximal for NH_3, maximal for CO_2 and minimal for NH_3, and maximal both CO_2 and NH_3. Minimum reserves correspond to the situation where growth has completely ceased, and the full catabolic flux from that reserve is spent on maintenance. The section on reserve kinetics for algae will explain why the minimum reserve density 1 and corresponding reserve density 2 are given by

$$m_{E_1A} = \frac{\dot{J}_{E_1A,\,MA}}{\dot{k}_{E_1A}} \quad \text{and} \quad m_{E_2A} = \frac{\dot{J}_{E_2A,\,A_2A,\,m} - \kappa_{R_2}\dot{J}_{E_2A,\,MA}}{(1 - \kappa_{R_2})\dot{k}_{E_2A}} \tag{7}$$

The role of reserves 1 and 2 can be interchanged, but the reserves at maximum CO_2 and NH_3 input must be obtained numerically. Growth is asymptotic, and the aging process makes sure that Expression (4) sets an upper bound on body volume that no individual can reach.

Apart from somatic maintenance costs, *Daphnia* invests in maturity maintenance, increases its state of maturity, and pays overhead costs for reproduction, which are all processes that do not directly relate to increase in mass, and should be combined with somatic maintenance to arrive at a dissipative flux. We will evaluate the dissipating flux in the section on reproduction.

3.5.3.3 Daphnia Reproduction and Aging

Stage transitions, from embryo to juvenile and from juvenile to adult, occur if investment into maturation exceeds some threshold value. If maturity maintenance equals κ times the somatic maintenance, these transitions occur at fixed amounts of structural body mass, or structural body volume, say, V_b and V_p (the indices refer to birth and puberty, conceived at point events). At other values for the maturity maintenance, the threshold values for structural mass depend on food history, and we have to make use of these relationships to estimate the maturity maintenance costs. We choose here the simple option, because we do not expect that small deviations affect the results.

The investment of reserves into maturity maintenance plus increase in maturity for embryos and juveniles, and into maturity maintenance plus reproduction for adults equals $(1 - \kappa)\dot{J}_{ED.CD}$.

The maturity maintenance costs for adults equals $\frac{1 - \kappa}{\kappa} \dot{J}_{ED,\,MD}V_p[M_{VD}]$, while a fraction $(1 - \kappa_R)$ of the investment into reproduction is dissipated as overhead costs. The total investment into reproduction equals the difference of $(1 - \kappa)$ times the catabolic flux, minus the maturity maintenance costs, while the flux to embryonic reserves equals $\dot{J}_{ED.ED} = \kappa_R\dot{J}_{ED.RD}$.

If the respiration ratio does not depend on the state of the individual, the oxygen flux that is not associated with feeding is proportional to the catabolic flux.[42] If oxygen is converted to free radicals with a fixed efficiency, DNA change is proportional to the oxygen flux, the accumulation rate of transformed proteins is proportional to the amount of changed DNA, and the hazard rate is proportional to the concentration of transformed proteins. The hazard rate equals for $a > a_b$. The aging acceleration $\ddot{h}_a$ and the maintenance rate coefficient $\dot{k}_{MD} = \dot{j}_{ED,\,MD}y_{VD,\,ED}$ are treated as fixed parameters. The approximative expression for the hazard rate in Table 3.12 holds for the situation where growth becomes negligibly small after some age a_x and the volume becomes arrested at V_x with $V_a = a_x^{-1}\int_0^{a_x}V(a)\,da$. If the growth period is short with respect to the life span, and mortality builds up after growth has been ceased, this relationship reduces to $\dot{h}(a) = 0.5\ddot{h}_a\dot{k}_{MD}a^2$, which is the Weibull model with shape paramter 3.

Growth of the embryo is important to evaluate its structural mass and reserves, which needs to be done to construct the full mass balance of the DAB community. As derived in Reference 41, the change in scaled length l and scaled reserve density e is given by

$$\frac{d}{da}e = -\dot{k}_{MD}g\frac{e}{l} \quad \text{and} \quad \frac{d}{da}l = \dot{k}_{MD}\frac{g}{3}\frac{e-l}{e+g} \tag{8}$$

where the conversions of e and l to masses is given in Table 3.4.

At steady state, where food density does not change, the volume at age a since birth can be obtained explicitly:

$$V(a)^{1/3} = V_\infty^{1/3} - (V_\infty^{1/3} - V_b^{1/3})\exp\{-\dot{r}_B a\} \tag{9}$$

where the von Bertalanffy growth rate is given by $\dot{r}_B = (3/\dot{k}_{MD} + 3V_\infty^{1/3}/\dot{v})^{-1}$.

Dilution by growth ensures that transformed proteins hardly build up during the embryonic period, which means that birth initializes the aging process. Derivation and backgrounds of the model for aging are given in Reference 41, together with tests against experimental data, including an evaluation of effects of energetics and mutagenic compounds on aging.

3.5.4 DAPHNIA POPULATION STRUCTURE

We now evaluate the fluxes to and from the population of *Daphnia*, assuming the age and size distributions are close to the stable ones.

At steady state, the easiest approach is to relate the states of the individuals to age. We introduce the relative density $\phi°(a) = \phi(a)/N$, where N denotes the number of juveniles plus adults, and $\phi_e°(a) = \phi_e(a)/N_e$ for embryos. These relative densities no longer depend on time at steady state, so we omit the reference to time. We will write $j_{*_1D.*_2D}(a)$ for the flux of compound $*_1$ linked to transformation $*_2$ in an individual of age a, where a_b is the age at birth and a_p the age at puberty (the transition from juvenile to adult). The DEB model obtains these ages from $V(a_b) = V_b$ and $V(a_p) = V_p$.

The characteristic equation applies at steady state:

$$M_{E0} = \int_{a_p}^{\infty} \exp\left\{-\int_{a_b}^{a} \dot{h}(a_1)\,da_1\right\} j_{ED,RD}(a)\,da \tag{10}$$

where M_{E0} refers to the reserves of an embryo of age 0. Given that reserve density at birth equals that of the mother, it can most easily be obtained from the backward integration of the $d/dt\,e$ and $d/dt\,l$, starting from $e = e_b$ and $l = l_b$. This integration has to be done anyway with to evaluate the expected masses of embryo reserves and structure, and the growth and dissipating fluxes for embryos.

The characteristic equation simply states that each individual is expected to replace itself exactly during its life span. This is only possible if $V_\infty > V_p$, but the difference between V_∞ and V_p should be really small if the life span is much larger than a_p. In practice, this difference is expected to be temporally larger, because of the intrinsic oscillations.

The constraint on the ultimate volume translates into a constraint on the specific assimilation, $\{j_{ED,MD}\} > V_p^{1/3}[M_{VD}]\,j_{ED,MD}/\kappa$, which can be related to a constraint on food densities. We use the characteristic equation to solve the required assimilation rate for the replacement of daphnids; given the assimilation rate, the trajectories of the state variables (reserves and structural mass) are fixed.

Since embryos do not feed (remove algae or bacteria, or produce feces) or die (produce dead biomass), we separate embryos from juveniles and adults. The number of individuals in the two groups are denoted by N_e and N. The embryonic period a_b is of no relevance, at steady state. Embryos are still important, because they represent a mass that plays a role in the mass balance, and they contribute to mineral fluxes. Their role is much more important in transient situations.

The age distributions of embryos and juveniles plus adults are given by

$$\overset{\circ}{\phi}_e(a) = a_b^{-1} \qquad \text{for } a \in [0, a_b] \tag{11}$$

$$\overset{\circ}{\phi}_e(a) = \frac{\exp\left\{-\int_{a_b}^{a} h(a_1)\, da_1\right\}}{\int_{a_b}^{\infty} \exp\left\{-\int_{a_b}^{a_2} h(a_1)\, da_1\right\} da_2} \qquad \text{for } a \in [a_b, \infty] \tag{12}$$

Since $\phi_e(a_b) = \phi(a_b)$, the number of embryos relates to that of juveniles plus adults as $N_e^{-1} N = a_b^{-1} \int_{a_b}^{\infty} \exp\{-\int_{a_b}^{a} h(a_1)\, da_1\}\, da$. The step from individuals to the population is most easily made via the concept of "randomly selected individual," and multiplication by the number of individuals. To this end, we introduce the expectation operators $\mathcal{E}_e$ and $\mathcal{E}$, i.e., $\mathcal{E}_e Z \equiv \int_0^{a_b} Z(a) \overset{\circ}{\phi}_e(a)\, da$ and $\mathcal{E}Z \equiv \int_{a_b}^{\infty} Z(a) \overset{\circ}{\phi}_e(a)\, da$, for any function $Z(a)$ of age.

At steady state, the dead biomass production must balance the consumption by bacteria.

$$j_{VD,H+} = -j_{PV,AB} \qquad \text{and} \qquad j_{ED,H+} = -j_{PE,AB}$$

3.5.5 ALGA

The process of growth in algae is complicated by the fact that two reserve fluxes $J_{E_iA,GA}$ that are allocated to growth must be merged to produce the structural biomass flux $j_{VA,GA}$, which implies the existence of rejected fluxes of reserves due to stoichiometric constraints. The rejected fluxes are partially returned to the originating reserves, and partially mineralized and excreted into the environment.

The growth of structural algal mass follows the kinetics of a fast SU, while the flux allocated to growth equals the difference between the catabolic flux, which is proportional to the reserve density, and the maintenance flux $j_{E_iA,MA}$. The parameters $\dot{k}_{E_iA}$ have the interpretation of the algal reserve turnover rates. The specific maintenance fluxes $j_{E_iA,MA}$ are treated as parameters, and combine both somatic and maturity maintenance. Since the increase in the state of maturity is combined with somatic growth, it leads to an increase of the value for $y_{E_iA,VA}$. The dissipating flux for algae equals the maintenance flux plus the excreted flux of mineralized reserves. The growth equation is implicit, due to the phenomenon of dilution by growth. The numerical evaluation of the specific growth rate from the reserve densities can be done following a Newton-Raphson scheme, starting from value 0. Two iterations already result in a high accuracy. Similar to the process of assimilation, the sink of reserves into growth is quantified via the specification of growth, and the (fixed) coupling between reserves and structural biomass.

At steady state, the reserve densities do not change, so $d/dt\, M_{E,A} = \dot{r}_{VA.\,GA}M_{E,A}$, which implies for $j_{E,A.\,A,A} = j_{E,A.\,A,A}/M_{VA}$

$$j_{E,A,A,A} = \dot{r}_{VA,GA}m_{E,A} + (1 - \kappa_{R_i})(\dot{k}_{E,A} - \dot{r}_{VA,GA})m_{E,A} + \kappa_{R_i}(J_{E,A,MA} + y_{E,A,VA}\dot{r}_{VA,GA}) \quad (13)$$

The removal of reserves by grazing is obviously directly coupled to the removal of structural mass, because cells are grazed, which combine the compounds. At steady state, the growth rate must equal the consumption rate by daphnids: $j_{VA.\,GA} = j_{VA.\,AD}$.

3.5.6   Bacterium

The growth of bacterial structural biomass is proportional to the difference between the catabolic rate and the maintenance costs. The dynamics of the reserve density is a first-order process, and again, the parameter k_{EB} represents the turnover rate of bacterial reserves. The specific maintenance flux $j_{EB.\,MB}$ is treated as a parameter, and the maintenance flux represents the dissipating flux.

At steady state, the reserve densities do not change, so $d/dt\, M_{EB} = \dot{r}_{VB.\,GB}M_{EB}$, which implies for $j_{EB.\,AB} = \Sigma_i j_{E,B.\,AB}/M_{VB}$

$$J_{EB,AB} = \dot{k}_{EB}m_{EB} \quad (14)$$

At steady state, the growth rate must equal the consumption rate by daphnids: $j_{VB.\,GB} = J_{VB.\,GB.\,AD}$, while the sink of bacterial reserves due to grazing is directly coupled to that of structural mass.

3.5.7   Mineral Fluxes and Mass Balance

The mineral fluxes follow from the organic fluxes and the uptake of minerals by algae, as a direct result of the conservation law for chemical elements. Since this law has to apply for each transformation, it is possible to decompose the total mineral fluxes into those involved in each of the transformations. It is even possible to decompose the mineral fluxes into contributions of each individual, but we will confine the analysis to the community level.

Let us partition the scheme matrix $\boldsymbol{J}$ into two submatices that correspond with the minerals and the organic compounds, respectively: $\dot{\boldsymbol{J}}_T = (\dot{\boldsymbol{J}}_M^T : \dot{\boldsymbol{J}}_O^T)$. The mineral fluxes $\boldsymbol{J}_M$ can be linked linearly to the organic fluxes $\boldsymbol{J}_O$, using the chemical indices $\boldsymbol{n}$ (see Table 3.5), by

$$\dot{\boldsymbol{J}}_M = -\boldsymbol{n}_M^{-1}\boldsymbol{n}_O\dot{\boldsymbol{J}}_O \quad (15)$$

The change in mass $\boldsymbol{M}$ can be compactly written as $d/dt\, \boldsymbol{M} = \boldsymbol{j}\boldsymbol{1}$. At steady state, we have by definition $d/dt\, \boldsymbol{M} = \boldsymbol{0}$. As for the organic compounds, all mineral fluxes must balance at steady state, so $0 = j_C^T\boldsymbol{1} = j_H^T\boldsymbol{1} = j_O^T\boldsymbol{1} = j_N^T\boldsymbol{1}$.

The equations that have been discussed above fully specify mass turnover and, indirectly, the structure in the DAB community. One way to obtain the masses at steady state is to evaluate the time trajectories, starting from appropriate initial values, and follow them until they stabilize. The total amounts of the elements do not change in time; only the distribution of the elements among the compounds changes, so this information about the initial condition is conserved.

3.6 MASS TURNOVER

Some interesting physiological quantities follow directly from the composition of reserves relative to structural mass. If reserves and structural mass have identical compositions in terms of element frequencies, it can be shown[42] that the Respiration Quotient (RQ), the Urination Quotient (UQ), and be Watering Quotient (WQ) are constant, i.e., they do not depend on the states of the individual. The RQ is defined as the ratio of the carbon dioxide production and the oxygen consumption, excluding the contributions from feeding and assimilation. The UQ and WQ are defined as similar ratios for ammonia (N-waste) and water production over oxygen consumption. These quotients are given by

$$RQ = \frac{1 - n_{NE}\frac{n_{CN}}{n_{NN}}}{1 + \frac{n_{HE}}{4} - \frac{n_{OE}}{2} - \frac{n}{4}\frac{n_{NE}}{n_{NN}}} = \left(\frac{n_{NN}}{n_{NE}} - n_{CN}\right)UQ \tag{16}$$

$$UQ = \frac{\frac{n_{NE}}{n_{NN}}}{1 + \frac{n_{HE}}{4} - \frac{n_{OE}}{2} - \frac{n}{4}\frac{n_{NE}}{n_{NN}}} \tag{17}$$

$$WQ = \frac{\frac{n_{HE}}{2} - n_{NE}\frac{n_{HN}}{2n_{NN}}}{1 + \frac{n_{HE}}{4} - \frac{n_{OE}}{2} - \frac{n}{4}\frac{n_{NE}}{n_{NN}}} = \left(\frac{n_{HE}}{2}\frac{n_{NN}}{n_{NE}} - \frac{n_{HN}}{2}\right)UQ \tag{18}$$

where $n = 4n_{CN} + n_{HN} - 2n_{ON}$. The interest in the RQ is its information about the relative contributions of carbohydrates, lipids, and proteins in the fueling of metabolism, while UQ and WQ are the logical counterparts of the RQ, which have not yet found their way into physiological applications. They can become valuable in the study of protein turnover and the measurement of respiration with the technique of double-labeled water. The RQ, UQ, and WQ are only constant if the elemental composition of reserves and structure are identical. For $n_{HE} = 1.8$, $n_{OE} = 0.5$ and $n_{NE} = 0.2$, and ammonia as N-waste, the quotients are RQ = 0.952, UQ = 0.19, and WQ = 0.571.

The Specific Dynamic Action (SDA) stands for the food-specific oxygen consumption that is associated with feeding, which is also known as the heat increment of feeding, although it has only an indirect relationship with heat flux. It can be shown[42] that the SDA amounts to

$$\frac{j_{OA}}{j_X} = (n_M^{-1})_{o*}\cdot n_o \begin{pmatrix} 1 \\ 0 \\ -y_{EX} \\ -y_{PX} \end{pmatrix} = \begin{pmatrix} 1 & \frac{1}{2} & \frac{3}{4} \end{pmatrix} \begin{pmatrix} -n_{CX} & n_{CE} & n_{CP} \\ n_{OX} & -n_{OE} & -n_{OP} \\ n_{NX} & -n_{NE} & -n_{NP} \end{pmatrix} \begin{pmatrix} 1 \\ y_{EX} \\ y_{PX} \end{pmatrix} \tag{19}$$

where $(n_M^{-1})_{o*}$ stands for the row of n_M^{-1} that corresponds with oxygen, which amounts to $(-1 \quad \frac{-1}{4} \quad \frac{1}{2} \quad \frac{3}{4})$ in the present case. The SDA can be calculated for daphnids feeding on algae and bacteria, but the variable reserves of these food items shows that the SDA is variable as well. This problem disappears when the SDA is calculated for the structural mass of algae and bacteria.

The dynamics of the DAB system has some interesting properties. If the consumers become extinct, the bacteria will follow, and the algae will grow to a density where assimilation just balances maintenance, which occurs when

$$\dot{J}_{E_iA, A_iA, m}f_{A_i} = (1 - \kappa_{R_i})\dot{k}_{E_iA}m_{E_iA} + \kappa_{R_i}\dot{J}_{E_iA, MA}.$$

Together with carbon and nitrogen balances, this equation defines the biomass and reserves of the algae. An infinite number of trivial steady states exist, consisting of distributions of elements over minerals and detritus, without living components.

The conservation law for elements makes sure that the Jacobian of the equilibrium ($d/dM^T j\mathbf{1}$), if it exists, has two eigenvalues that are zero; a property known as neutral stability: a (small) shift in the state results in a (small) shift in the point or cyclic attractor, because the total amount of elements is conserved a closed system. This property can only be removed by supplying the system with a mass input and leak. These two eigenvalues can also be removed by omitting the equations for change in carbon dioxide and ammonia, and expressing their amounts as the difference between total carbon and nitrogen and the carbon and nitrogen locked in biota and detritus.

The Jacobian of the equilibrium that has been selected in Tables 3.2 and 3.3, and Figure 3.5 has a very small positive eigenvalue. This means that the equilibrium is not stable. The system moves very slowly toward extinction for daphnids and bacteria after perturbation in the direction of the corresponding eigenvector and for extinction of all biota in the opposite direction. The equilibrium dose attract trajectories a large part of the state space toward its direct neighborhood, however, and extinction is really slow, which makes the unstable equilibrium still of interest for the analysis of the temporary behavior of the DAB community. Several parameters can be used to decrease the value of the real positive eigenvalue, but when it becomes almost zero, the equilibrium point disappears. Algae can escape grazing and force consumers into extinction if the conversion from structural mass of algae to reserves of consumer is too low to pay consumers' maintenance costs, in combination with alga's ability to survive on low reserves.

A simple experiment with a DAB community in a real bottle in the window shows that such a community can function for quite some time, but ultimately goes extinct. Another reason for ultimate extinction is that nutrient recycling is not as perfect as modeled here, and organic matter builds up that cannot be readily degraded by bacteria that managed to survive; although bacteria in a rich sample from the field are metabolically extremely versatile, each species can use a very much restricted set of substrates. Intraspecific competition among bacteria leads to a decline of species diversity in spatially homogeneous experimental units. Nitrogen locked into poorly degrad-able organic matter is found to be the largest nitrogen pool in oceans, next to N_2 gas. Peptidoglycan consitutes the major component of ultrafiltrated dissolved organic matter, and originates from cell walls of Gram-negative bacteria,[53] with a C/N ratio of 17.

Trying to understand the steady-state behavior of the community in the bottle, it helps to realize that, when we change light, or total carbon of nitrogen, while all other parameters are not changed, we know a priori that, at steady state, the consumer grows at a rate that is not affected by these three variables. This is because growth must balance death, which is set by aging. The reserve density of the consumer directly relates to growth, and is also not affected by the three variables.

Figure 3.4 shows the effect of maintenance and reserves, when compared to Figure 3.5 which shows the way the stable steady state for masses depends on changes in total carbon, total nitrogen, and light. The Jacobian in this steady state has two eigenvalues zero, two pairs of conjugated complex ones with negative real parts, and seven real negatives ones. This implies that the steady state is stable, indeed. A detailed comparison with the Monod model is complicated by freedom in the choice of parameter values. Bacteria can now also feed on reserves of dead daphnids, algae have two nutrient uptake routes, and reserve kinetics can be changed via turnover rates and recovery fractions. For this reason, the presented results can only be preliminary, and this presentation just

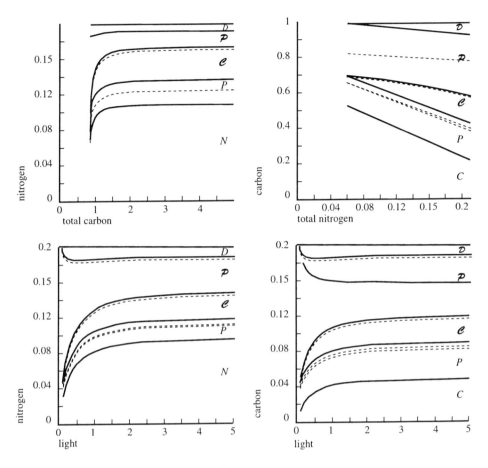

FIGURE 3.4 The steady-state distribution of carbon and nitrogen in the DAB community while increasing the total amount of carbon (upper left), nitrogen (upper right), or light (lower panels), using the DEB model. The nonchanged amounts are 1 unit for carbon, 0.2 units for nitrogen, and 5 for light. The amounts of carbon and nitrogen are plotted cumulatively, from bottom to top, across the minerals (carbon dioxide, C, or ammonia, N), detritus (P, three types), daphnids (i.e., consumers $\mathcal{C}$), algae (i.e., producers $\mathcal{P}$), and bacteria (i.e., decomposers $\mathcal{D}$). Parameters: $j_{VA,AD,m} = 1.5$, $j_{VB,AD,m} = 1.25$, $X_{K,VA} = 2$, $X_{K,VB} = 5$, $j_{E_1A,AA,m} = 6$, $j_{E_2A,AA,m} = 4$, $X_{K,L1} = 5$, $X_{K,C1} = 1$, $X_{K,L2} = 2$, $X_{K,C2} = 2$, $X_{K,N2} = 2$, $j_{PA,AB,m} = 3$, $j_{PB,AB,m} = 3$, $j_{PV,AB,m} = 2$, $j_{PEAB,m} = 3$, $X_{K,PA} = 2$, $X_{K,PB} = 2$, $X_{K,PV} = 2$, $X_{K,PE} = 1$, $j_{ED,MD} = 0.1$, $\dot{k}_{ED} = 1.5$, $j_{E_1A,MA} = 0.05$, $j_{E_2A,MA} = 0.025$, $\dot{k}_{E_1A} = 6$, $\dot{k}_{E_2A} = 6$, $\kappa_{R1} = 0.8$, $\kappa_{R2} = 0.7$, $\kappa_{EB} = 3$, $j_{EB,MB} = 0.15$, $h = 0.01$, $g = 1$, $y_{ED,VD} = 0.3$, $y_{ED,E_1A} = 0.8$, $y_{ED,E_2A} = 0.8$, $y_{VD,VB} = 0.3$, $y_{VD,EB} = 0.8$, $y_{PA,VA} = 0.7$, $y_{PA,VB} = 0.7$, $y_{C,E_1A} = 1$, $y_{C,E_2A} = 1$, $y_{L,E_1A} = 0.5$, $y_{L,E_2A} = 0.2$, $y_{N,E_2A} = 1$, $y_{EB,PA} = 0.7$, $y_{EB,PB} = 0.7$, $y_{EB,PV} = 0.7$, $y_{EB,PE} = 0.9$, $y_{ED,VD} = 2.5$, $y_{E_1A,VA} = 1.5$, $y_{E_2A,VA} = 1.5$, $y_{EB,VB} = 1.25$.

aims to identify the processes that need to be considered in the study of the dynamics of closed systems.

Apart from the stable steady state at low total carbon and nitrogen, an interesting unstable steady state exists at high C and N levels. Figure 3.5 illustrates that an increase of total nitrogen, starting from a situation where nitrogen is limiting, shifts carbon proportionally from detritus and algae to carbon dioxide, consumers, and decomposers, until it ceases to be limiting. A similar increase in carbon also results in a proportional increase in biomass of all three living components, but ammonia is linearly decreasing, until it hits a threshold at which the community goes extinct. An increase of light has a more complex effect on biomass. It results in a peak for the consumers and the decomposers, and a dip for the producers, while an increase beyond the level where light

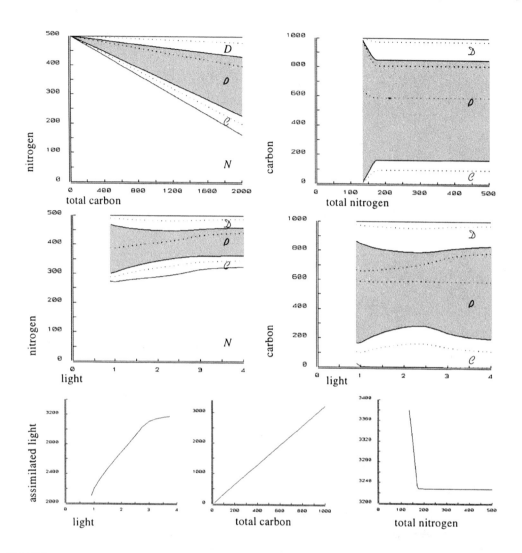

FIGURE 3.5 The steady state distribution of carbon and nitrogen in the DAB community while increasing the total amount of carbon (upper left), nitrogen (upper right), or light (middle panels), using the DEB model. The lower panels present the amounts of assimilated light (by the algae), which is proportional to the amount of dissipating heat. The nonchanged amounts are 1000 units for carbon, 500 units for nitrogen, and 1000 for light. The amounts of carbon and nitrogen are plotted cumulative, from bottom to top, across the minerals (carbon dioxide, C, or ammonia, N; the amount of carbon dioxide is very small with respect to carbon in living matter), detritus (four types, but the band is too narrow to be visible), daphnids (i,e., consumers $\mathcal{C}$, structure, and reserve), algae (i.e., producers $\mathcal{P}$; structure, C- and N,C-reserves), and bacteria (i.e., decomposers $\mathcal{D}$, structure, and reserve). The algae have three carbon components, and two for nitrogen, because one reserve lacks nitrogen. An increase of light above 4 units has no effect (so all lines proceed horizontally).

ceases to be limiting has no effect at all. Assimilated light, in the lower panels of Figure 3.5 quantifies "the rate of living." It is curious to note that it is decreasing for increasing nitrogen, as long as nitrogen is limiting. Coupled to this phenomenon is that ammonia is practically absent if nitrogen is strongly limiting; all nitrogen is then fixed into the biota. This corresponds well with widely known qualitative observations; nitrogen minerals are extremely low in oligotrophic systems (lakes, oceans as well as rain forests).

3.7   ENERGY TURNOVER AND DISSIPATION HEAT

We used fixed stoichiometries in all chemical transformations. It is only fair to point to some implicit assumptions here, which are basic to all complex biochemical transformations, such as the transformation of reserves into structural mass.

Some of the reserve "molecules" are used as building blocks for structure, and part are used to generate power to drive the synthesis through combustion, which relates to the production of carbon dioxide and ammonia. The power that is required to drive this transformation depends on the chemical potentials of the compounds that are involved, while the chemical potentials themselves depend, generally, on the local chemical environment, i.e., the set of concentrations of *all* compounds. If the chemical potentials of any compound involved in the transformation from reserve "molecules" to structural mass "molecules" plus combustion products would vary, the stoichiometric coefficients will vary, as a consequence of the dual role of reserves (fuel *and* building blocks). The use of fixed stoichiometries, and, therefore, constant chemical potentials, translates into the assumption that the chemical environment does not change, despite the transformation. This odd assumption can be justified by the realistic assumption that the actual transformation is via monomers, which concentrations are kept low and constant by the cells in which this transformation occurs, while the bulk of mass in reserves and structure is present in the form of polymers that do not partake in transformations in that form. As a consequence, the entropy of reserves and structure is assumed to be negligibly small, which implies, via Gibbs' relationship between free energy and enthalpy, that their numerical values become almost the same. This results when pushing the implications of homeostasis to the extreme. It is obviously a very simplified and idealized point of view, which nevertheless seems to work.

The empirical evidence for this point of view is in the success of the empirical method of indirect calorimetry, which relates dissipation heat linearly to carbon dioxide production, oxygen consumption, and N-waste production. Mathematically, this method is based on $\dot{p}_T = J_M^T \mu_T$, where

$$\mu_T = \begin{pmatrix} \mu_{TC} & \mu_{TH} & \mu_{TO} & \mu_{TN} \end{pmatrix}^T = \begin{pmatrix} 60 & 0 & -350 & -590 \end{pmatrix} \quad \text{kJ mol}^{-1}$$

is the vector of coefficients that are obtained via multiple regression.[9,11] The fact that these coefficients are directly fixed supports the assumption that the chemical potentials are constant, which implies that the entropy is zero (or at least small enough to be neglected). The empirical weight coefficients could, in principle, be species dependent, but practice learns that they vary little. The method has been well tested for a wide variety of animals over the many years of use, but (as far as we know) not for autotrophic systems.

Thus the DEB model offers a theoretical underpinning of the method of indirect calorimetry,[42] which rests on the work of Crawford[15] and Lavoisier and de Laplace,[47] in the 18th century.[55] All mass fluxes in the DEB model for feeding on a single food source can be written as weighted sums of three powers: assimilation, dissipation, and growth, so can the dissipating heat, as an implication of the chemical potentials being constant. The consequence is that dissipating heat can be written as a sum of three mass fluxes (linear functions of linear functions are again linear functions). The selection of oxygen, carbon dioxide, and ammonia is very convenient, because they can be easily measured. The fact that the empirical method of indirect calorimetry is based on three mineral fluxes directly relates to the three organizing powers behind the DEB structure: assimilation, dissimilation, and growth.

Reproduction power is *not* part of the short list of organizing powers because reserves of the mother are converted into reserves of the embryo in the egg, which must have an identical chemical composition due to the strong homeostasis assumption. From a strict chemical perspective, there is no transformation, only a reduction of reserves that is related to the overhead costs for reproduction, which is included in the dissipating power. The transformation involved in reproduction has big kinetic consequences, but not for the relationship between mass and energy. Development power is also not part of the list of organizing powers, because it does not involve the synthesis of compounds that are abundant enough to affect the whole body. The state of maturity should be treated as information rather than matter, although this cannot be quantified in the entropy of biomass, because it is zero. Development power should, therefore, be treated as a dissipating flux. The class of models that is consistent with the method of indirect calorimetry is quite small.

The addition of assimilation fluxes, as in Table 3.8, should not affect the weight coefficients, because we could repeat the thought experiment for different food sources, one at a time, while the different experiments have growth and dissipating powers in common, that should not depend on food composition, if the organism uses the same reserves. The coefficients can be species dependent, but we will not implement that for simplicity's sake.

The respiration ratio (RQ), i.e., the ratio between carbon dioxide production and oxygen consumption, usually varies little, while ammonia production is small. This has led to the common practice of taking respiration (i.e., carbon dioxide production or oxygen consumption) proportional to dissipating heat, and considering respiration as a measure for metabolic work. If the composition of reserves, relative to structural mass, obeys certain constraints, the RQ is constant indeed, within the context of the DEB model.[42] The simplified relationship between respiration and heat will not be used here, because the flux of ammonia is essential.

The dissipating heat $\dot{P}_T$ directly follows from the energy balance equation. The reasoning is as follows. Let $\dot{J}_L$ denote the vector of photon fluxes that are involved in the various transformations, which is zero for all transformations, except for the two algal assimilation processes. So

$$\dot{J}_L = (0 \ \ 0 \ \ 0 \ \ 0 \ \ 0 \ \ \dot{J}_{LA_1A} \ \ \dot{J}_{LA_2A} \ \ 0 \ \ 0 \ \ 0 \ \ 0 \ \ 0 \ \ 0 \ \ 0 \ \ 0)^T \qquad (20)$$

and $\dot{p}_L = \mu_L \dot{J}_L$ are the powers that are associated with the photon fluxes, where μ_L denotes the exergy of light, i.e., the exergy: energy that is extracted by the photosynthetic system from a mole of useful photons. The energy content of a photon with wavelength L_λ is $h\dot{v}_L / L_\lambda$, where $h = 6.625 \ 10^{-34}$ Js is Planck's constant, and $\dot{v}_L = 3 \ 10^8$ ms^{-1} is the velocity of photons in vacuum. A photon of 720 nm (the maximum length that can be used for by photo pigments) has an enthalpy of $1239.5/720 = 1.72$ eV (a 1239.5 nm photon has an energy of 1 eV). Photosystem II, using pigment P_{680}, increases about 1.2 eV in potential upon absorbing of a photon, Photosystem I, using pigment P_{700}, makes a jump of 1.7 eV.[21] (The pigments are named by the wavelength of the maximum absorption of the complex in which they are embedded.) Since both P_{680} and P_{700} have to be excited to trigger a reaction, the mean exergy of a photon in photosynthesis is either zero, for photons with wavelengths outside the interval between 550 and 720 nm, or 1.45 eV (which is also very close to the potential increase of bacterial photo pigments P_{870} and P_{840} upon capturing a useful photon), which amounts to 140 kJ per mole of useful photons (the number of Avogadro is $6.02 \ 10^{23}$, while 1 eV = $1.6 \ 10^{-19}$ J). The quantity $Y_{L,E_iA} = \dot{J}_{L.A,A} / \dot{J}_{E.A.A,A}$ stands for the quantum requirement of reserve i (in moles of photons per C-mole of reserve i).

The dissipating heat that is associated with the various transformations is

$$\dot{p}_T = \dot{p}_L - \dot{J}^T \mu \qquad (21)$$

where μ denotes the enthalpies of the various compounds. The balance equation excludes the heat flux that results from absorbed light that is not involved in photosynthesis, as well as the energy of used photons that exceeds the amount that is extracted. The balance equation just states that the heat that dissipates in association with a transformation equals the difference between the energy in the light that is used for photosynthesis and the energy that is locked in the compounds. Endothermic reactions in terms of overall transformations cannot occur here because the entropy of biomass is zero, which implies that all elements of $\dot{p}_T$ should be nonnegative, while the Second Law of Thermodynamics implies that all are positive.

We can use the method of indirect calorimetry as follows. From (15) we know that $\mu^T \dot{J} = \mu_O^T \dot{J}_O + \mu_M^T \dot{J}_M = (\mu_O^T - \mu_M^T n_M^{-1} n_O) \dot{J}_O$, where the vector of enthalpies is partitioned into contributions from the minerals and from the organic compounds, i.e., $\mu^T = (\mu_M^T : \mu_O^T)$. Using $\dot{p}_T = \dot{J}_{M\mu_T}^T$ and (21) gives

$$\dot{p}_L^T = (\mu_O^T - (\mu_M^T + \mu_T^T) n_M^{-1} n_O) \dot{J}_O \qquad (22)$$

In dark situations, where $\dot{p}_L = 0$, it follows that $\mu_O^T = (\mu_{OM}^T - \mu_T^T) n_M^{-1} n_O$. In a combustion frame of reference we have, by definition $\mu_M = 0$, and the chemical potentials of the organic compounds reduces to $\mu_O^T = \mu_T^T n_M^{-1} n_O$. In our numerical example, this amounts in kJ mol^{-1} to

μ_{PA}	395	μ_{PB}	395	μ_{PV}	310	μ_{PE}	310	μ_{VD}	310	μ_{ED}	310
μ_{VA}	310	$\mu_{E_1 A}$	410	$\mu_{E_2 A}$	139	μ_{VB}	310	μ_{EB}	139		

For comparison, Heijen and van Dijken report the Gibbs energy of formation of (microbial) biomass $CH_{1.8}O_{0.5}N_{0.2}$ to be 474.6 kJ/ C-mol in a combustion frame of reference,[26,27] while these element frequencies would lead to 310 kJ C-mol on the basis of calorimetric coefficients (obtained from animals). Wacasey and Atkinson report combustion values scattering around 580 kJ/C-mol for a wide variety of invertebrates.[86]

The total dissipating heat amounts to $P_T = \dot{p}_T^T 1$, while at steady state we must have that the total dissipating heat must balance assimilated light, so $\dot{p}_T = -\dot{p}_L^T 1$, since $\mu^T \dot{J} 1 = 0$, because $\dot{J} 1 = 0$ at steady state.

3.8 TOP-DOWN VS. BOTTOM-UP CONTROL

A popular issue in the analysis of food web dynamics is the question of top-down or bottom-up control[8,13,31,34,56,57,68,75]: Are predators in control of prey densities or vice versa? Simple models of linear food chains predict that if competition among consumer species is mediated only through the effects of the consumers on their resources, then increases in primary productivity lead to increases in steady-state population and biomass at the top trophic level and at even-numbered trophic levels below it, whereas the steady states at odd-numbered levels below the top level are unaffected by enrichment.[1,2,39,40,66,70,71] These simple results become invalid for more complex food webs; in particular, omnivory complicates the predictions significantly. Thus even in the "Monod" DAB model, the steady-state density of algae is not "controlled" by the *Daphnia*, since they also eat bacteria.

The situation is yet more complicated if the consumers are represented by the full DEB model. The effect of an increase in prey density is not on growth, but on assimilation. Figure 3.6 illustrates how *Daphnia* assimilation reacts on an incremental increase in algae and bacteria. Since *Daphnia* nutrition mainly depends on the reserves of algae and bacteria, the incremental increase in algae and bacteria is not only in their structural mass, but also in the reserves; we add an extremely small amount of algal and bacterial cells, with the same composition as the ones already present in the bottle. The sensitivity coefficients show that an increase of total nitrogen leads to an increase in the effect of food additions on consumers assimilation, until nitrogen ceases to be limiting. This holds for both the scaled and the unscaled sensitivities. An increase of total carbon, however, leads to a decrease in the assimilation sensitivity for food additions, until a plateau level, and an increase in the scaled sensitivity. An increase in light leads to a peak for the assimilation sensitivity.

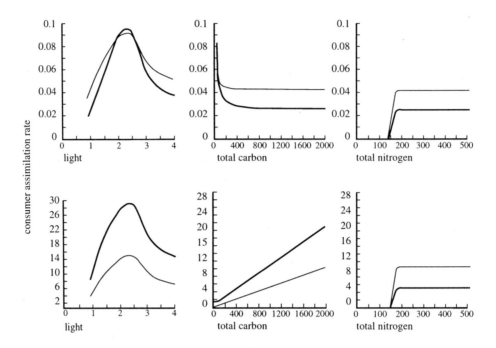

FIGURE 3.6 The steady-state control of consumer assimilation by producer (—) and decomposer (...) densities in the DAB community while increasing the total amount of light (left), carbon (middle), or nitrogen (right). The upper panels give the unscaled values, lower panels the scaled ones. The nonchanged amounts are 1000 units for carbon, 500 units for nitrogen, and 1000 for light. The reserve densities of producers and decomposers are kept fixed, e.g.,

$$\frac{d}{dM_A}\dot{J}_{ED.A_1D} \overset{def}{=} \lim_{dM_{VA}\downarrow 0} \frac{\dot{J}_{ED.A_1D}(M_{VA} + dM_{VA}, M_{E_1A} + dM_{VA}m_{E_1A}, M_{E_2A} + dM_{VA}m_{E_2A}) - \dot{J}_{ED.A_1D}(M_{VA}, M_{E_1A}, M_{E_2A})}{dM_{VA}}.$$

A potentially useful piece of theory to address the control question quantitatively is Metabolic Control Analysis (MCA),[28,29,37] which aims to evaluate the amount of control each enzyme exercises on the flux of metabolites in a metabolic pathway. One source of interest in this evaluation is to identify the most limiting enzymes for the overall flux through the pathway, in order to increase the amount of that enzyme, or set of enzymes, in one way or another, and so increase the overall flux. MCA is built on two sets of control coefficients: flux control coefficients, which give the fractional change in flux resulting from a fractional change in concentrations of enzyme, and metabolite control coefficients, which give the fractional change in concentrations of substrate (metabolites), resulting from a fractional change in concentrations of enzyme. MCA is particularly useful, because of two relationships held under certain conditions: the sum of all flux control coefficients amounts to unity, and the sum of all metabolite control coefficients amounts to zero. These relationships, known as summation theorems, only hold under rather restrictive conditions, including the assumption that enzyme concentrations are considered as model parameters, rather than part of the dynamic system; they are independent from the reactions that they catalyze.[3] This hampers the application of MCA in ecological contexts. Allison et al.[3] showed that the flux control coefficients for substrate on the dilution rate and the maximum specific growth rate in a chemostat with a microorganism that follows the Monod model, does obey the summation relationships. The feedbacks and the existence of structural mass and reserves in our bottle seem too complex for an easy application of MCA, pointing to the need for further work in this area.

3.9 FROM CANONICAL TO NATURAL COMMUNITIES

Although the numerical values that we used in this contribution are biologically realistic, the selected parameter values are only meant to illustrate the reasoning practically, and are not based on carefully executed estimation procedures with real empirical data.

Indeed, there are many difficult issues involved in parameter estimation and model testing for DEB models.[64] Rather, our aim has been to describe principles of the systematic formulation of models that describe the mass and energy balances in the canonical community, as these principles carry over to systems with additional features. Furthermore, we have only touched on the dynamical analysis. This work obviously needs expansion, and we think that this can only be done in a meaningful way by comparing each model aspect in a family of related simplified models. We now describe some possible extensions to the theory, but reemphasize that every additional feature in a model carries with it new parameters whose values have to be estimated.

The canonical community has properties that are relevant to most aquatic and terrestrial ecosystems, but it requires considerable system-specific extensions to acquire a minimum degree of realism in any given context. As an example, for planktonic communities, these extensions might include:

- *Nutrients.* Since ammonia is rare relative to nitrate, the inclusion of nitrate is obvious, but such an extension comes with ammonia-nitrate interactions in terms of uptake by algae[22] and conversion by oxidation, which involves bacteria. The generation of nitrate from (atmospheric) nitrogen by blue-green bacteria also seems essential. The next step is to include phosphate and iron, because they are frequently found to be limiting. Phosphate is present as ion, but also as polyphosphate. Iron occurs in several chemical species, and $Fe^{2+} - Fe^{3+}$ interactions need to be considered.
- *Species diversity.* The inclusion of one species of producer and one species of consumer is close to a caricature of the real situation.

  The quantitative effects of the replacement of the consumer population by food web of consumer populations is hard to evaluate. Body-size scaling relations for feeding, reproduction, characteristic times, and spatial scales become important in this extension. The DEB theory offers valuable entries into this complex matter.[41]

  The realistic inclusion of competing algae has to account for differences in their properties. Diatoms are the first group to appear in the season, probably because of their large surface area relative to the maintenance-requiring mass (they have huge vacuoles that probably require little maintenance). The occurrence of diatoms cannot be understood without the inclusion of silica, which they use to build their cell walls. Diatom blooms are followed by various other groups, such as haptophyceans and, in summer, dinophyceans. The increase of concentrations of organic compounds during the season comes with a general tendency among "algal" groups to changes from rather pure autotrophy to mixothrophy, by using organic compounds and prey on bacteria and microzooplankton. Of particular importance are changes in the *size structure* of the algal community, since size is a major determinant of edibility by zooplankton, and inedible algal species may constitute "refuges" for substantial quantities of elemental matter.[46]

  The quantitative importance of (viral) diseases in plankton dynamics is still an open question.[79] Blooming makes plankton susceptible, because of the short time required to travel from one host to the other.
- *Products and export.* There elements of plankton dynamics are of major importance to the oceans ecosystem.

  The export of organic material travels to layers below the euphotic zone. Marine snow serves as an elevator, or conveyer belt if you wish, and represents food for communities below the euphotic zone. Fecal pellets play a role, but also microbial floc formation, and appendicularian feeding houses. This export implies an uptake of carbon dioxide from the atmosphere (and from the layers below the euphotic zone). The production of organic compounds by algae, its decomposition by bacteria, and the consumption by zooplankton and mixotrophs might be of importance, as indicated in the introduction.

The export of carbonate to deeper layers is a process where coccolithophores play an important role.[88,89] This transport enhances uptake from the atmosphere and might affect global climate. The residence time of carbonate in deep layers is really long, and burial in sediments might extend the characteristic time to geological time scales.

The export of dimethyl sulfide (DMS) to the atmosphere is of importance due to its oxidation to sulfuric acid, which serves as condensation kernel for water (cloud formation), affecting the albedo, and, therefore, the global heat balance.[12,48] The discussion about its importance is far from completed, and its inclusion into modeling requires extensions involving the dynamics of excretion of DMS precursors and their microbe-mediated transformation.

- *Spatial structure.* The above-mentioned export processes cannot be understood fully without a vertical spatial structure, that acknowledges that light comes from above, while nutrients are more abundant in the layers below the photic zone. Wind-induced turbulence is essential to quantify the availability of light and nutrients. The role of intracellular reserves becomes even more pronounced in this context. It is not yet obvious to what extent horizontal structure can be avoided for a basic quantitative understanding of the plankton system. The description of ocean currents requires, apart from effects of wind, details of basin morphology to evaluate effects of earth (and moon) rotation.

This list can easily be extended; we only mention the effects of UV radiation on the physiological performance of plankton and the production and effects of toxicants by certain algal groups, which recently became of interest, in connection with reductions of ozone in the upper atmosphere and eutrophication problems in coastal areas, respectively.

Although the complexity of real-world plankton systems is well known by people working in plankton dynamics, it can easily become of depressive complexity in the eyes of modelers and quantitative analysts. A lot has been done already, but it is still little with respect to what is necessary for a quantitative understanding. We think that a few elements in the art of modeling are essential: to make explicit use of mass and energy conservation, in order to exclude weird unrealistic behavior and reduce degrees of freedom; to model organisms using Dynamic Energy Budgets, in order to include some sound biology; to simplify considerably in order to avoid frightening amounts of variables and parameters. Although modern computational methods solved many problems in the numerical evaluation, complex models usually contribute little to our basic understanding because each parameter value and each relationship comes with considerable uncertainty. Work with simplified models implies a modular setup of model structure and comparisons of models with simplified versions to tell details apart from main relationships. We are still far away from a quantitative understanding of real-world ecosystem dynamics and still have a long way to go and develop theoretical explorations before more detailed descriptions and predictions are feasible, if ever.

3.10 DISCUSSION

We showed how dynamical mass and energy balances can be used to constrain the dynamics of a simplified community, leading to a dynamics that differs substantially from models that leave part of the compounds that are involved in the "black box." An example is the popular class of logistic models, where the carrying capacity relates to the ratio of intrinsic food production by the environment and the maintenance costs by individuals, while the amount of food is not modeled explicitly. The use of balances is still rare in population dynamics models, while the more successful models for weather and climate are all based on energy balances.[54] We expect to see a parallel development for population dynamical models in the future.

We made extensive use of the linear nature of mass and energy fluxes, by deriving them by addition of fluxes to and from individuals. Although we pushed simplicity to the extreme, many more realistic features complicate the dynamics, not the incorporation of dynamic balances for mass and energy. If individuals would show complex forms of interaction, for instance, this would primarily and directly affect feeding, but most of the other formulations can be left unchanged. Opening of the bottle does not give additional problems, and the application of balances should by no means be confined to closed systems.

The DEB model treats several mass fluxes different from many other models and physiological texts. Energy in nitrogen waste, for instance, is frequently subtracted from food intake to evaluate production fluxes. This procedure originates from evaluations of static energy budgets, which are frequent in physiological and ecological texts. In the DEB model, assimilation, maintenance, and growth can contribute to nitrogen waste, which is included into overhead costs for these processes. The same holds for respiration (i.e., oxygen consumption or carbon dioxide production), which is frequently converted conceptually into energy using fixed conversion coefficients, and interpreted as maintenance costs: so energy use, as reflected by respiration, is subtracted from assimilation to evaluate production. In the DEB model all three basic energy fluxes contribute to respiration (which is treated as two different mass fluxes), while the ratio between oxygen consumption and carbon dioxide production need not be constant (and is, in fact, found to be varying, especially among microorganisms). The basic difference is that the DEB model makes a sharp distinction between energy allocated to growth, and energy fixed into new tissue. (The same holds for reproduction.) This distinction can only be made in dynamic models, because the overhead costs for growth can only be evaluated from the relationship between changes in food intake and growth. A full discussion of similarities and differences between the DEB and other models is beyond the scope of this text.

A remarkable observation is the existence of stable steady states at low carbon and nitrogen levels, and unstable ones for high levels. Maximum assimilation, in the latter case, is achieved at low nitrogen levels, which requires further investigation.

ACKNOWLEDGMENTS

We thank Bob Kooi, Martin Boer, Fleur Kelpin, Erik Muller, and James Kay for their very constructive suggestions. This study has been supported by Grant 013/1204.10 to SALMK from the Dutch Government, National Research Programme on global air pollution and climate change, and by the U.S. Environmental Protection Agency (Grant R82-3588-01-2 to RMN). It has also been supported by the National Center for Ecological Analysis and Synthesis (NCEAS), through a sabbatical fellowship to RMN and support for a working group on ecotoxicology.

REFERENCES

1. Abrams, P. A., The responses of unstable food chains to enrichment. *Evolut. Ecol.,* 8:150–171, 1994.
2. Abrams, P. A. and Roth, J. D., The effects of enrichment on three-species food chains with nonlinear functional responses. *Ecology,* 75:1118–1130, 1994.
3. Allison, S. M., Small, J. R., Kacser, H., and Prosser, J. I., Control analysis of microbial interactions in continuous culture: a simulation study. *J. Gen. Microbiol.,* 139:2309–2317, 1993.
4. Andersen, T. R., *Pelagic nutrient cycles—Herbivores as sources and sinks.* Springer-Verlag, Berlin, 1997.
5. Azam, F., Fenchel, T., Field, J. G., Meyer-Reil, L. A., and Thingstad, F., The ecological role of water-column microbes in the sea. *Mar. Ecol. Prog. Ser.,* 10:257–263, 1983.

6. Bader, F. G., Frederickson, A. G., and Tsuchiya, H. M., Dynamics of an algal-protozoan grazing interaction. In R. P. Canale, editor, *Modelling Biochemical Processes in Aquatic Ecosystems*. Ann Arbor Science, Ann Arbor, MI, 1976.

7. Baretta-Bekker, J. G., Baretta, J. W., and Rasmussen, E. K., The microbial food web in the European regional seas ecostem model. *Neth. J. Sea Res.*, 33:363–379, 1995.

8. van den Berg, H. A., Propagation of permanent perturbations in food chains and food webs. *Ecol. Model.*, 107:225–235, 1998.

9. Blaxter, K., *Energy Metabolism in Animals and Man*. Cambridge University Press, Cambridge, 1989.

10. Boraas, M. E., A chemostat system for the study of rotifer-algal-nitrate interactions. In Kerfoot, W. C. editor, *Evolution and Ecology of Zooplankton Communities*. Ann Arbor Science University Press of New England, Hanover, NH, 1980.

11. Brafield, A. E. and Llewellyn, M. J., *Animal Energetics*. Blackie, Glasgow, 1982.

12. Bürgermeister, S., Zimmerman, R. L., Georgh, H.-W., Bingemer, H. G., Kirst, G. O., Janssen, M., and Ernst, W., On the biogenic origin of dimethylsulfide: relation between chlorophyll, {ATP}, organismic {DMSP}, phytoplankton species, and DMS distribution in {A}tlantic surface water and atmosphere. *J. Geophys. Res. D.*, 95(D12):20607–20615, 1990.

13. Carpenter, S. R., Kitchell, J. R., Hodgson, J. F., Cochran, P., Elser, J. J., Elser, M. M., Lodge, D. M., Kretchmer, D., He, X., and van Ende, C. N., Regulation of lake primary productivity by food web structure. *Ecology*, 68:1863–1876, 1987.

14. Corner, E. D. S. and Davies, A. G., Plankton as a factor in the nitrogen and phosphorous cycles in the sea. *Adv. Mar. Biol.*, 9:101–204, 1971.

15. Crawford, A., Experiments and observations on animal heat, and the inflammation of combustible bodies, 1979.

16. DeAngelis, D. L., Energy flow, nutrient cycling, and ecosystem resilience. *Ecology*, 61:764–771, 1980.

17. DeAngelis, D. L., *Dynamics of nutrient cycling and food webs*. Chapman & Hall, London, 1992.

18. Ducklow, H. W., Purdie, D. A., Williams, P. J. L., and Davies, J. M., Bacterioplankton: a sink for carbon in a coastal marine plankton community. *Science*, 232:865–867, 1986.

19. Dugdale, R. C. and Wilkerson, F. P., Silicate regulation of new production in equatorial Pacific upwelling. *Nature*, 391:270–273, 1998.

20. Elliott, E. T., Castanares, L. G., Perlmutter, D., and Porter, K. G., Trophic-level control of production and nutrient dynamics in an experimental planktonic community. *Oikos*, 41:7–16, 1983.

21. Falkowski, P. G. and Raven, J. A., *Aquatic photosynthesis*. Blackwell Science, Oxford, 1997.

22. Flynn, K. J., Fasham, M. J. R., and Hipkin, C. R., Modelling the interactions between ammonium and nitrate in marine phytoplankton. *Philos. Trans. R. Soc. Lond. B*, 352:1625–1645, 1997.

23. Goldman, J. C., Oceanic nutrient cycles. In M. J. R. Fasham, editor, *Flows of Energy and Materials in Marine Ecosystems: Theory and Practice*. 137–170. Plenum Press, New York, 1984.

24. Gurney, W. S. C. and Nisbet, R. M., *Ecological Dynamics*. Oxford University Press, New York, 1998.

25. Hecky, R. E. and Kilham, P., Nutrient limitation of phytoplankton in freshwater and marine environments: a review of recent evidence on the effects of enrichment. *Limnol. Oceanogr.*, 88:796–822, 1988.

26. Heijnen, J. J., A new thermodynamically based correlation of chemotrophic biomass yields. *Antonie van Leeuwenhoek*, 60:235–256, 1991.

27. Heijnen, J. J. and van Dijken, J. P., In search of a thermodynamic description of biomass yields for the chemotrophic growth of micro organisms. *Biotechnol. Bioeng.*, 39:833–858, 1992.

28. Heinrich, R. and Rapoport, S. M., Utility of mathematical models for the understanding of metabolic systems. *Biochem. Soc. Trans.*, 11:31–35. 1974.

29. Heinrich, R. and Schuster, S., *The regulation of cellular systems*. Chapmann & Hall, London, 1996.

30. Henry III, R. L., The impact of zooplankton size structure on phosphorus cycling in field enclosures. *Hydrobiologia*, 120:3–9, 1985.

31. Herendeen, R. A., A unified quantitative approach to trophic cascade and bottom-up:top-down hypotheses. *J. Theor. Biol.*, 176:13–26, 1995.

32. Hirata, H., A model of hierarchical ecosystems with utility efficiency of mass and its stability. *Int. J. Sys. Sci.*, 11, 1980.

33. Hirata, H. and Fukao, T., A model of mass and energy flow in ecosystems. *Math. Biosci.*, 33:321–334, 1977.

34. Hunter, M. D. and Price, P. W., Playing chutes and ladders: heterogeneity and the relative roles of bottom-up and top-down forces in natural communities. *Ecology,* 73:724–732, 1992.
35. Johannes, R. E., Nutrient regeneration in lakes and oceans. *Microb. Sea,* 1:203–213, 1968.
36. Jørgensen, S. E., *Integration of ecosystem theories: a pattern.* Kluwer Academic Publishers, Dordrecht, 1997.
37. Kacser, H. and Burns, J., The control of flux. *Symp. Soc. Exp. Biol.,* 27:65–104, 1973.
38. Koch, A. L., Mathematical modeling in microbial ecology. In Koch, A. L., Robinson, J. A. and Milliken, G. A. editors, *The Monod Model and Its Alternatives.* Chapman & Hall Microbiology Series, Chapman & Hall, London, 1998.
39. Kooi, B. W. and Kooijman, S. A. L. M., Existence and stability of microbial prey-predator systems. *J. Theor. Biol.,* 170:75–85, 1994.
40. Kooi, B. W. and Kooijman, S. A. L. M., The transient behavior of food chains in chemostats. *J. Theor. Biol.,* 170:87–94, 1994.
41. Kooijman, S. A. L. M., *Dynamic Energy and Mass Budgets in Biological Systems.* Cambridge University Press, Cambridge, 2000.
42. Kooijman, S. A. L. M., The stoichiometry of animal energetics. *J. Theor. Biol.,* 177:139–149, 1995.
43. Kooijman, S. A. L. M., The synthesizing unit as model for the stoichiometric fusion and branching of metabolic fluxes. *Biophys. Chem.,* 73:179–188, 1998.
44. Kooijman, S. A. L. M., Kooi, B. W., and Hallam, T. G., The application of mass and energy conservation laws in physiologically structured population models of heterotrophic organisms. *J. Theor. Biol.,* 2000, in press.
45. Kooijman, S. A. L. M. and Zonneveld, C., Multiple nutrient limitation of algal growth modelled by synthesizing units., 2000, in press.
46. Kretzschmar, M., Nisbet, R. M., and McCauley, E., A predator-prey model for zooplankton grazing on competing algal populations. *Theor. Pop. Biol.,* 44:32–66, 1993.
47. Lavoisier, A. and de Laplace, P. S., Memoire sur la chaleur. *Mem. Acad. R. Sci.,* 1780:355–408, 1780.
48. Lawrence, M. G., An empirical analysis of the strength of the phytoplankton-dimethylsulfide-cloud-climate feedback cycle. *J. Geophys. Res. D.,* 98:20663–20673, 1993.
49. Mann, H. H., *Ecology of coastal waters: a systems approach.* Blackwell Scientific Publications, Oxford, 1982.
50. Mann, K. H. and Lazier, J. R. N., *Dynamics of marine ecosystems.* Blackwell Science, Oxford, 1996.
51. Martin-Jézéquel, V., Zonneveld, C., Kooijman, S. A. L. M., and Quéguiner, B., Growth of diatoms as determined by si, n and c metabolism, 2000, in press.
52. May, R. M., Mass and energy flow in closed ecosystems: a comment. 39:155–163, 1973.
53. McCarthy, M. D., Hedges, J. I., and Benner, R., Major bacterial contribution to marine dissolved organic nitrogen. *Science,* 281:231–234, 1998.
54. McGuffie, K. and Henderson-Sellers, A., *A climate modelling primer.* Wiley, New York, 1997.
55. McNab, B. K., Mammalian energetics. In T. E. Tomasi and T. H. Horton. editors, *Energy Expenditure: A Short History,* pages 1–15. Comstock Publishing Associates, Ithaca, NY, 1992.
56. McQueen, D. J., Jojannes, M. R. S., Post, J. R., Stewart, T. J., and Lean, D. R. S., Bottom-up and top-down impacts on freshwater pelagic community structure. *Ecol. Monogr,* 59:289–309, 1989.
57. Menge, B. A., Community regulation: under what conditions are bottom-up factors important on rocky shores? *Ecology,* 73:755–765, 1992.
58. Metz, J. A. J. and Diekmann, O., *The dynamics of physiologically structured populations,* volume 68 of *Lecture Notes in Biomathematics.* Springer-Verlag, Berlin, 1986.
59. Murdoch, W. W., Nisbet, R. M., McCauley, E., de Roos, A. M., and Gurney, W. S. C., Plankton abundance and dynamics across nutrient levels: tests of hypotheses. *Ecology,* 79:1339–1356, 1998.
60. Nisbet, R. M., McCauley, E., Gurney, W. S. C., Murdoch, W. W., and de Roos, A. M., Simple representations of biomass dynamics in structured populations. In Othner, H. G., Adler, F. R., Lewis, M. A., and Dallon, J. C. editors, *Case Studies in Mathematical Modeling — Ecology, Physiology, and Cell Biology,* pages 61–79. Prentice Hall, Upper Saddle River, NJ, 1997.
61. Nisbet, R. M. and Gurney, W. S. C., Model of material cycling in a closed ecosystem. *Nature,* 264:633–634, 1976.
62. Nisbet, R. M., McCMuley, E., de Roos, A. M., Murdoch, W. W., and Gurney, W. S. C., Population dynamics and element recycling in an aquatic plant-herbivore system. *Theor. Pop. Biol.,* 40:125–147, 1991.

63. Nisbet, R. M., McKinistry, J., and Gurney, W. S. C., A 'strategic' model of material cycling in a closed ecosystem. *Math. Biosci.*, 64:99–113, 1983.

64. Noonburg, E. G., Nisbet, R. M., McCauley, E., Gurney, W. S. C., Murdoch, W. W., and de Roos, A. M. Experimental testing of dynamic energy budget models. *Funct. Ecol.*, 12:211–222, 1998.

65. Nyholm, N., A simulation model for phytoplankton growth and nutrient cycling in eutrophic, shallow lakes. *Ecol. Modell.*, 4:279–310, 1978.

66. Oksanen, L., Fretwell, S. D., Arruda, J., and Niemela, P., Exploitation ecosystems in gradients of primary productivity. *Am. Nat.*, 118:240–261, 1981.

67. Parker, R. A., Nutrient recycling in closed ecosystem models. *Ecol. Modell*, 4:67–70, 1978.

68. Power, M. E., Top-down and bottom-up forces in food webs: do plants have primacy? *Ecology*, 73: 733–746, 1992.

69. Reder, C., Metabolic control theory: a structural approach. *J. Theor. Biol.*, 135:175–201, 1988.

70. Rosenzweig, M. L. and MacArthur, R. H., Graphical representation and stability conditions for predator/ prey interactions. *Am. Nat.*, 97:209–223, 1963.

71. Smith, F. E., Effects of enrichment in mathematical models. In *Eutrophication: causes, consequences, correctives,* pages 631–645. National Academy of Sciences, Washington, D.C., 1969.

72. Smith, S. V., Phosphorus versus nitrogen limitation in the marine environment. *Limnol. Oceanogr.*, 29:1149–1160, 1984.

73. Steele, J. H., *The Structure of Marine Ecosystems*. Oxford University Press, Oxford, 1993.

74. Sterner, R. W., The ratio of nitrogen to phosphorus resupplied by herbivores-zooplankton and the algal competitive arena. *Am. Nat.*, 136:209–229, 1990.

75. Strong, D. R., Are trophic cascades all wet? Differentiation and donor-control in speciose ecosystems. *Ecology*, 73:747–754, 1992.

76. Taylor, A. H., Harbour, D. S., Harris, R. P., Burkill, P. H., and Edwards, E. S., Seasonal succession in the pelagic ecosystem of the north Atlantic and the utilization of nitrogen. *J. Plankton Res.*, 15: 875–891, 1993.

77. Taylor, A. H. and Joint, I., A steady-state analysis of the "microbial loop" in stratified systems. *Mar. Ecol. Prog. Ser.*, 59:1–17, 1990.

78. Thingstad, T. F., Modelling the microbial food web structure in pelagic ecosystems. *Arch. Hydrobiol. Beih. Ergebn. Limnol.*, 37:111–119, 1992.

79. Thingstad, T. F., Heldal, M., Bratbak, G., and Dundas, I., Are viruses important partners in pelagic food webs? *Tree*, 8:209–213, 1993.

80. Thingstad, T. F. and Pengerud, B., Fate and effect of allochthonous organic material in aquatic microbial ecosystems. An analysis based on chemostat theory. *Mar. Ecol. Prog. Ser.*, 21:47–62, 1985.

81. Tilman, D., *Resource Competition and Community Structure*. Princeton University Press, Princeton, 1982.

82. Trégure, P., Nelson, D. M., van Bennekom, A. J., DeMaster, D. J., Leynaert, A., and Quéguiner, B., The silica balance in the world ocean: a reestimate. *Science*, 268:375–379, 1995.

83. Tuljapurkar, S. and Caswell, H., *Structured Population Models in Terrestrial, Marine and Freshwater Systems*. Chapman & Hall, New York, 1997.

84. Ulanowicz, R. E., Mass and energy flow in closed ecosystems. *J. Theor. Biol.*, 34:234–253, 1972.

85. Vanni, M. J. and Layne, C. D., Nutrient recycling and herbivory as mechanisms in the "top-down" effect of fish on algae in lakes. *Ecology*, 78:21–40, 1997.

86. Wacasey, J. W. and Atkinson, E. G., Energy values of marine benthic invertebrates from the Canadian Arctic. *Mar. Ecol.*, 39:243–250, 1987.

87. Weers, P. M. M. and Gulati, R. D., Growth and reproduction of *daphnia* galeata in response to changes in fatty acids, phosphorus, and nitrogen in *chlamydomonas* reinhardtii. *Limnol. Oceanogr.*, 42: 1584–1589, 1997.

88. Westbroek, P., *Life as a Geological Force*. W. W. Norton & Company, New York, 1991.

89. Westbroek, P., Brown, C. W., van Bleijswijk, J., Brownlee, C., Brummer, G. J., Conte, M., Egge, J., Fernandez, E., Jordan, R., Knappertsbusch, M., Stefels, J., Veldhuis, M., van der Wal, P., and Young, J. R., A model approach to biological climate forcing. The example of *emiliania huxleyi*. *Global and Planetary Change*, 8:1–20, 1993.

90. Williams, G. R., *The Molecular Biology of Gaia*. Columbia University Press, Columbia, 1966.

91. Williams, G. R., The coupling of biogeochemical cycles of nutrients. *Biogeochemistry*, 4:61–75, 1998.

*The sun: the ultimate
energy source for all
activities on earth*

CHAPTER 4

This chapter is devoted to the concept of emergy, introduced in ecology and ecological economics by H. T. Odum. Emergy expresses the cost of a process or a product in solar energy equivalents. The basic idea is that solar energy is our ultimate energy source and by expressing the value of products in emergy units, it becomes possible to compare apples and pears.

A number of other useful concepts are defined in the chapter: (1) Transformity of an output flow, as the total emergy driving the process divided by the free energy of the output. (2) Emergy Yield Ratio, as the ratio of output emergy to input emergy. (3) The Environmental Loading Ratio, ELR, measuring the impact of a process to the local ecosystem, defined as the ratio of purchased emergy + nonrenewable indigenous emergy to free environmental emergy. (4) The Empower Density, ED, which is the emergy flow per unit of time and unit of area. (5) The Energy Sustainability Index, ESI = EYR/ELR. The latter three indices are important in environmental management context.

The chapter gives a practical procedure on emergy accounting and the appendix to this chapter contains a long list of emergy for various products. This practical approach makes it possible to utilise emergy in ecological economics. It is, for instance, possible to make emergy balances of entire countries and give, on basis of the results, some environmental management recommendations. This application of emergy is illustrated by several examples in the chapter.

H. T. Odum, who was previously a strong advocate for Lotka's Maximum Power Principle, has reformulated the principle to the Maximum Empower Principle: the survival and growth of systems depends on their ability to self-organise to maximise the inflow of resources through reinforcement actions. This is consistent with "capture as much solar radiation as possible" (= maximise the inflow of resources (see Chapter 13) to build as much self-organisation as possible to obtain reinforcement actions (obtain as much biomass, structure, and information as possible ≈ measured by exergy which measures the amount of work the system can perform when it is brought into thermodynamic equilibrium with its environment (also Chapters 12 to 14).

Emergy measures the cost of a product in solar energy while exergy (see Chapter 7 for rigorous definition) measures the result in form of work capacity. It is therefore not surprising that the emergy/exergy ratio is a strong indicator of the efficiency of a system. Man-made systems, for instance, a biological treatment system, have high emergy/exergy ratios in contrast to well-balanced natural systems that have low emergy/exergy ratios.

Those interested in learning more about the emergy analysis are referred to *Environmental Accounting — Emergy and Environmental Decision Making,* by H. T. Odum (John Wiley & Sons, New York, 1996).

4 Emergy Accounting of Human-Dominated, Large-Scale Ecosystems

Sergio Ulgiati and Mark T. Brown

CONTENTS

4.1 Introduction 63
4.2 Emergy Concepts and Definitions 64
    4.2.1 Procedure for Emergy Accounting 67
    4.2.2 The "Energy Memory" 69
    4.2.3 The Emergy Algebra 69
4.3 Maximum Empower 70
4.4 Hierarchies 70
    4.4.1 The Thermodynamics of Hierarchies 71
    4.4.2 Hierarchies in Ecosystems Including Humans 71
4.5 Evaluating Ecosystems for Economic Questions 73
    4.5.1 Valuing Environmental Work 74
    4.5.2 Evaluating Renewability and Replacement of Resources 75
    4.5.3 Assessing Sustainability 76
4.6 Environmental and Human-Dominated Processes 76
    4.6.1 Emergy Evaluation of Agricultural Production 77
        4.6.1.1 The Case of Italian Agriculture 77
        4.6.1.2 The Case of Sugarbeet Production 87
        4.6.1.3 The Case of Biofuels Production 100
    4.6.2 Emergy Evaluation of a National Economic System 101
        4.6.2.1 Monitoring the Performance of the Italian Economic System 104
        4.6.2.2 Emergy as a Basis for Policies 105
4.7 Thermodynamics of Oscillating Systems 107
4.8 Concluding Remarks 108
Acknowledgments 109
References 109

4.1 INTRODUCTION

The biosphere as a whole and individual component ecosystems are the product of a continuous and never completed process of self-organization, driven and constrained by resources availability. Ecosystems are dynamic systems, maintained far from equilibrium by resource input flows and by their ability to discharge outside the system the entropy produced during the process. The commonly used tools of classical thermodynamics and chemistry do not allow for complete understanding of the dynamic equilbria that are fundamental for systems to survive or prevail in competition with other systems. Innovative points of view and new methodologies of analysis are needed to deal appropriately with a system's complexity and with its interaction with the environment.

One of the goals of this chapter is to provide a short introductory, unambiguous assessment of the emergy accounting methodology, by means of selected case studies. We will also show that the emergy can be used as a tool for the investigation of natural systems, systems at the interface of nature and human society, and human society itself. A recent contribution to clarify the fundamentals of emergy theory has been provided by Odum (1996).

We feel that time is ripe to recognize that complementary approaches may be needed to provide a more complete description and understanding of the behavior of dynamic, complex systems and ecosystems, including those under human control (Jørgensen, 1997; Ulgiati et al., 1998). For this to be possible, the assumptions underlying each approach as well as their field of applicability should be better clarified and known. This is what we are trying to do in this chapter with the emergy procedure. Analysts should be able to jointly apply different approaches to a case, according to the problem they are investigating and the goals they are trying to reach.

4.2   EMERGY CONCEPTS AND DEFINITIONS

The solar emergy, U, of a flow coming out of a given process is defined as the solar energy that is directly or indirectly required to drive the process itself.

$$U = \sum_i Tr_i E_i \qquad i = 1,...,n \qquad (1)$$

where E_i is the available energy (or free energy) content of the ith independent input flow to the process itself (Figure 4.1), and Tr_i is the solar transformity of the ith input flow, defined as follows:

$$Tr_i = U_i / E_i \qquad i = 1,...,n \qquad (2)$$

In Equation (2), U_i is the total solar emergy driving the ith process. This circular definition is made operational by putting Tr_S, the solar transformity of direct solar radiation, equal to 1.

The energy and matter inputs E_i to a process can be locally renewable, R_i, locally nonrenewable, N_i, and imported from outside the system, F_i (feedbacks from outside to reinforce the process). Therefore, an equivalent form for Equation (1) is:

$$U = \sum_i Tr_i R_i + \sum_j Tr_j N_j + \sum_k Tr_k F_k \qquad i = 1,...,n; \quad j = 1,...,n'; \quad k = 1,...,n'' \qquad (1a)$$

By definition the total solar emergy, U, driving a process is assigned to the output (see emergy algebra, Section 4.2.3), as a measure of the resources investment that is supporting it. Therefore, we can say that U is the emergy of the output. Sometimes it is indicated with Y (Yield, $U = Y$).

The transformity of the output flow from a process is, therefore:

$$Tr_{out} = \frac{\text{Total emergy } U \text{ driving the process}}{\text{Available energy of the output}} = \frac{\sum_i Tr_i R_i + \sum_j Tr_j N_j + \sum_k Tr_k F_k}{E_{out}}.$$

When a process is directly driven by solar energy, its transformity clearly appears as a measure of the convergence of solar energy to originate the product flow. When a set of transformities $\{Tr_i\}$ has been calculated for a certain number of flows or products originated by direct solar energy (wind, rain, etc.), it is possible to evaluate the indirect solar energy requirements as well as the transformities of other processes where the input flows are known, according to Equations (1) and (2).

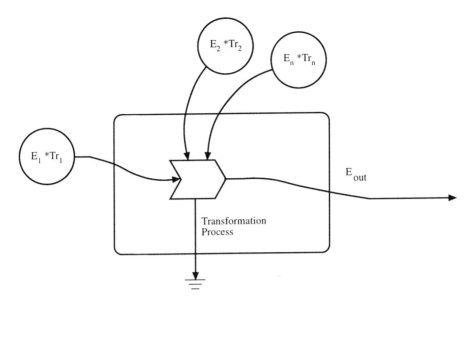

$$Em_{out} = \Sigma E_n \, *Tr_n$$

$$Tr = Em_{out} \, / \, E_{out}$$

FIGURE 4.1 Calculating a new transformity requires that the energy inputs, E_i, to a process be evaluated, then multiplied by their transformities, Tr_i, resulting in each input expressed as emergy, Em_i. The sum of the emergy inputs, Em, is the emergy required for the output, and its transformity is calculated as the quotient of the total emergy divided by the energy, E, of the output.

Typical values of transformities are listed in Table 4.1. As will appear evident in the following sections, increasing transformity is associated with longer turnover time, increasing range and territory, and more information processing.

Energy and resource inputs to a process are usually quantified by using their content of available energy relative to the environment (= exergy*). In this chapter, available energy (relative to the environment) and exergy will be used as synonyms. Sometimes other measures are used (combustion enthalpy, mass, hours, dollars, etc.), provided that the appropriate transformation coefficient (emergy per unit Item) is known, instead of transformities (emergy per unit energy). Measuring inputs by means of their calorific value may be an easier approximation instead of calculating their free energy. For example, the exergy of direct solar radiation is about 93.4% of its actual energy (Petela, 1964). Fossil fuels also show very small differences (Szargut et al., 1988).

Solar emergy is usually measured in solar emergy joules (sej), while solar transformity is measured in solar emergy joules per joule of product (sej/J). By measuring all inputs in emergy units, they can be evaluated on a common basis and their relative importance (contribution to the process) can be accordingly compared. This procedure appears capable of assessing the total environmental input required to support a process or growth of a system.

---

* Exergy can be calculated according to Szargut et al. (1988). Jørgensen (1997) calculates exergy in living matter (physical and chemical exergy + information embodied) by means of convergence of genetic information. In so doing, he built an accounting methodology that is very close to emergy and could offer a complementary point of view. We discussed elsewhere interesting analogies and parallels with other biophysical theories, also accounting for converging flows and processes that drive the growth and the evolution of complex systems and ecosystems (Ulgiati and Bianciardi, 1997).

TABLE 4.1
Selected Values of Solar Transformities and Emergies per Unit Mass

Flow or Product	Transformity (sej/J)	Emergy/Mass (sej/g)
Solar energy	1	[1]
Heat, at temperature 30°C	1,179	[1]
Surface wind	1,500	[1]
Heat, at temperature 100°C	7,660	[1]
Rain on land, physical potential	1.05 E4	[1]
Heat, at temperature 500°C	2.22 E4	[1]
Earth cycle	3.44 E4	[1]
Coal	3.98 E4	[1]
Natural gas	4.80 E4	[1]
Oil	5.40 E4	[1]
Gasoline	6.60 E4	[1]
Sugarcane	3.40 E4–5.53 E4	[5]
Plantation wood	8.0 E3–6.94 E4	[1,5]
Corn	4.52 E4–6.03 E4	[1,5]
Sugarbeet	6.14 E4	[2]
Soybeans	7.65 E4	[5]
Sunflower	7.72 E4	[5]
Rapeseed	8.88 E4	[5]
Electricity	6 E4 ÷ 2 E5	[1,3]
Bioethanol	9.50 E4–2.63 E5	[1,5]
Biodiesel	1.41 E5–2.41 E5	[2,5]
Methanol from wood	1.51 E5	[5]
Potassium fertilizer (sej/g K)	1.74 E9	[1]
Nitrogen fertilizer (sej/g N)	4.6 E9	[1]
Phosphate fertilizer (sej/g P)	1.78 E 10	[1]
Iron ore (sej/g)	1 E9	[1]
Cement (sej/g)	1.2 E9	[1,6]
Steel (sej/g)	1.25 E9–2.27 E9	[4,6]
Machinery (sej/g)	6.7 E9	[1]

References for transformities:
1. Odum, 1996.
2. Ulgiati and Brown, 1998.
3. Ulgiati, 1996.
4. Ulgiati et al., 1994.
5. Ulgiati, unpublished manuscript.
6. Haukoos, 1994.
7. Lapp, 1991.

Once solar transformities of the main input flows of the biosphere have been calculated (Odum, 1996), processes on a smaller spatial and temporal scale are evaluated by means of a similar procedure. The same flow or product can be originated through different pathways, so that many different transformities may be available for processes under human control, according to the specific time, location, and technological development. The same is not true for natural systems, where it is a common rule that processes in a given environment, operating under natural selection, develop organizations that achieve maximum performance. The result for natural systems is an optimum value of the transformity, selected over a long pathway, to achieve the present throughput and operating efficiency.

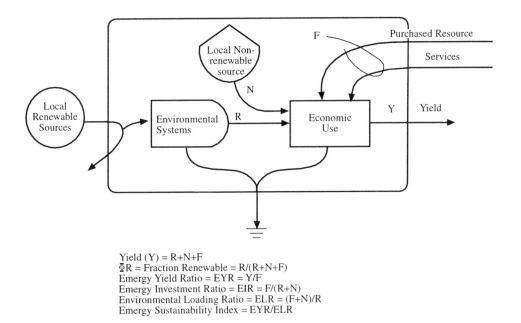

Yield (Y) = R+N+F
%R = Fraction Renewable = R/(R+N+F)
Emergy Yield Ratio = EYR = Y/F
Emergy Investment Ratio = EIR = F/(R+N)
Environmental Loading Ratio = ELR = (F+N)/R
Emergy Sustainability Index = EYR/ELR

FIGURE 4.2 Aggregated diagram of renewable, nonrenewable and purchased emergy inputs to an economic system, illustrating computation of some emergy based indices and use ratios.

In addition to the above-defined transformity, some of the most used indices and ratios are defined as follows (Figure 4.2):

1. The Emergy Yield Ratio (EYR) is the ratio of the emergy of the output Y divided by the emergy of those inputs F to the process that are fed back from outside the system under study. It is an indicator of yield compared to inputs other than local and gives a measure of the ability of the process to exploit local resources.
2. The Environmental Loading Ratio (ELR) is the ratio of purchased emergy F plus non-renewable indigenous emergy N to free environmental emergy R. It is an indicator of the pressure of the process to the local ecosystem and may be considered a measure of the ecosystem stress due to production activity.
3. The Empower Density (ED) is the emergy flow per unit time and unit area, defined as the ratio of total emergy U used in the process to the area involved and the time required by the process to develop. It is a measure of area and time convergence of emergy used.
4. The Emergy Sustainability Index (ESI) is defined as the ratio of the EYR to the ELR (Brown and Ulgiati, 1997), globally indicating if a process is providing a suitable contribution to the user at the cost of a low environmental pressure.

4.2.1 PROCEDURE FOR EMERGY ACCOUNTING

The general methodology for emergy analysis is a "top-down" systems approach. It can be organized in three steps.

The first step is to draw a detailed energy systems diagram, as a way to gain an initial network overview, combine information, and organize data-gathering efforts. These diagrams should be considered as a means of organizing thinking and relationships between components and pathways of exchange and resource flow. The following are the steps in the initial diagramming of a system to be evaluated:

1. The boundary of the system is defined.
2. A list of principal component parts believed important considering the scale of the defined system is made.
3. List of processes (flows, relationships, interactions, production, and consumption processes, etc.) is made. Included in these are flows and transactions of money believed to be important.
4. The diagram is drawn, by means of the symbols of the energy systems language, each having a rigorous energy and mathematical meaning (examples in Figures 4.2 to 4.9). A second system diagram is often drawn that represents an aggregated overview of the system under study. Processes and storages are aggregated to reduce complexity, but to retain overall system integrity and aggregation.

The second step is to construct emergy evaluation tables directly from the diagrams, to facilitate calculations of main sources and contributions of the system. Raw data on inflows are converted into emergy units, and then summed into a total emergy inflow driving the system. A table for storage reservoirs is also often constructed to place in perspective the emergy content of major system components.

An emergy analysis table can be prepared by using a spreadsheet on a personal computer. Its columns usually have the following headings:

1	2	3	4	5	6	7
Note	Item	Raw Data	Units	Transformity	Ref. Transf.	Emergy

If the table is for flows, it represents flows per unit time (usually per year). If the table is for reserve storages, it includes those storages with a turnover time longer than 1 year.

- Column number 1 is the line Item number, which is also the number of the footnote in the table where the raw data source is cited and calculations shown.
- Column number 2 is the name of the Item, which is also shown on the aggregated diagram.
- Column number 3 is the raw data in joules, grams, dollars, or other units, that are shown in column 4.
- Column number 5 is the transformity, or the emergy per unit, used for calculations, in solar emergy joules per unit of raw input (sej/J; sej/g; sej/$). These are obtained from previous studies cited in literature or calculated for the system under study. If transformities from other authors are used, source reference should be shown in column number 6.
- Column number 7 is the solar emergy of a given flow, calculated as raw input times its transformity (column 3 times column 5).

Finally, when the emergy tables have been completed, a third step involves calculating several emergy indices that relate emergy flows of the economy with those of the environment, and allow the evaluation of a system's performance as well as predictions of economic viability and carrying capacity. Additionally, using the results of the emergy analysis, comparisons between the emergy costs and benefits of proposed developments as well as insights related to international flows of money and resources can be made.

To make it clearer how the procedure is applied, a closer look to the footnotes of Table 4.6 may be useful. Footnote 1 calculates the annual amount of solar energy reaching the soil (1 hectare of agricultural land) as incoming radiation minus the outcoming fraction due to albedo. References of data used in these calculations as well as physical units are also supplied. The resulting annual energy flow, expressed in J/yr, is then transferred up in the table, line 1, and multiplied by the appropriate transformity to give the corresponding emergy flow. When this has been done for all of the main inputs to the process, emergy flows are summed in partial totals (renewable inputs, local inputs, imported inputs, etc.) as well as in final total, avoiding double counting inputs that are by-products of the same source (see Section 4.2.3, emergy algebra). Partial totals and final total are then used to calculate the transformity of the product and other emergy-based indicators.

The spreadsheet should be organized in such a way that changing one of the inputs will automatically change the calculated transformity as well as the other indicators.

4.2.2 THE "ENERGY MEMORY"

It might be useful to recall that *emergy is not energy*. The emergy of a given flow or product is, by definition, the total amount of available solar energy that is directly or indirectly required to generate (and maintain) the flow or the production process. This much solar energy is provided by the continuous work of self-organization of the planet as a whole. Part of this work has been performed in the past (for instance, the production of fossil fuels) over a million years. Part is the present work of self-organization of the geo-biosphere. The larger the flow required, the larger the present and past environment that is exploited to support a given process. Thus emergy can be considered like an "energy memory," i.e., the memory of the available energy that has been and is being used during the whole process. According to the efficiency of the process along a given pathway, more or less emergy might have been required to reach the same result. Natural processes have been selected over long biological times, in accordance with the available flow and turnover rate of resources, for maximum empower output (see Section 4.3).

4.2.3 THE EMERGY ALGEBRA

Emergy is the amount of a source exergy it takes to make another form of exergy. It can also be defined as "embodied exergy of one kind." There are definite rules that are followed to assign emergy to flows of exergy. The sum total of these rules can be termed Emergy Algebra (Scienceman, 1987; Brown and Herendeen, 1996). Scienceman (1996, personal communication) also defined it as a "memory algebra" as opposed to the ordinary "conservation algebra." In short, the main rules of emergy algebra are:

1. All source emergy to a process is assigned to the process output(s).
2. By-products from a process (i.e., product Items showing different physicochemical characteristics, and which can only be produced jointly) have the total emergy assigned to each pathway.
3. When a pathway splits (originating flows showing the same physicochemical characteristics), the emergy is assigned to each "leg" of the split based on their percent of the total exergy flow on the pathway.
4. Emergy cannot be counted twice within a system:
   a. Emergy in feedbacks should not be double counted;
   b. By-products, when reunited, cannot be summed.

Giannantoni (1997, 1998, 1999) has given a mathematical formulation of the emergy algebra and emergy balance, by means of coinjection coefficients, source functions, accumulation, and coproduction terms.

As a consequence of the above rules, when a process results in the output of two different products (i.e., by-products) the input emergy is assigned to both outputs, since each cannot be made without the other and all emergy is required to make each. This fact creates much confusion since, at first glance, it appears that more emergy is output from a process than is input, and it is thus a violation of the Conservation Law of Thermodynamics. However, under no circumstances should the emergy outputs from a process be added together. It would be a violation of rule 4, a double counting of emergy.

4.3 MAXIMUM EMPOWER

System self-organization is the way to maximize resources through-flow and reinforce production. The Maximum Power Principle (Lotka, 1992a, b; 1945), recently restated as Maximum Empower Principle (Odum H. T., 1983a, b; 1996), suggests that the survival and growth of systems depends on their ability to self-organize to maximize the inflow of resources through reinforcement actions, thus displacing systems with less reinforcement of their productive basis. Generally, this means the system organization that can develop uses for the largest amount of resources in the shortest time will displace other patterns that do not use resources as effectively. According to this principle, "natural selection tends to make the energy flux through the system a maximum, so far as compatible with the constraints to which the system is subject" (Lotka, 1992a). Systems self-organize for survival: under competitive conditions, systems prevail when they develop designs that allow for maximum flow of available energy. The flow circuit "learns" to reinforce such energy flow through feedbacks; structure which promotes these kinds of feedbacks is encouraged (rewarded). It has been underlined that "these systems perform at an optimum efficiency for maximum power output, which is always less than maximum efficiency" (principle of time speed regulator, Odum and Pinkerton, 1955).

Ecosystems and economic systems tend to provide more of whatever is limiting to maximum production and useful consumption. Thus, in different situations they may build biomass, store nutrients, develop variety, become complex, earn profit, generate information, export products, etc. Redirecting resources to eliminate limiting factors is a common mechanism of self-organization for maximum power output.

4.4 HIERARCHIES

While Lotka (1922a) and Odum (1996) suggested the Maximum Power Principle as a possible Fourth Law of Thermodynamics,* the hierarchical organization of systems may be recognized as a Fifth Law of Thermodynamics (Odum, 1996). Hierarchy can be thought of in two ways. First as a set of Chinese boxes of systems within systems (the holarchies described by some authors, see O'Neill et al., 1986) where each successively more complex system contains all lower order systems, for instance, atoms, compounds, chemicals, organelles, cells, organs, organisms, ecosystems, biomes, biosphere. A second way of describing hierarchies is as a chain of exergy transformations, such as the food chain, where many lower order transformation processes converge exergy to fewer and fewer higher order processes. In reality, exergy transformation chains are really webs like that illustrated in Figure 4.3. When each of the parallel processes is aggregated, a hierarchical series of transformations results. It is this second way of viewing hierarchies that we are interested in.

In any real process, available energy is transformed resulting in many joules of energy of one kind being used up, or degraded, to produce fewer joules of another kind. This "loss of available energy" is a consequence of the second law. It takes many units at a lower level to support the growth and development of one unit at a higher level. Organization patterns that require many small components to support fewer and fewer larger components are hierarchical (Figure 4.3). Since there is energy in everything, including information, and since there are energy transformations in all processes on earth and possibly in the universe as well, all energy processes can be regarded as part of an energy hierarchy. All these energy transformations are observed to be connected to others so as to form a network of energy transformations. Self-organizing systems are always organized in hierarchies.

Recognizing and measuring hierarchies is fundamental. This is because the stability of the whole cannot be based on the stability of an individual component of the system, but should be founded on

---

* Other, not contrasting ways of defining a tentative Fourth Law of Thermodynamics can be found in Landsberg (1961), Wright (1979), Jorgensen (1997).

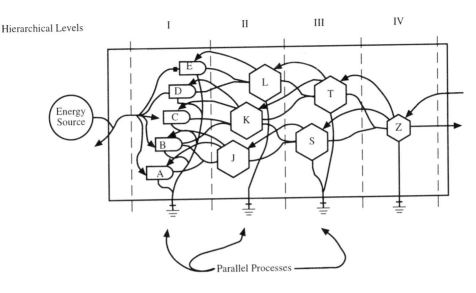

FIGURE 4.3 Hierarchical organization of systems, with many small units supporting few large units at higher levels. Feedbacks from units at higher levels to units at lower levels are also shown (Odum, 1996).

the overall simultaneous dynamical stability at all the levels of the hierarchy (Giampietro, 1994). Moreover, structure and hierarchy depend upon input flows, which can vary over time. Thus, structure will also vary, increase or decrease, following the evolution of resources inflows. Evaluating the time course of resource inflows and storages, by means of their emergy content and transformities, can be the way to forecast possible changes in the system structure and performance.

4.4.1   THE THERMODYNAMICS OF HIERARCHIES

Implicit in the formulation of the emergy theory is that:

1. Energy is conserved within each hierarchical level, but its availability (available energy, exergy) is used up, according to the Second Law of Thermodynamics.
2. Emergy is conserved in a different way than energy: all emergy converging to the upper level is assigned to this level itself. When emergy is used up, it only means that it is transferred to other levels of the hierarchy.
3. Quality of energy at each level is measured by its transformity.
4. Emergy and exergy of a given flow or Item are linked to each other by the transformity of the Item itself (emergy = exergy × transformity).

By determining transformities in units of one kind, in this case solar emjoules, it is possible to show and compare the position of all the components of ecosystems in the global energy hierarchy.

4.4.2   HIERARCHIES IN ECOSYSTEMS INCLUDING HUMANS

The hierarchical organization of natural systems has been clearly pointed out by many authors (including Odum, E. P., 1983; Odum, H. T., 1983a; O'Neill et al., 1986). More recent is the increasing interest for measures of hierarchical organization in man-made systems, like, say, a city or a nation. When analyzing the city structure and its parts, we realize that understanding the dynamics of the general design of the system is crucial in order to assess a sustainability criterion for the whole system itself. A city is organized in hierarchies where many different sectors are linked in a web (Figure 4.4). Many kinds of hierarchies

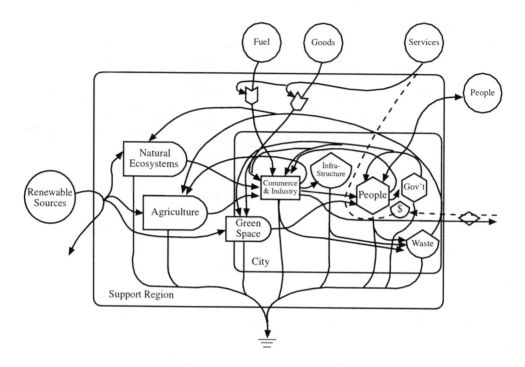

FIGURE 4.4 Hierarchical organization of sectors within a city.

can be pointed out, all contributing to the global stability of the city. Examples are

- Individuals, families, condominiums, districts, the whole city;
- Small artisan shops, small industries, commerce sector, consumers;
- Kindergarten, elementary schools, middle schools, high schools, colleges, universities.

Cities are points of convergence in the landscape that represent large concentrations of people, structure, and information (Odum et al., 1996). Energies and materials inflow from surrounding regions producing large volumes of wastes in air, water, and solid waste dumps. It has long been demonstrated that landscapes of cities are organized hierarchically, where there are many small rural towns, fewer small cities, and fewer and fewer larger cities. It has been suggested (Christaller, 1966; Losch, 1954) that the hierarchy results from the distribution of goods having varying market regions. It was found that market region increased with city size and the array of goods increased because of the larger market areas from which to draw demand.

Probably just as important is the convergence of energies and materials into cities. Environmental support of cities must be converged from larger and larger support regions as city size increases. The larger the city, the greater the area of support required to produce necessary inputs or from which inputs are extracted. The hierarchy of cities results from the interplay of both market regions and support regions. Energy and materials are concentrated in pathways of convergence, while information and goods feedback in diverging pathways of control and amplifier actions. Table 4.2 lists several characteristics of classes of cities in Florida, U.S.A. Class 1 cities are the largest in the region, serving as central places and having populations of over one half million people. Class 5 cities are the smallest incorporated towns found scattered throughout the landscape having typical populations of about 2000 people. Annual empower is the total inflow of emergy per year consumed within the city. Empower density is the flow of emergy per unit area of the city per year. Annual use of emergy varies from 30.6 E21 sej/yr to only 0.1 E21 sej/yr for the class 1 and class 5 cities, respectively.

TABLE 4.2
Characteristics of Urban Systems in North Central Florida, U.S.A

Urban Class	Population (E3 people)	Area (km²)	Annual Empower (E21 sej/yr)	Empower Density (E12 sej/m²·yr⁻¹)	Support Region (km²)
Class 1	504	399.9	30.6	76.4	9855.1
Class 2	99	254.5	14.9	58.6	4778.3
Class 3	37	55.5	2.8	50.4	896.6
Class 4	13	18.6	0.8	44.2	263.1
Class 5	1.8	2.8	0.1	35.8	32.2

Source: Brown, M. T., Ph.D. dissertation, University of Florida, Gainesville, 1980.

TABLE 4.3
Characteristics of Selected Land Uses in Florida Cities, U.S.A

Land Use Type	Empower Density (E12 sej m⁻² yr⁻¹)	Emergy of Structure (E12 sej m⁻²)	Structural Mass (E3 kg m⁻²)	% of City Area
Single family residential (med. density)	20.7	149.1	181.5	81.3
Multifamily residential (avg. 4 floors)	126.6	1135.4	1170.0	4.5
Commercial strip	46.4	517.1	720.4	3.5
Commercial mall	220.7	1248.7	1429.4	0.9
Central business district (avg. 4 floors)	294.2	2026.8	2067.0	0.8

Source: Brown, M. T., Ph.D. dissertation, University of Florida, Gainesville, 1980.

The most intensely developed cities are the class 1 central places where commercial and industrial uses make up a greater proportion of the total city area than in the smaller cities. Intensity of activity can be measured by empower density (emergy flow per unit area, sej/m² yr⁻¹). The empower density of the Florida cities ranges from about 76 El2 sej/m² yr⁻¹ to about 36 E12 sej/m² yr⁻¹ as given in Table 4.2. Within cities spatial organization is hierarchical from the low intensity rural fringe to the high concentrations of information and business in the Central Business District. The central city, where buildings and populations are largest, is surrounded by rings of decreasing intensity. When structure and land use "metabolism" (energy use) are expressed as emergy, the increasing intensity of activity is obvious. Given in Table 4.3 are characteristics of several typical urban land uses in Florida cities and the percent of city area that is devoted to these uses. The empower density and emergy in structure increase with increasing intensity of activity while area decreases.

4.5   EVALUATING ECOSYSTEMS FOR ECONOMIC QUESTIONS

Traditional economic or energy analyses usually do not take into account inputs that they cannot evaluate on an actual monetary or energy basis. As economies rely upon very large inputs from the environment (natural capital), if these inputs are not accounted for and given a value, misuse of resources can follow and future prospects for the system cannot be adequately inferred. An economy is vital when it has abundant goods and resources and uses them to reinforce and maintain productivity. Energy, material, and information are the real wealth. It takes energy to concentrate the minerals needed by an economy. It takes energy to maintain and process information. When resources are abundant and cheap, there is abundant wealth and a high standard of living. If resources and basic products are imported cheaply, abundant wealth is imported and raises the standard of living.

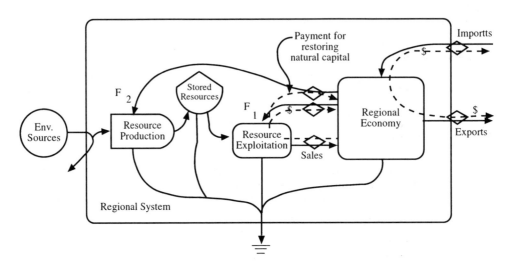

FIGURE 4.5 Model of resource building and use. Resources provided and stored by nature's work are then exploited for economic uses. A reinforcing feedback from consumers is needed to avoid the depletion of natural capital.

4.5.1 VALUING ENVIRONMENTAL WORK

A storage of environmentally generated resource is shown in Figure 4.5. Environmental work builds and maintains this storage, that is in turn exploited by human economies by investing work, fuels, goods and services, and information. Natural resources exploited by human economies are taken for free: money only pays for the work done to extract and process the resource, not just the work performed by nature to store it. Abundant resources therefore have a low market value, due to the small investment that is required for their exploitation.

When a resource is used up by the growth of the demand, it becomes scarce and its price rises. The higher price makes the resource more valuable in economic terms, encouraging further depletion. If a feedback is not provided to maintain or restore the environmental storage, the resource is pulled down and the economic activity based on it will ultimately follow the same trend. A sustainable mechanism requires the economic system to invest money and work in maintaining the natural capital, so that the real source of wealth is maintained (Odum, 1994b).

Microeconomic theory uses market value to measure the contribution of goods, resources, and services to an economy. The market value of something is the amount humans are willing to pay for it. Microeconomic theory works fine at the grocery store, but it is quite inadequate when evaluating natural resources because it does not work where markets do not exist. For instance, there is not a market for the services of water storage and filtration provided by wetlands. It is very common that large debates develop on economic costs and benefits of processes affecting natural environment or programs for restoration of environmental assets, yet, it still remains difficult to make decisions, because some benefits and costs are evaluated in dollars and others are environmental costs and benefits that cannot be given a dollar value. Thus society is left with making decisions based on only half the information it needs.

Conventional energy analysts only account for inputs under human control, without including natural inputs, that they take for free. Furthermore, inputs accounted for are not given a quality factor, and input joules are assumed to be all equal, no matter the pathway that was followed to supply a given resource. On the contrary, at the basis of emergy, accounting approach is the recognition that even inputs that are small in energy terms may have huge weight in the process. This weight is acknowledged by assigning an appropriate transformity to each input. Under Maximum Empower constraints, the higher the transformity of an input, the higher the environmental support that has been required to produce it, and the higher the potential driving or control ability of the input in the process.

By evaluating resources in emergy units a more complete analysis of human systems and their trends can be performed. It allows the comparison among resource uses in different systems (subsistence and industrialized agriculture, developed and developing countries, different regions within the same nation, and so on) and underlines the existence of hierarchies among them. Also, by means of transformities, a quality assessment of emergy concentration per unit flow or product within a system can be made, thus moving a step beyond the monetary, market evaluation of commodities (Ulgiati and Brown, 1998a).

4.5.2 Evaluating Renewability and Replacement of Resources

Human societies have learned how to exploit resources at a higher rate than they are replaced; this raises the problem of their sustainability according to the biosphere carrying capacity and available storages (Brown and Ulgiati, 1999). The renewability concept is very relative. If we define slowly renewable as a source whose turnover time occurs at rates slower than our exploitation rate (and fast renewable the opposite case), we realize that only sources like sun, rain, wind, waves, tides, and maybe geothermal heat can be considered renewable. All fossil energy sources and other minerals, and fertile topsoil itself (approximately 500 years are needed to replace 2.5 cm of topsoil by natural processes under normal agricultural conditions, Pimentel et al., 1988) should be considered nonrenewable. Underground waters, when drained faster than they are replaced by natural processes of water cycle, should also be considered nonrenewable. Forests that we exploit at rates faster than their replacement time are nonrenewable.

The ratio ϕ_R of fast renewable resources (usually called renewables) to the total emergy used (renewable plus nonrenewable inputs) is therefore an important measure of sustainability of the process itself. The ratio of renewable to total emergy, ϕ_R, is linked to the above defined Environmental Loading Ratio by the equation ELR $= (1/\phi_R) - 1$.

Transformities may also be indicators of renewability and turnover time. As defined above, the transformity is a measure of the convergence of emergy flows to yield a given product. Therefore, if a large activity from the (past and present) environment is required to provide the inputs, it follows that,

1. A similarly large activity would be required to replace them, when excessively exploited. Larger turnover times are usually linked to higher transformities (Doherty, 1995). Replacement of Items supported by a long chain of transformation processes takes time, making output Items less renewable than input Items driving the process.
2. If two processes A and B yield the same kind of product (say electricity) and A requires a lower emergy input per unit output than B, the resulting lower transformity of A is a measure of its higher thermodynamic efficiency. This comparison in terms of efficiency is obviously meaningless when outputs are different. Ulgiati and Brown (1998a) have stressed the meaning of transformities as quality, hierarchy, and efficiency indicators, suggesting that through their use quality of matter and energy flows is accounted for in resources policymaking.

Environment is a scarce resource, in comparison to exploitation by the human economy. Some Items (i.e., topsoil, clean water, etc.) are already scarce; some (sand, gravel, other minerals) do not seem to be scarce at present, but their long replacement time puts an upper limit to the amount that can be exploited without damage to the environment as a whole. Just think of the environmental and economic problems caused by the excessive withdrawal of sand and gravel from riverbeds; the apparent low economic value of these minerals results from their large availability, but availability would no longer be such, if environmental constraints were applied. If these constraints are not taken into account, the overall self-organization of the system would be disturbed and the economy would also be affected.

4.5.3 ASSESSING SUSTAINABILITY

The amount of the investment that is required to exploit a local renewable or nonrenewable resource is a key parameter when evaluating the feasibility of a process and its relationship with the outside environment and economy. The ratios of N and R to the investment F (N/F and R/F) can be used as measures of this relationship (Figure 4.2). Let it be given that: $\eta = R/F$ and $\vartheta = N/F$. In this way, the three independent variables give rise to two functions $\eta(R,F)$ and $\vartheta(N,F)$, that we have called *exploitation functions*. These functions define a measure of the locally available renewable or nonrenewable emergy flows that are exploited using an emergy investment from outside the system.

We have introduced elsewhere (Brown and Ulgiati, 1997; Ulgiati and Brown, 1998b) an Emergy Sustainability Index, ESI, defined as the ratio of a yield emergy indicator (EYR) to an environmental loading indicator (ELR). It can be shown that the ESI is an aggregated measure of economic performance and ecological sustainability. Its mathematical definition is

$$\text{ESI} = \text{EYR/ELR} = \eta + [\,\eta^2/(1 + \vartheta)]\qquad\qquad(3)$$

where it clearly appears to be a function of η and ϑ. This index shows a different sensitivity to variations of the components of the emergy inputs. It decreases when ϑ (the nonrenewable component) is increasing. An increase of the function occurs at increasing values of η (the renewable component), with a parabolic trend at low ϑ values, tending to a linear one when ϑ becomes very large. It is important to keep in mind that η and ϑ are ratios of locally renewable and nonrenewable inputs to feedbacks from outside: the variables are three, not just two. Sustainability is not provided by a low requirement of outside investments F, but by a huge renewable input in comparison to the investment itself, that may also be large. Therefore, a huge investment from outside the process can also be useful and sustainable, provided that it allows the exploitation of a large amount of emergy from locally renewable sources, and therefore an increase in the value of ESI. Detailed examples and calculation of ESI in selected case studies are reported in Brown and Ulgiati (1997).

A sustainability assessment requires the simultaneous evaluation of the transformity as well as the ESI. High transformities can be due both to renewable and nonrenewable inputs: the higher the transformity, the higher the share of environmental work converging into the output. Transformities are not sensitive to the nature of the inputs, yet they can indicate the efficiency of the process of converting inputs into the final product (above Section 4.5.2, point 2). On the contrary, the higher the ESI function, the higher the fraction of locally renewable emergy that is exploited in comparison to nonrenewable as well as imported: if the yield is supported by a high contribution from locally renewable flows of resources, the process is in a better equilibrium with the surrounding environment. Therefore, the same transformity may characterize processes with the same conversion efficiency, but a different global sustainability.

A general rule that may be suggested when assessing the sustainability of a process is that high transformity of a process yield (i.e., demand of large environmental support, high turnover time, low renewability of the product) should be coupled with high values of the ESI (large contribution from locally renewable sources, equilibrium with the surrounding environment).

4.6 ENVIRONMENTAL AND HUMAN-DOMINATED PROCESSES

The total emergy driving a process becomes a measure of the self-organization activity of the surrounding environment, that is converged to make that process possible. It is a measure of the environmental work necessary to provide a given resource. In the case of stored resources, part of the work of past and present ecosystems is used to make a given resource available (be it the present oxygen stock in the atmosphere or the present stock of gold or oil deep in the planet). Such a measure may provide the maximum size allowed for system's growth, i.e., represents the upper limit to the carrying capacity of the system under study. This can be better understood if we think of an electric hydropower plant. Plant size and performance must be tuned to the availability of water flow; a larger plant would not be supported by a small flow. The same happens with biosphere, completely supported and constrained by a flow of (solar) emergy from outside and the flow of stored resources from within. The same also can be said for all ecosystems, whose growth occurs within the limits of available emergy. Giampietro and Pimentel (1991) developed a similar approach, introducing a Biosphere Space-Time Activity (BIOSTA) indicator as a measure of environmental life-supporting work. Rees and Wackernagel (1994) recently proposed a similar approach, called Ecological Footprint.

Man-made systems at different levels can be evaluated by using this methodology:

1. Economic systems as a whole at the national or local level;
2. Production sectors, like agriculture or the energy sector of a country;
3. Individual industrial processes, like production of electricity, biofuels, or a given commodity from raw material, focusing on a given plant or production methodology.

Different time and space scales characterize systems 1 to 3. Of course, their evaluation requires the accounting methodology to be adapted to the subject under study; also, each system may require the calculation and discussion of specific indices, while others may be meaningless or scarcely useful. We have analyzed different systems that could be classified as above. Among them are (1) Italian agriculture as a whole (Ulgiati et al., 1993); (2) production of biofuels from different biomass substrates (Bastianoni et al., 1995; Ulgiati and Bastianoni, 1995a,b); (3) the economy of Italy (Ulgiati et al., 1994b; Ulgiati et al., 1995b). Results confirm that emergy-based indices may help to give a deeper insight into a system's dynamics and its relationships with the natural environment. We believe that emergy indices are able to account for pollution factors (Ulgiati et al., 1995a) or to indicate if a production pathway is sustainable or not (Brown and Ulgiati, 1997). In the following section we present and discuss selected case studies.

4.6.1   EMERGY EVALUATION OF AGRICULTURAL PRODUCTION

Photosynthesis and plant growth driven by environmental input occur at rates that have been selected over millions of years of natural evolution. The introduction of agriculture, 10,000 years ago, has deeply modified this pattern. Wild habitats have been partially substituted by domestic and simplified habitats, characterized by a lower biodiversity, lower standing biomass, and a higher productivity of selected crops of interest to humans. Agriculture, a production process interacting with many other sectors of the biosphere (atmosphere and oceans, surrounding ecosystems, storages of mineral underground reserves, human societies), is also driven by inputs under human control.

4.6.1.1   The Case of Italian Agriculture

As a nation becomes highly developed with urban civilization, what kinds of agriculture are compatible? A spatial hierarchy develops with cities as the centers. Lands near cities are either removed from agricultural production or require intensive, high value crops to compete with alternative economic uses, pay the high taxes, and continue as part of the high density urban economy. An emergy analysis of Italian agricultural and livestock system has been performed to evaluate the role of energy quality and environmental inputs (Ulgiati et al., 1993). Natural as well as purchased inputs from main economy, all converging to drive the global system of food production at the national level have been evaluated together with yielded agricultural and livestock products.

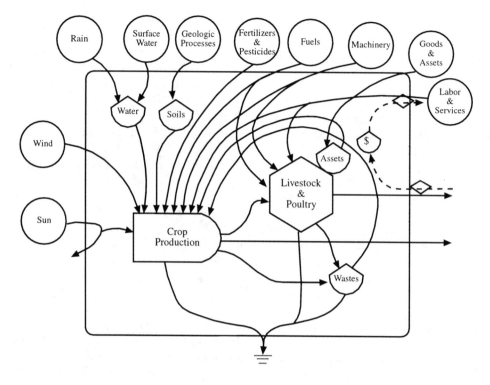

FIGURE 4.6 Emergy diagram of Italian agricultural and livestock system (Ulgiati et al., 1993).

Figure 4.6 shows an aggregated energy systems diagram of Italian agriculture and livestock sector, as detailed in Table 4.4. The reader may like to refer to the published paper for a detailed discussion of data, calculations, and results. Here, we only point out, from Table 4.4,

1. Italian agriculture-livestock sector relies for about two-thirds upon nonrenewable imported emergy flows, that enhance production on available land (mostly fossil substrates for fertilizers and fuels manufacturing, plus other fossil inputs supporting cattle feedstock production and human labor). Carrying capacity of Italy would be lower if imports of primary raw materials would not provide, in a hidden form, a substitute for land, making evident the friction between what is needed and what is locally available.
2. The emergy value of each input flow measures the different amount of environmental work embodied in the Item itself. Some results are very unexpected: only think of the huge input of emergy embodied in the process of making labor available in a developed country like Italy. Fossil fuels and chemicals inputs are largely lower than labor input, in emergy terms.
3. Average transformies of global food production from agricultural and livestock sectors are calculated as the ratio of total emergy input of that sector to total energy output. Livestock production differs by more than one order of magnitude from crop production, thus indicating a larger convergence of environmental work and a higher hierarchical level.
4. When individual crop productions are evaluated, the total emergy input splits into different, parallel processes, from which a hierarchy of transformies can be calculated according to Equation (3) and the emergy algebra, Section 4.2.3. Table 4.5 shows some of the emergy-based indices that can be calculated as measures of solar concentration, size, and environmental loading of the different crop production systems, compared to the average values of Italian agriculture and Italy as a whole. The possibility of a comparative overview of

TABLE 4.4
Emergy Analysis of Resources Basis for Italian Agriculture, 1989 (data per hectare per year)

No.[a]	Item	Amount	Units	Solar Transform. (sej/unit)	Ref. for Transf.[b]	Solar Emergy (E20 sej)
RENEWABLE RESOURCES						
1	Sunlight	6.17E+20	J	1	[1]	6.17
2	Rain chem. potential	3.59E+17	J	18199	[1]	65.33
3	Rain geopotential	3.15E+17	J	10488	[1]	33.07
4	Earth cycle	5.07E+17	J	34377	[1]	174.29
NONRENEWABLE SOURCES FROM WITHIN THE SYSTEM						
5	Net loss of topsoil	2.12E+16	J	62500	[1]	13.26
APPLIED ENERGY AND LABOR						
6	Electricity, crop prod.	1.05E+16	J	2.00E+05	[1]	21.02
7	Electricity, livestock	2.94E+15	J	2.00E+05	[1]	5.89
8	Lubricants	9.04E+14	J	66000	[1]	0.60
9	Diesel, crop prod.	9.53E+16	J	66000	[1]	62.88
10	Diesel, livestock	2.94E+15	J	66000	[1]	1.94
11	Gasoline	1.07E+16	J	66000	[1]	7.04
12	Labor, crop prod.	4.27E+15	J	6.23E+06	[2]	266.04
13	Labor, livestock	3.37E+15	J	6.23E+06	[2]	209.70
GOODS AND ASSETS FOR CROP PRODUCTION						
14	Potash fertilizers, K_2O	4.37E+11	g	1.10E+09	[1]	4.81
15	Nitrogen fertilizers, N	9.23E+11	g	4.62E+09	[1]	42.64
16	Phosphate fertil., P_2O_5	6.86E+11	g	3.90E+09	[1]	26.75
17	Pesticides	1.29E+16	J	6.60E+04	[1]	8.52
18	Mechanical equipment	6.42E+11	g	6.70E+09	[1]	43.01
19	Seeds (oil equiv.)	8.93E+15	J	66000	[1]	5.89
20	Assets, crop prod. (oil equiv.)	6.56E+15	J	66000	[1]	4.33
GOODS AND ASSETS FOR LIVESTOCK						
21	Assets, livestock (oil equiv.)	3.57E+15	J	66000	[1]	2.36
22	Industrial fodder (oil equiv.)	4.72E+16	J	66000	[1]	31.15
23	Forage	3.16E+17	J	63282	[3]	199.93
24	Self-produced fodder (oil equiv.)	3.00E+15	J	66000	[1]	1.98
PRODUCTS OF AGRICULTURAL AND LIVESTOCK SECTORS						
25	Total crop production	8.16E+17	J	8.34E+04		681.08
26	Total livestock production	9.42E+16	J	4.81E+05		452.94

[a] numbers in first column refer to footnotes of the table, following.

[b] **References for transformities:**

[1] Odum, 1996.

[2] Ulgiati et al., 1994.

[3] Ulgiati et al., 1993.

Performance Indicators	
Emergy Yield Ratio	1.38
Environmental Loading Ratio	2.91
Emergy Sustainability Index	0.47

(Continued)

TABLE 4.4
Emergy Analysis of Resources Basis for Italian Agriculture, 1989 (data per hectare per year) (continued)

Footnotes of Table 4.4

1 SOLAR ENERGY

Land area	=	1.69E+11 m²	[ISTAT, 1990a,b]
Insolation	=	1.09E+02 (kcal/cm²)/yr	(ISTAT, 1991]
Albedo land	=	0.20 (% given as decimal)	[Henning, 1989]
Energy (J/yr) = (land area) (avg. insolation)(1-albedo)	=	6.17E+20 J/yr	

2 RAIN CHEMICAL POTENTIAL

Land area	=	1.69E+11 m²	[ISTAT, 1990a,b]
Rain (average)	=	0.99 m/yr	[ISTAT, 1991]
Evapotransp. rate	=	0.43 m/yr (43.6% of total rainfall)	[ISTAT, 1991; Henning, 1989]
Energy on land = (Area)(Evapo-transpired rainfall)(Water density) (free energy of water)	=	3.59E+17 J/yr	

3 RAIN, GEOPOTENTIAL ENERGY

Area	=	1.69E+11 m²	[ISTAT, 1990a,b]
Rainfall	=	0.99 m/yr	[ISTAT, 1991]
Average elevation	=	340.00 m	[IGDA, 1975]
Runoff rate	=	0.56 m/yr (56.4% of total rainfall)	[ISTAT, 1991]
Energy = (area)(runoff rate) × (water density)(avg. elevation) × (gravity)	=	3.15E+17 J/yr	

4 EARTH CYCLE (steady state uplift balanced by erosion)

Heat flow per area	=	3.00E+06 (J/m²)/yr	[Odum, 1994a]
Land area	=	1.69E+11 m²	[ISTAT, 1990a,b]
Energy (J/yr) = (land area) × (heat flow per area)	=	5.07E+17 J/yr	

5 NET LOSS OF TOPSOIL

Farmed area		1.69E+11 m²	[ISTAT, 1990a,b]
Erosion rate	=	2.00E+02 (g/m²)/yr	[Triolo, 1989]
% organic in soil	=	0.03	[Odum, 1992]
Energy cont./g organic	=	5.00E+00 kcal/g	[Odum, 1992]
Net loss = (farmed area) × (erosion rate)	=	3.38E+13 g/yr	
Energy of net loss (J/yr) = (net loss)(% org. in soil) × (5.4 kcal/g)(4186 J/kcal)	=	2.12E+16 J/yr	

6 ELECTRICITY USED FOR CROP PRODUCTION

Total use	=	2.92E+09 kWh/y	[ISTAT, 1990b, 1991]
	=	1.05E+16 J/yr	

7 ELECTRICITY USED FOR LIVESTOCK

Total use	=	8.18E+08 kWh/y	[ISTAT, 1990b, 1991]
	=	2.94E+15 J/yr	

8 LUBRICANTS, CROP PRODUCTION

Total use	=	1.20E+07 kg/yr	[ISTAT, 1990b]
Energy content per kg	=	7.53E+07 J/kg	[Triolo et al., 1984]
Energy (J/yr) = (total use) × (Energy content per kg)	=	9.04E+14 J/yr	

9 DIESEL FOR CROP PRODUCTION (included fodder production for livestock)

Total use	=	1.85E+09 kg/yr	[ISTAT, 1990a,b]
Energy content per kg	=	5.15E+07 J/kg	[Biondi et al., 1989]
Energy (J/yr) = (total use) × (Energy content per kg)	=	9.53E+16 J/yr	

10 DIESEL FOR LIVESTOCK (fodder production is not included)

Total use	=	5.71E+07 kg/y	[ISTAT, 1990a,b; Triolo et al., 1984]
Energy content per kg	=	5.15E+07 J/kg	[Biondi et al., 1989]
Energy (J/yr) = (total use) × (Energy content per kg)	=	2.94E+15 J/yr	

11 GASOLINE

Total use	=	1.93E+08 kg/yr	[ISTAT, 1990a,b]
Energy content per kg	=	5.53E+07 J/kg	[Biondi et al., 1989]
Energy (J/yr) = (total use) × (Energy content per kg)	=	1.07E+16 J/yr	

12 LABOR FOR CROP PRODUCTION

Energy input:			[ISTAT, 1990a,b; Triolo et al., 1984; Biondi et al., 1989]
Total man-days applied	=	3.40E+08 working days (mostly not trained labor)	
Daily metabol. energy	=	3.00E+03 kcal/day per person	[Odum, 1992]
Total energy applied			
per person per year	=	8.55E+05 kcal/person/yr (285 working days/year)	
	=	3.58E+09 J/yr/person	
Total energy input = (total metabolic energy/person/day)(total man-days applied)(4186 J/kcal)	=	4.27E+15 J/yr	

(Continued)

TABLE 4.4
Emergy Analysis of Resources Basis for Italian Agriculture, 1989 (data per hectare per year) (continued)

Emergy per person (Italy, 1989)	=	2.23E+16 sej/yr	[Ulgiati et al., 1994]
Solar transformity of labor = (Total emergy/yr/person)/ (Total applied energy/yr/person)	=	6.23E+06 sej/j	

13 LABOR FOR LIVESTOCK

Energy input:			[ISTAT, 1990a,b; Triolo et al., 1984; Biondi et al., 1989]
Total man-days applied	=	2.68E+08 working days (mostly not trained labor)	
Daily metabol. energy	=	3.00E+03 kcal/day per person	[Odum, 1992]
Total energy input = (total metabolic energy/person/ day) × (total work-days applied) × (4186 J/kcal)	=	3.37E+15 J/yr	

14 POTASH FERTILIZER

K_2O content	=	4.37E+11 g/yr	[ISTAT, 1991]

15 NITROGEN FERTILIZER

N content	=	9.23E+11 g/yr	[ISTAT, 1991]

16 PHOSPHATE FERTILIZER

P_2O_5 content	=	6.86E+11 g/yr	[ISTAT, 1991]

17 PESTICIDES/Commercial Products

Anticryptogamics	=	1.06E+08 kg/yr	[ISTAT, 1991]
Production energy	=	5.60E+07 J/kg	[Biondi et al., 1989]
Herbicides	=	2.88E+07 kg/yr	[ISTAT, 1991]
Production energy	=	9.10E+07 J/kg	[Biondi et al., 1989]
Insecticides	=	3.59E+07 kg/yr	[ISTAT, 1991]
Production energy	=	5.30E+07 J/kg	[Biondi et al., 1989]
Fytohormones	=	2.47E+07 kg/yr	[ISTAT, 1991]
Production energy	=	1.00E+08 J/kg	[Biondi et al., 1989]
Total use	=	1.95E+08 kg/yr	
Total energy	=	1.29E+16 J/yr (oil equivalent)	

18 MECHANICAL EQUIPMENT

Total equipment used	=	6.42E+11 g/yr	[Biondi et al., 1989]
Energy for prod. of machinery	=	9.20E+07 J/kg	[Biondi et al., 1989]
Total energy for machinery	=	5.91E+19 J/yr (oil equivalent)	

19 SEEDS

Cereal seeds	=	3.25E+08 kg/yr	[ISTAT, 1990b]
Potato	=	7.65E+07 kg/yr	[ISTAT, 1990b]
Vegetables	=	1.11E+07 kg/yr	[ISTAT, 1990b]
Oilseeds	=	1.52+06 kg/yr	[ISTAT, 1990b]
Sugarbeet	=	1.15E+06 kg/yr	[ISTAT, 1990b]
Tobacco	=	7.70E+03 kg/yr	[ISTAT, 1990b]
Forage	=	3.11E+07 kg/yr	[ISTAT, 1990b]
Total use of seeds	=	4.46E+08 kg/yr	
Average energy production of seeds	=	2.00E+07 J/kg	[Biondi et al., 1989]
Total energy for seeds = (total use) (Energy for production)	=	8.93E+15 J/yr (oil equivalents)	

20 ASSETS FOR CROP PRODUCTION

(Total assets and energy embodied for production and maintenance)   [Biondi et al., 1989]

Greenhouses	=	2.00E+04 ha
Unit energy cost	=	2.50E+11 J/ha/yr
Plastic mulch	=	2.00E+04 ha
Unit energy cost	=	7.80E+10 J/ha/yr
Total energy in assets	=	6.56E+15 J/yr (oil equivalents)

21 ASSETS FOR LIVESTOCK

(Total assets and energy embodied for production and maintenance)

Stables	=	8.50E+03 ha	[Biondi et al., 1989]
Unit energy cost	=	4.20E+11 J/ha/yr	
Total energy in assets	=	3.57E+15 J/yr (oil equivalents)	

22 FORAGE

Forage crops	=	1.04E+11 kg/yr	[ISTAT, 1990b]
Pasture	=	8.68E+09 kg/yr	[ISTAT, 1990b]
Total forage	=	1.13E+11 kg/yr	
Energy content/unit	=	7.25E+02 kcal/kg	[Triolo et al., 1984]
Total energy content = (Total fodder) × (Energy content per unit) =		3.16E+17 J/yr	

23 INDUSTRIAL FODDER

Total used	=	1.18E+13 g/yr	[ISTAT, 1990b]
Energy for production	=	4.00E+03 J/g	[Biondi et al., 1989]
Total energy required = (total used) (Energy requirement)	=	4.72E+16 J/yr (oil equivalents)	

24 SELF-PRODUCED FOODER (production in the farm)

Total used	=	7.49E+11 g/yr	[ISTAT, 1990b]
Energy for production	=	4.00E+03 J/g	[Biondi et al., 1989]
Total energy required = (total used) × (Energy requirement)	=	3.00E+15 J/yr (oil equivalents)	

(Continued)

TABLE 4.4
Emergy Analysis of Resources Basis for Italian Agriculture, 1989 (data per hectare per year) (continued)

25 CROP PRODUCTION

RICE

Total production	=	1.25E+09 kg/yr	[ISTAT, 1990b]
Energy content per kg	=	3.00E+03 kcal/kg	[Triolo et al., 1984]
Total energy content	=	3.74E+12 kcal/yr	
	=	1.56E+16 J/yr	

FORAGE

Total production	=	1.49E+11 kg/yr	[ISTAT, 1990b]
Energy content per kg	=	7.25E+02 kcal/kg	[Triolo et al., 1984]
Total energy content	=	1.08E+14 kcal/yr	
	=	4.52E+17 J/yr	

SUGARBEET

Total production	=	1.70E+10 kg/yr	[ISTAT, 1990b]
Energy content per kg	=	6.67E+02 kcal/kg	[Triolo et al., 1984]
Total energy content	=	1.13E+13 kcal/yr	
	=	4.74E+16 J/yr	

CORN

Total production	=	6.44E+09 kg/yr	[ISTAT, 1990b]
Energy content per kg	=	3.50E+03 kcal/kg	[Triolo et al., 1984]
Total energy content	=	2.25E+13 kcal/yr	
	=	9.44E+16 J/yr	

WHEAT

Total production	=	7.88E+09 kg/yr	[ISTAT, 1990b]
Energy content per kg	=	3.30E+03 kcal/kg	[Triolo et al., 1984]
Total energy content	=	2.60E+13 kcal/yr	
	=	1.09E+17 J/yr	

FRUITS (Apples, pears, peaches, plums, and apricots)

Total production	=	4.34E+09 kg/yr	[ISTAT, 1990b]
Energy content per kg	=	5.50E+02 kcal/kg	[Triolo et al., 1984]
Total energy content	=	2.39E+12 kcal/yr	
	=	9.99E+15 J/yr	

VINEYARD

Total production	=	9.64E+09 kg/yr	[ISTAT, 1990b]
Energy content per kg	=	6.80E+02 kcal/kg	[Triolo et al., 1984]
Total energy content	=	6.55E+12 kcal/yr	
	=	2.74E+16 J/yr	

ORANGES AND LEMONS

Total production	=	2.82E+09 kg/yr	[ISTAT, 1990b]
Energy content per kg	=	4.40E+02 kcal/kg	[Triolo et al., 1984]
Total energy content	=	1.24E+12 kcal/yr	
	=	5.19E+15 J/yr	

OLIVES

Total production	=	3.07E+09 kg/yr	[ISTAT, 1990b]
Energy content per kg	=	1.70E+03 kcal/kg	[Triolo et al., 1984]
Total energy content	=	5.22E+12 kcal/yr	
	=	2.18E+16 J/yr	

SUNFLOWERS

Total production	=	2.78E+08 kg/yr	[ISTAT, 1990b]
Energy content per kg	=	6.10E+03 kcal/kg	[Triolo et al., 1984]
Total energy content	=	1.69E+12 kcal/yr	
	=	7.09E+15 J/yr	

ALMONDS

Total production	=	1.02E+08 kg/yr	[ISTAT, 1990b]
Energy content per kg	=	1.60E+03 kcal/kg	[Triolo et al., 1984]
Total energy content	=	1.63E+11 kcal/yr	
	=	6.80E+14 J/yr	

TOTAL AGRICULTURAL PRODUCTION   [ISTAT, 1990b; Triolo
                                                 et al., 1984]

Energy (J/yr)	=	1.95E+14 kcal/yr* (4186 J/kcal)	
	=	8.16E+17 J/yr	

26 LIVESTOCK PRODUCTION (meat, eggs, milk)

Total meat	=	3.45E+09 kg/yr	[ISTAT, 1990b]
Total milk and cheese	=	1.05E+10 kg/yr	[ISTAT, 1990b]
Total eggs	=	6.52E+09 kg/yr	[ISTAT, 1990b]
a. Meat:		3.45E+09 kg/yr	[ISTAT, 1990b]

  Total protein content = (Total prod.)(0.22 organic)   [Odum, 1996]

  Energy (J/yr) = (Total production)(0.22)(1000 g/kg)(5.0 kcal/g)

	=	(4186 J/kcal)(3.45E+9 kg/yr)(1E+03 g/kg)
	=	(5.0 kcal/g)(4186 J/kcal)(0.22)=1.59E+16 J/yr

  b. Milk and Cheese (produced 1.05E+10 kg/yr in the farm):   [ISTAT, 1990b]

  Total protein content = (Total prod.)(0.22 organic)   [Odum, 1996]

  Energy (J/yr) = (Total production)(0.22)(1000 g/kg)(5.0 kcal/g)(4186 J/kcal)

	=	(1.05E+10 kg/yr)(1E+03 g/kg)(5.0 kcal/g)
	=	(4186 J/kcal)(0.22) = 4.83E+16 J/yr

(Continued)

TABLE 4.4
Emergy Analysis of Resources Basis for Italian Agriculture, 1989 (data per hectare per year) (continued)

c. Eggs:	6.52E+09 kg/yr	[ISTAT, 1990b]

Total protein content = (Total prod.)(0.22 organic)   [Odum, 1996]

Energy (J/yr) = (Total production)(0.22)(1000 g/kg)(5.0 kcal/g)(4186 J/kcal)

$$= \quad (6.52E+9 \text{ kg/yr}) \ (1E+03 \text{ g/kg})(5.0 \text{ kcal/g})$$

$$= \quad (4186 \text{ J/kcal})(0.22) = 3.00E+16 \text{ J/yr}$$

d. Total production   =   2.05E+10 kg/yr   [ISTAT, 1990b]

Total protein content = (Total prod.)(0.22 organic)   [Odum, 1996]

Energy (J/yr) = (Total production)(0.22)(1000 g/kg)(5.0 kcal/g)(4186 J/kcal)

$$= \quad (1.54E+10 \text{ kg/yr})(1E+03 \text{ g/kg})(5.0 \text{ kcal/g})$$

$$= \quad (4186 \text{ J/kcal})(0.22) = 9.42E+16 \text{ J/yr}$$

Source: Ulgiati, S. et al., 1993, revised by authors.

TABLE 4.5
Emergy Yield Ratio and Other Indices for Selected Crops in Italian Agriculture

Item	Crop	Solar Transformity (E4 sej/J)	Environmental Loading Ratio	Emergy Yield Ratio	EYR/ELR
1	Rice	5.18	2.53	1.44	0.57
2	Forage	5.66	1.38	1.82	1.32
3	Corn	6.03	4.57	1.70	0.37
4	Sugarbeet	6.14	5.75	1.41	0.25
5	Soybeans	7.65	3.14	1.77	0.56
6	Sunflower	7.72	3.55	1.82	0.51
7	Rapeseed	8.88	3.49	1.84	0.53
8	Wheat	9.26	2.51	1.44	0.57
9	Fruits	22.70	10.26	1.11	0.11
10	Oranges and lemons	25.40	10.74	1.10	0.10
11	Grapes	31.60	7.06	1.15	0.16
12	Olives	40.90	4.74	1.23	0.26
13	Almonds	70.50	3.71	1.30	0.35
14	Tot. Crop Prod.	8.34	2.91	1.38	0.47

Source: Modified from Ulgiati, S. et. al., 1993.

the different production systems helps to evaluate the efficiency of each conversion process, on a global basis.

5. The Environmental Loading Ratios from Table 4.5 allow an evaluation of the matching of invested to local resources. The amount of outside and nonrenewable emergy invested to exploit a unit of locally available resources is an indirect measure of environmental

equilibrium and eventual disturbance of local dynamics (sustainable agriculture) by human controlled emergy flows.

Appropriate land use as well as incentives to technical innovation and other financial tools may follow according to these and other indicators, to increase the matching of investments to locally available resources.

4.6.1.2   The Case of Sugarbeet Production

A detailed analysis of Italian sugarbeet production is shown in Tables 4.6 and 4.7, where this crop is evaluated as a substrate for biofuels production by means of a comparative energy, emergy, and carbon balance. The process is diagrammed in Figure 4.7. We will discuss in the following section the feasibility of biofuel from sugarbeets. Here we only evaluate the process of producing the crop to show the differences and the potential integration between emergy and conventional approaches.

Inputs to the process are calculated in the footnotes of Tables 4.6 and 4.7. The are evaluated on a matter and energy basis, and then oil equivalents, carbon equivalents, and emergy content related to each input are calculated by means of suitable transformation coefficients. References are provided for each input and transformation coefficient. Finally, total energy and emergy inputs as well as total carbon dioxide emissions have been calculated, in order to provide a set of indices for the final assessment.

If we only account for energy and carbon flows (Table 4.7), the energy cost of sugarbeet production can be calculated by the total amount of energy invested into the agricultural phase, 0.90 Mg/kg of fresh sugarbeet. The energy input/output ratio can be calculated by means of the ratio of total energy value of fresh sugarbeet production, 1.14 E5 MJ, to total energy invested into the agricultural phase, 3.60 E4 MJ, yielding an E.R. = 3.16. The CO_2 emissions can be evaluated as 1.80 kg CO_2 per kg of available sugar yield.

TABLE 4.6
Emergy Analysis of Ethanol Production from Sugarbeet in Italy (1984 nationwide average values per hectare per year)

No.[a]	Item	Unit	Amount	Solar Transformity (sej/unit)	Ref. for Transf.[b]	Solar Emergy (E14 sej/ha/yr)
ENVIRONMENTAL INPUTS						
1	Sunlight	J	4.41E+13	1.00E+00	[2]	0.44
2	Rain chemical potential	J	4.45E+10	1.82E+04	[2]	8.09
3	Wind	J	8.82E+10	1.50E+03	[2]	1.32
4	Earth cycle	J	3.00E+10	3.44E+04	[2]	10.32
AGRICULTURAL PRODUCTION PHASE						
5a	Loss topsoil, resid. in field	J	1.36E+10	7.38E+04	[1]	10.01
5b	Loss topsoil, resid. harvested	J	2.71E+10	7.38E+04	[1]	20.02
Inputs assuming that residues are left in field						
6	Nitrogen fertilizer (N)	g	1.36E+05	3.80E+09	[2]	5.17
7	Phosphate fertilizer (P_2O_5)	g	1.87E+05	3.90E+09	[2]	7.29
8	Potash (K_2O)	g	7.28E+04	1.10E+09	[2]	0.80
9	Insecticides and pesticides	g	4.07E+04	1.48E+10	[1]	6.02
10	Herbicides	g	1.52E+04	1.48E+10	[1]	2.25
11	Diesel	J	1.33E+10	6.60E+04	[1]	8.75
12	Lubricants	J	2.53E+08	6.60E+04	[1]	0.17
13	Gasoline	J	4.42E+08	6.60E+04	[1]	0.29
14	Human labor	J	1.26E+08	7.38E+06	[3]	9.27
15	Agric. machinery	g	8.37E+04	6.70E+09	[1]	5.61

(Continued)

TABLE 4.6

Emergy Analysis of Ethanol Production from Sugarbeet in Italy (1984 nationwide average values per hectare per year) (continued)

No.[a]	Item	Unit	Amount	Solar Transformity (sej/unit)	Ref. for Transf.[b]	Solar Emergy (E14 sej/ha/yr)
16	Electricity	J	5.86E+08	2.00E+05	[1]	1.17
17	Seeds	J	5.58E+07	8.94E+04	[3]	0.05
18a	Surface water for irrigation	J	6.17E+09	4.10E+04	[1]	2.53
18b	Fuel for irrigation[c]	J		6.60E+04	[1]	
Additional inputs if 70% residues are harvested						
19	Nitrogen loss with erosion	g	4.50E+04	3.80E+09	[2]	1.71
20	Phosph. loss with erosion	g	2.25E+04	3.90E+09	[2]	0.88
21	Potash loss with erosion	g	1.50E+05	1.10E+09	[2]	1.65
22a	Additional water demand	J	2.47E+09	4.10E+04	[1]	1.01
22b	Fuel for additional water demand	J	2.39E+09	6.60E+04	[1]	1.57
23	Nitrogen harv. in residues	g	3.50E+04	3.80E+09	[2]	1.33
24	Phosphorus harv. in resid.	g	2.10E+04	3.90E+09	[2]	0.82
25	Potash harv. in residues	g	1.40E+05	1.10E+09	[2]	1.54
26	Diesel for residues	J	2.23E+09	6.60E+04	[1]	1.47
27	Labor for residues	J	4.38E+06	7.38E+06	[3]	0.32
28	Machinery for residues	g	2.46E+03	6.70E+09	[1]	0.16
Products of the agricultural phase						
29	Sugarbeet produced	J	1.14E+11	6.14E+04		69.70
30	Sugar available in sugarbeet	J	1.05E+11	6.66E+04		69.70
31	Residues in field as such[d]	J	4.67E+10	1.49E+05		69.70
32	70% harvested agric. resid.[d]	J	n.a.	n.a.		92.19
INDUSTRIAL PRODUCTION PHASE						
33	Plant machinery	g	7.24E+03	6.70E+09	[1]	0.48
34	Diesel for transport	J	1.77E+09	6.60E+04	[1]	1.17
35	Diesel for process heat	J	3.72E+10	6.60E+04	[1]	24.58
36	Electricity	J	1.87E+10	2.00E+05	[1]	37.49
Product of industrial phase						
37a	Ethanol, without residues	J	9.42E+10	1.42E+05		133.42
37b	Ethanol, with residues use	J	9.42E+10	1.65E+05		155.91

Note: Legend: R = renewable resource; N = nonrenewable resource; L = local resource,
F = feedback, resource purchased from outside, n/a = not applicable.

[a] Numbers of each item refer to footnotes below.

[b] *References for transformities:*
[1] Brown M.T. and Arding J., 1991.
[2] Odum H.T., 1996.
[3] Estimated from Ulgiati et al., 1994.

[c] Already accounted for as diesel and electricity consumption.

[d] Energy content of residues in field is assumed to be their gross calorific value.
(higher heating value); energy content of harvested and treated residues is assumed to be their "usable energy"
(see discussion in Item 31).

Aggregated input data for emergy analysis

(A) Sum of local renew. inputs without double counting (maximum among Items 1 to 4)	10.32

In case residues are left in field [hypothesis (a)]:

(B,a) Sum of all local inputs without double counting (sum of maximum among Items 1 to 4, and 5a)	20.33
(C,a) Sum of imported inputs (sum of Items 6 to 18)	49.37
(D,a) Total emergy used in the agricultural phase (sum of B,a and C,a)	69.70
(E,a) Sum of imported inputs to industrial phase (Items 33 to 36)	63.72
(F,a) Total emergy used, in case residues are left in field (sum of Items D,a and E,a)	133.42

In case residues are harvested [hypothesis (b)]:

(B,b) Sum of all local inputs without double counting (maximum among Items 1 to 4 and 5b)	30.34
(C,b) Sum of imported inputs (sum of Items 6 to 28)	61.85
(D,b) Total emergy used in the agricultural phase (sum of B,b and C,b)	92.19
(E,b) Sum of imported inputs to industrial phase (Items 33 to 36)	63.72
(F,b) Total emergy used, in case residues are harvested (sum of Items D,b and E,b)	155.91

Emergy based indices for sugarbeet, sugar, agricultural residues, and ethanol

Sugarbeet, sugar, and agricultural residues

Transformity of sugarbeet (fresh weight)	6.14E+04 sej/J	1.75E+08 sej/g
Transformity of sugar available in sugarbeet	6.66E+04 sej/J	1.09E+09 sej/g
Transformity of residues in field (as such)	1.49E+05 sej/J	
Emergy yield ratio of fresh sugarbeet (Item 29/Item C)		1.41
Environmental loading ratio of fresh sugarbeet (Items (29-A)/A)		5.75
Emergy density of fresh sugarbeet (Item 29/area)		6.97E+11 sej/m^2
Investment ratio (Items C/B) of fresh sugarbeet		2.43
EYR/ELR of fresh sugarbeet		0.25

Ethanol

Transformity	1.42E+05 sej/J	3.74E+09 sej/g
Emergy yield ratio (Items 37/(C+E))		1.18
Environmental loading ratio (Items (37-A)/A)		11.93
Emergy density (Item 37/area)		1.33E+12 sej/m^2
Investment ratio (Items (C+E)/B)		5.56
EYR/ELR, without residues		0.10

(Continued)

TABLE 4.6
Emergy Analysis of Ethanol Production from Sugarbeet in Italy (1984 nationwide average values per hectare per year) (continued)

Footnotes of Tables 4.6 and 4.7

RENEWABLE RESOURCES

1 SOLAR ENERGY

Land area	=	10000 m^2	
Insolation	=	132 (kcal/cm^2)/yr	[ENEA, 1989]
Albedo of land	=	0.20 (reflected radiation)	[Henning, 1989]
Energy (J/yr) = (land area)(avg. insolation)(1-albedo)	=	4.41E+13 J/yr	

2 RAIN CHEMICAL POTENTIAL

Land area	=	10000 m^2	
Rain (average)	=	0.90 m/yr	[ISTAT, 1993]

A percent of rainfall, depending upon local factors, is evapotranspired, allowing the growth of the crop. Irrigation, see below, also contributes to evapotranspiration.

Gibbs free energy of water	=	4.94 J/g	[Odum, 1996]
Energy on land = (Area)(total rainfall)(Water density)(free energy per gram of water)	=	4.45E+10 J/yr	

3 WIND KINETIC ENERGY

Wind energy on land	=	2.45E+04 kWh/yr	[Himschoot, 1988]
	=	(Wind energy on land) (3.6E6 J/kWh)	
	=	8.82E+10 J/yr	

4 EARTH CYCLE (steady state uplift balanced by erosion)

Heat flow per area	=	3.00E+06 J/m^2/yr	[Loddo et al., 1978/79]
Land area	=	10000 m^2	
Energy (J/yr) = (land area) (heat flow per area)	=	3.00E+10 J/yr	

AGRICULTURAL PRODUCTION PHASE

5a Net Loss of Topsoil, When Residues Are Left in the Field

Land area	=	10000 m^2	
Erosion rate	=	1500 g/m^2/yr	[our estimate for 2 to 4% slope soils]

Other estimates: 3966 g/m^2/yr, our evaluation from Magaldi et al., 1981.

% organic in soil	=	0.04	[Pimentel et al., 1995, p. 1118]

Energy cont./g organic	=	5.40 kcal/g

Net loss of topsoil =
  (farmed area)(erosion rate)   =   1.50E+07 g/yr

Organic matter in topsoil used up =
  (total mass of topsoil)(% organic)   =   6.00E+05 g/yr

Energy loss = (loss of organic matter)
  (5.4 kcal/g)(4186 J/kcal)   =   1.36E+10 J/yr

5b Net Loss of Topsoil, When Residues Are Harvested

Harvesting of 100% residues from sugarbeet production on land with a 2 to 4% slope can more than double the erosion of topsoil, that might increase up to 30 to 40 t/ha/yr. A conservative estimate for 70% residues harvesting could be 30 t/ha/yr.

Erosion rate   =   3000 g/m^2/yr   [our estimate for 2% to 4% slope soils]

Net loss of topsoil = (farmed area)
  (erosion rate)   =   3.00E+07 g/yr

Organic matter in topsoil used up   =   1.20E+06 g/yr

Energy loss = (loss of organic
  matter) × (5.4 kcal/g)(4186 J/kcal)   =   2.71E+10 J/yr

6 NITROGEN

N content   =   1.36E+05 g/yr   [Triolo et al., 1984, p. 40]

Other estimates: 143 kg/ha (Biondi et al., 1989, pp. 127–129); 145 kg/ha (CCPCS, 1991, France nationwide average); 180 kg/ha (European Union average, ERL, 1991); 180 kg/ha (IEA, 1994, p. 27); nitrogen demand is reported in the range 100–180 kg/ha in SERI (U.S.A., 1986).

7 PHOSPHATE FERTILIZER

P_2O_5 content   =   1.87E+05 g/yr   [Triolo et al., 1984, p. 40]

Other estimates: 175 kg/ha (Biondi et al., 1989, pp. 127–129); 130 kg/ha (CCPCS, 1991, France nationwide average); 140 kg/ha (European Union average, ERL, 1991); 140 kg/ha (IEA, 1994, p.27); phosphate demand is reported in the range 40–80 kg/ha in SERI (U.S.A., 1986)

8 POTASH FERTILIZER

K_2O content   =   7.28E+04 g/yr   [Triolo et al., 1984, p. 40]

Other estimates: 60 kg/ha (Biondi et al., 1989, pp. 127–129); 250 kg/ha (CCPCS, 1991, France nationwide average); 240 kg/ha (European Union average, ERL, 1991); 240 kg/ha (IEA, 1994, p.27); potash demand is reported in the range 145–336 kg/ha in SERI (U.S.A., 1986).

9 INSECTICIDES AND PESTICIDES

Total used   =   4.07E+04 g/yr   [Triolo et al., 1984, p. 40]

Other estimates: 9–17 kg/ha (including herbicides; Biondi et al., 1989, pp. 127–129);

10 HERBICIDES

Total used   =   1.52E+04 g/yr   [Triolo et al., 1984, p. 40]

11 DIESEL

Total used   =   298 kg/yr   [Triolo et al., 1984, p. 40]

(Continued)

TABLE 4.6
Emergy Analysis of Ethanol Production from Sugarbeet in Italy (1984 nationwide average values per hectare per year) (continued)

Other estimates: 541 kg/ha (Biondi et al., 1989, pp. 127–129); 176 kg/ha (CCPCS, 1991, France nationwide average); 184 kg/ha (European Union average, ERL, 1991).

Lower heating value	=	4.45E+07 J/kg	[Ellington et al., 1993, p. 408]
Energy (J/y) = (total use) (Energy content per kg)	=	1.33E+10 J/yr	

12 LUBRICANTS

Total used	=	5.50 kg/yr	[Triolo et al., 1984, p. 40]

Other estimates: 6 kg/ha (Biondi et al., 1989, pp. 127–129);

Lower heating value	=	4.61E+07 J/kg	[Ellington et al., 1993, p. 408]
Energy (J/y) = (total use)(Energy content per kg)	=	2.53E+08 J/yr	

13 GASOLINE

Total used	=	11 l/yr	[Triolo et al., 1984, p. 40]
Lower heating value	=	4.07E+07 J/kg	[Ellington et al., 1993, p. 408]
Total energy	=	4.42E+08 J/yr	

14 HUMAN LABOR

Total applied labor	=	8.00E+01 h/yr	(Estim. from Biondi et al., 1989)
	=	1.00E+01 working days per year (mostly not trained labor; 8 h/day)	
Daily metabol. energy	=	3.00E+03 kcal/day per person	
Total energy applied per year (total daily metabolic energy) (total applied labor)	= =	3.00E+04 kcal/yr	
	=	1.26E+08 J/yr	

15 AGRICULTURAL MACHINERY (assuming a 10-year life span)

Total machinery used	=	8.37E+04 g/yr	[Triolo et al., 1984, p. 40]

Other estimates: 39 kg/ha (Biondi et al., 1989, pp. 127–129);

Embodied energy machin.	=	8.00E+07 J/kg	(Biondi et al., 1989, p. 68)
Total energy demand	=	6.70E+09 J/yr	

16 ELECTRICITY

Total used	=	163 kWh/yr	[Triolo et al., 1984, p. 40]
	=	5.86E+08 J/yr	

Other estimates: 8 kWh/ha (Biondi et al., 1989, pp. 127–129, assuming energy demand completely met by diesel use.

17 SEEDS

Total used	=	2.00E+04 g/yr	[Biondi et al., 1989, pp. 127–129]
Energy cont. of seeds	=	2792 J/g	[estim. from Triolo et al., 1984, p. 29]
Energy of seeds = (seeds used) (unit energy content of seeds)	=	5.58E+07 J/yr	

18 SURFACE WATER FOR IRRIGATION

Water demand of sugarbeet (water transpired per kg fresh beet)	=	257 l/kg beet produced	[Stuani et al., 1986, p. 792]
Transpired water	=	10250 m³/yr	
Water from rainfall	=	9000 m³/yr	
Water from irrigation = (Total transpired water-water from rainfall)	=	1250 m³/yr	
Gibbs free energy of water	=	4.94 J/g	[Odum, 1996]
Total free energy in water = (volume of water)(water density) (Gibbs free energy per gram)	=	6.17E+09 J/yr	

The fossil energy cost in surface water irrigation is about:

	=	4.77E+06 J/m³	[Pimentel, 1980]
Total fossil energy cost	=	5.96E+09 J/yr	

ADDITIONAL INPUTS IF 70% RESIDUES ARE HARVESTED

19 NITROGEN (N) LOSS WITH EROSION

(Average estimate, based on values in Pimentel et al., 1995, p. 1118)

Nitrogen per unit soil	=	0.003 g Nitrogen/g soil used up
Total N loss = (additional soil loss)(Nitrogen per unit of soil)	=	4.50E+04 g/yr

20 PHOSPHATE (P_2O_5) LOSS WITH EROSION

(Average estimate, based on values in Pimentel et al., 1995, p. 1118)

Phosphate per unit soil	=	0.0015 g Phosphate/g soil used up
Total P_2O_5 loss = (additional soil loss)(Phosphate per unit of soil)	=	2.25E+04 g/yr

21 POTASH LOSS WITH EROSION

(Average estimate, based on values in Pimentel et al., 1995, p. 1118)

Potash per unit soil	=	0.01 g Potash/g soil used up
Total K_2O loss = (additional soil loss)(Potash per unit of soil)	=	1.50E+05 g/yr

(*Continued*)

TABLE 4.6
Emergy Analysis of Ethanol Production from Sugarbeet in Italy (1984 nationwide average values per hectare per year) (continued)

22 ADDITIONAL SURFACE WATER FOR IRRIGATION, DUE TO INCREASED RUNOFF FROM SOIL EROSION

Pimentel et al., 1995, estimate a 75 mm increased runoff when soil erosion increases by 17 t/ha. In the case under study, erosion increases by 15 t/ha. Therefore, an estimate of 50 mm increased runoff seems to be reasonable.

Quantity of water lost	=	0.05 m/yr	[Estim. from Pimentel et al., 1995, p. 1119]
	=	500 m³/yr	

This amount is not "used" by the crop, it is only a measure of increased supply required.

It runs off soon. Therefore it should not be added to emergy inputs, if we look for the actual plant performance (plant thermodynamic efficiency). As we are here evaluating the global process efficiency, which is affected by erosion, we add this value into the total. There is no doubt that it must be accounted for in the total input for energy and carbon balance.

Gibbs free energy of water	=	4.94 J/g	[Odum, 1996]

Total free energy in water = (volume of water)(water density)(Gibbs free energy per gram)

	=	2.47E+09 J/yr	

The fossil energy cost of surface water irrigation is about:

	=	4.77E+06 J/m³	[Pimentel, 1980]
Total fossil energy cost	=	2.39E+09 J/yr	

23 NITROGEN HARVESTED IN RESIDUES

Unit Nitrogen content of residues	=	0.003 g N/gram wet residues	[CNR-PFMA-1981, p. 22]
70% residues harvested	=	1.12E+07 g dry matter/yr	

Total nitrogen harvested from soil, to be replaced for a sustainable culture

	=	3.50E+04 g/yr	

24 PHOSPHORUS HARVESTED IN RESIDUES

Unit Phosphorus content of residues	=	0.002 g P/gram wet residues	[CNR-PFMA-1981, p. 22]
70% residues harvested	=	1.12E+07 g wet matter/yr	

Total phosphorus harvested from soil, to be replaced for a sustainable culture

	=	2.10E+04 g/yr	

25 POTASH HARVESTED IN RESIDUES

Unit Potash content of residues	=	0.013 g k/gram wet residues	[CNR-PFMA-1981, p. 22]
70% residues harvested	=	1.12E+07 g wet matter/yr	

Total potash harvested from soil, to be replaced for a sustainable culture

	=	1.40E+05 g/yr	

26 ADDITIONAL DIESEL used for 70% of residues (collection, transport, and treatment)

Pellizzi (1986, p. 321) globally estimated at least 1.25 MJ/kg d.m., also including the embodied energy in the biomass burner.

We will therefore assume the lower estimate of 1.00 MJ/kg d.m.

Total energy demand	=	2.23E+09 J/yr	
Diesel requirement	=	50 kg/yr	

27 ADDITIONAL LABOR for residues harvesting and storage

Estimated from CNR-PFMA, 1981, p. 24]	=	2.79 hours (mostly not trained labor)
	=	0.35 working days (8 hours per day)

Applied energy for residues = (daily metabolic energy)(days labor applied)(4186 J/kcal)

	=	1.05E+03 kcal/yr
	=	4.38E+06 J/yr

28 ADDITIONAL MACHINERY FOR RESIDUES (assuming a 10-year life span)

Machinery demand was estimated from embodied energy per ton of dry matter collected (CNR-PFMA, 1981, p. 24)

Unit demand of machin. (J)	=	8.82E+07 J/t d.m.	[estim. from CNR-PFMA, 1981, p. 24]
Embodied energy machin.	=	8.00E+10 J/t machin.	[Biondi et al., 1989, p. 68]
Unit demand of machin. (g)	=	0.001 t machin/t d.m.	
Total harvested residues	=	2.23 tons d.m.	[see Item 32]
Total demand of machinery	=	2462 g/yr	
Total energy demand	=	1.97E+08 J/yr	

PRODUCTS OF THE AGRICULTURAL PHASE

29 SUGARBEET PRODUCED

Total wet sugarbeet	=	3.99E+07 g/yr	[Triolo et al., 1984, p. 40]

Other estimates: 42 tons/ha (Biondi et al., 1989, pp. 127–129); 45 Tons/ha (CNR, 1981, p. 23); 66 tons/ha (CCPCS, 1991, France average); 60 tons/ha (ERL, 1990, European Union average); 60 tons/ha (IEA, 1994, p. 27);

Lower heating value of wet sugarbeet	=	2.85E+03 J/g	[Bartolelli et al., 1982, p. 22033;
Total energy in sugarbeet	=	1.14E+11 J/yr	CNR-PFMA, 1981, p. 9]

30 SUGAR AVAILABLE

Average sugar content	=	0.16	[Biancardi, 1984, p. 128]
Total sugar available	=	6.38E+06 g/yr	
Lower heating value of sugar	=	1.64E+04 J/g	
	=	1.05E+11 J/yr	

31 RESIDUES IN FIELD AS SUCH

Total agric. wet residues	=	1.60E+07 g/yr (w.m.)	(estimated from CNR-PFMA, 1981, p. 23)
Residues dry matter	=	3.19E+06 g/yr d.m.	(20% of total wet residues)

Higher heating value (gross anhydrous calorific value) of sugarbeet residues is reported as 14.6 E6 J/kg of dry matter (CNR-PFMA, 1981, p. 5). If we account for 80% moisture content (12.8 E3 kg water/ha), this calorific value must be reduced to 0.20 * 14.6 E6 = 2.92 E6 J/kg w.m.

Therefore, the global gross calorific value of wet residues is 2.92 E6 * 1.60E4 = 4.67E10 J. Following the accounting procedure developed by Lyons et al. (1985), about 2.44 E6 J are required to evaporate 1 kg of moisture: moisture removal therefore requires 2.44 E6 * 12.8 E3 = 3.12 E10 J.

(Continued)

TABLE 4.6
Emergy Analysis of Ethanol Production from Sugarbeet in Italy (1984 nationwide average values per hectare per year) (continued)

Another heat loss comes from the need to evaporate the water formed in the reaction of fuel's hydrogen content with oxygen, during combustion of dry matter. We assume an average estimate of 5% of dry weight for hydrogen in beet residues (OTA, 1980), while it is about 6% in wood (Lyons et al., 1985, p. 285). Hydrogen produces nine times its own weight of water when burned.

Therefore, water formed is calculated as mass of water = 9 * 0.05 * (dry mass of residues) = 1435 kg water.

The consequent vaporization heat loss is heat loss = (mass of water) * 2.44 E6 = 3.50 E9 J.

Other losses (globally accounting for approximately 2 E10 J) come from energy subtracted by superheated moisture vapour, dry flue gases, and air exceeding stoichiometric requirement, plus other losses associated to uncomplete combustion and humidity of combustion air. If all these energy requirements and losses are accounted for, it clearly appears that sugarbeet residues have no usable energy content at all.

Usable energy of residues in field	=	0.00E+00 J/yr

The energy that is spent to collect and transport them is completely wasted, without any real advantage.

32 RESIDUES HARVESTED (70%)

70% harvested wet residues	=	1.12E+07 g/yr (w.m.)
Harvested residues (dry)	=	2.23E+06 g/yr (d.m.)
Usable energy in harvested residues	=	0.00E+00 J/yr

INDUSTRIAL PRODUCTION PHASE

33 MACHINERY MANUFACTURE

Energy invested per litre ethanol	=	1.30E+05 J/l	[CCPCS, 1991, France average]

Other estimates: 1.33 MJ/l (U.K. average, Marrow, 1987); 2.06 MJ/l (IEA, 1994, p. 29); 4.2 MJ/l (Sachs, 1980, also including miscellaneous Item).

Ethanol produced	=	4454 l/yr	[See Item 37]
Total energy demand	=	5.79E+08 J/yr	
Embodied energy machin.	=	8.00E+07 J/kg	[Biondi et al., 1989, p. 68]
Mass of machinery	=	7238 g/yr	

34 DIESEL FUEL FOR TRANSPORT OF BEETS TO PLANT

Material transported each trip	=	2.00E+04 kg/trip	[Average out of different trucks]
Average kms each round trip	=	100 km/trip	[Our assumption]
Total materials transported	=	3.99E+04 kg w.m. (total sugarbeet produced)	
Total trips required	=	1.99E+00 trips	
Total kilometers	=	1.99E+02 km	
Diesel per kilometer	=	0.20 kg/km	[Average out of different trucks]
Total diesel demand	=	3.99E+01 kg	
Lower heating value	=	4.45E+07 J/kg	[Ellington et al., 1993, p. 408]
Total energy in used diesel	=	1.77E+09 J/yr	
Transport energy demand per litre of ethanol	=	3.98E+05 J/l	[From above calculations]
	=	0.40 MJ/l	

Other estimates: 0.58 MJ/1 (CCPCS, 1991, France average); 0.74 MJ/1
(European Union average, ERL, 1990);

35 DIESEL FOR PROCESS HEAT DEMAND

| Process energy demand per litre of ethanol produced | = | 8.36E+06 J/1 | [CCPCS, 1991, France average] |

Other estimates: 12.10 MJ/1 (European Union average, ERL, 1990); 5.55 MJ/1
(U.K. average, Marrow, 1987);

15.34 MJ/1 (IEA, 1994, also including energy for water removal in the different phases of the process;
10.9 MJ/1 (Sachs, 1980).

Lower heating value of diesel	=	4.45E+07 J/kg	[Ellington et al., 1993, p. 408]
Total energy demand	=	3.72E+10 J/yr	
Total diesel demand	=	838 kg/yr	

36 ELECTRICITY FOR WATER REMOVAL FROM ETHANOL

| Electricity demand/litre ethanol | = | 1.17 kWh/1 | [CCPCS, 1991, France average] |

Other estimates: 1.64 kWh/1 (European Union average, ERL, 1990); 2.65 kWh/1 (U.K. average, Marrow, 1987);
0.8 kWh/1 (Sachs, 1980).

| Total energy demand | = | 5207 kWh/yr | |

37 ETHANOL PRODUCED (as 90% of energy in sugar available)

| Total ethanol produced | = | 4454 1/yr | [our estimate from sugar content] |

Other estimates: 6692 1/ha (CCPCS, 1991, France average); 5580 1/ha
(ERL, 1990, European Union average); 5820 1/ha (IEA, 1994, p. 27),

| Lower heating value | = | 2.12E+07 J/1 | [Wyman et al., 1993, p. 807] |
| Total energy in ethanol | = | 9.42E+10 J/yr | |

TABLE 4.7
Energy and Carbon Balance of Ethanol Production from Sugarbeet in Italy
(1984 nationwide average values per hectare per year)

No.[a]	Item	Unit	Amount	Oil Equivalent (g/unit)	Ref. for Equiv.[b]	Oil Used Up (g/yr)	CO$_2$ Released (g/yr)
AGRICULTURAL PRODUCTION PHASE							
5a	Organic in topsoil used up, when residues are left in field[c]						
		g	6.00E+05				2.94E+05
5b	Organic in topsoil used up, when 70% residues are removed[c]						
		g	1.20E+06				5.88E+05
Inputs assuming that residues are left in field							
6	Nitrogen	g	1.36E+05	1.75	[1]	2.38E+05	8.14E+05
7	Phosphate	g	1.87E+05	0.32	[1]	5.99E+04	2.05E+05
8	Potash	g	7.28E+04	0.22	[1]	1.60E+04	5.47E+04
9	Insecticides and pesticides	g	4.07E+04	1.60	[1]	6.51E+04	2.23E+05
10	Herbicides	g	1.52E+04	1.60	[1]	2.43E+04	8.31E+04
11	Diesel	g	2.98E+05	1.23	[1]	3.67E+05	1.26E+06

(Continued)

TABLE 4.7
Energy and Carbon Balance of Ethanol Production from Sugarbeet in Italy
(1984 nationwide average values per hectare per year) (continued)

No.[a]	Item	Unit	Amount	Oil Equivalent (g/unit)	Ref. for Equiv.[b]	Oil Used Up (g/yr)	CO₂ Released (g/yr)
12	Lubricants	g	5.50E+03	2.00	[1]	1.10E+04	3.76E+04
13	Gasoline	g	8.70E+03	1.32	[1]	1.15E+04	3.93E+04
14	Human labor	h					
15	Agricultural machinery	g	8.37E+04	0.19	[1]	1.60E+04	5.47E+04
16	Electricity	kWh	1.63E+02	250.84	[1]	4.08E+04	1.40E+05
17	Seeds	g	2.00E+04	0.48	[1]	9.56E+03	3.27E+04
18	Fossil fuels needed to pump surface water for irrigation[d]	g		1.23	[1]	0.00E+00	0.00E+00

Additional inputs if 70% residues are harvested

No.[a]	Item	Unit	Amount	Oil Equivalent (g/unit)	Ref. for Equiv.[b]	Oil Used Up (g/yr)	CO₂ Released (g/yr)
19	Nitrogen loss with erosion	g	4.50E+04	1.75	[1]	7.88E+04	2.69E+05
20	Phosph. loss with erosion	g	2.25E+04	0.32	[1]	7.20E+03	2.46E+04
21	Potash loss with erosion	g	1.50E+05	0.22	[1]	3.30E+04	1.13E+05
22	Fossil fuels needed to provide the additional water demand	g	5.70E+04	1.23	[1]	7.01E+04	2.40E+05
23	Nitrogen harv. in residues	g	3.50E+04	1.75	[1]	6.13E+04	2.10E+05
24	Phosphorus harv. in resid.	g	2.10E+04	0.32	[1]	6.72E+03	2.30E+04
25	Potash harv. in residues.	g	1.40E+05	0.22	[1]	3.08E+04	1.05E+05
26	Additional diesel for residues	g	5.02E+04	1.23	[1]	6.18E+04	2.11E+05
27	Labor for residues	h					
28	Additional machinery for residues	g	2.46E+03	0.19	[1]	4.71E+02	1.61E+03

Products of the agricultural phase

No.[a]	Item	Unit	Amount	Oil Equivalent (g/unit)	Ref. for Equiv.[b]	Oil Used Up (g/yr)	CO₂ Released (g/yr)
29	Sugarbeet produced (w.m.)	g	3.99E+07				
30	Sugar available for ethanol	g	6.38E+06				
31	Residues in field as such (w.m.)	g	1.60E+07				
32	70% harvested wet resid.	g	1.12E+07				

INDUSTRIAL PRODUCTION PHASE

No.[a]	Item	Unit	Amount	Oil Equivalent (g/unit)	Ref. for Equiv.[b]	Oil Used Up (g/yr)	CO₂ Released (g/yr)
33	Plant machinery	g	7.24E+03	0.19	[1]	1.38E+03	4.73E+03
34	Diesel for transport	g	3.99E+04	1.23	[1]	4.91E+04	1.68E+05
35	Diesel for process heat	g	8.38E+05	1.23	[1]	1.03E+06	3.52E+06
36	Electricity for water removal	kWh	5.21E+03	250.84	[1]	1.31E+06	4.47E+06

Product of the industrial phase

37	Ethanol	J	9.42E+10

[a] Numbers of each Item refer to the footnotes in Table 4.6. Items 1 to 4, environmental inputs, and Items 14 and 27, labor, are not accounted for in energy and carbon balance.

[b] **References for equivalents:**

[1] Biondi et al., 1989

[c] CO_2 emissions from oil are calculated according to Desmarquest, 1991 (3.42 g CO_2/g oil used).

[d] Water content in organic matter is assumed 70%. Average C content in dry organic matter is assumed 45%. Therefore, conversion coefficient of organic matter to CO_2 is (1-0.70)(0.45)(molecular mass of CO_2/atomic mass of C)=0.30*0.45*3.66=0.49.

[e] Already accounted for as diesel and electricity consumption.

Indices based on energy and carbon balance

Energy demand in agricultural phase, without residues	3.60E+10 J/ha/yr
Energy demand in agric. phase, harvesting residues	5.06E+10 J/ha/yr
Additional energy demand due to residues collection and use	1.47E+10 J/yr
Energy demand in industrial phase, without residues	9.99E+10 J/ha/yr
Energy demand in industrial phase, harvesting residues	9.99E+10 J/ha/yr
Total applied energy, without residues	1.36E+11 J/ha/yr
Total applied energy, harvesting residues	1.51E+11 J/ha/yr
CO_2 released per ha, without residues	1.14E+07 g/ha/yr
CO_2 released per ha, harvesting residues	1.29E+07 g/ha/yr
CO_2 avoided, by replacing gasoline	1.02E+07 g/ha/yr
Energy ratio (out/in) without residues	0.69
Energy ratio (out/in) harvesting residues	0.63
Released/avoided CO_2 ratio, without residues	1.12
Released/avoided CO_2 ratio, harvesting residues	1.27

When the traditional energy accounting methodology is integrated with emergy evaluations (Table 4.6) and results from other crop production processes are compared (Table 4.5), we get a deeper understanding of the process under study.

Sugarbeet belongs to a group of crops (herbaceous crops) showing a relatively low transformity, in comparison to woody crops. Within this group, its emergy yield ratio is the lowest and its environmental loading ratio the highest, so that its emergy sustainability index (ESI = EYE/ELR) is only 0.25, one of the lowest within the whole set of crops investigated. This means that even if the conversion efficiency of emergy inputs into output energy is very high, the fraction of nonrenewable emergy inputs is much larger than the locally renewable fraction that is exploited. The process has a low ability to exploit the locally renewable resource and is therefore less sustainable than competing production alternatives with higher EYR/ELR.

The average transformity of purchased inputs is 13.73 E4 sej/J, calculated according to Equation (7b). This means that the invested inputs to the process globally have a higher hierarchical quality than the final product (6.14 E4 sej/J). This can be accepted if food production is the goal, for sugarbeet meets our food demand better than the fuels and goods input to the process. On the contrary, it is absolutely meaningless to use the sugar yield as a substrate for biofuel production; it would be better to use the inputs directly as fuels, instead of producing the sugar.

Different economic policies can be based upon the above energy, carbon, and emergy indices, to favour and stimulate production processes according to specific policy goals. Processes with high CO_2 emissions, low energy ratios, and low sustainability indices should not be encouraged. As these characteristics may not occur simultaneously, a policymaker will have to choose according to an integrated approach to the problem.

4.6.1.3 The Case of Biofuels Production

Fuels from biomass have been proposed as substitutes for fossil fuels, whose scarcity and cost are likely to increase in the future. Also environmental concerns related to an increasing greenhouse effect have been raised, to encourage the development of cleaner energy technologies. Competition for available land has sometimes been considered, on the basis of fertile land becoming a limiting factor in both food and energy production. Thus, it is possible to write down a list of needed requisites for a biomass fuel to be feasible.

1. The fuel should provide more energy than it is required for its production.
2. The fuel should be renewable and its long-run availability should be assured, in order to provide a solution to possible scarcity of fossils.
3. Release of carbon dioxide into the environment due to biofuel production should be lower than CO_2 release from an equivalent amount of energy from fossil fuels.
4. The biomass fuel production process should not provide stress to the environment (soil erosion, air and water pollution, etc.).
5. Land requirement should not be too high, in order to avoid competition with food production and to preserve wild areas supporting the biosphere's activity.
6. A clear assessment of fuel production costs should be possible. Not only the advantage to the user should be considered, but also its cost of production from the point of view of the unpaid environmental work needed to provide it.

We evaluated many different biofuels processes from selected crops (Giampietro et al., 1997). Figure 4.7 shows the energy diagram of a typical biofuel production process, while detailed data about bioethanol from sugarbeet are reported in Tables 4.6 and 4.7. Energy output/input ratio and released/uptaken CO_2 ratio are calculated in Table 4.7, to evaluate the feasibility of ethanol fuel from biomass and eventual contribution to decreasing carbon dioxide levels in the atmosphere. Emergy, transformities, and other emergy indices are calculated in Table 4.6 for fresh sugarbeet, available sugar, agricultural sugarbeet residues, and bioethanol.

An unfavorable output/input energy ratio (0.70) results when careful calculations are made and all main inputs are accounted for. This is mainly due to the energy demand in the industrial phase (process heat and water removal). Such results mean that ethanol from sugarbeet cannot be considered a source of energy for the society. The use of residues does not improve the performance of the process.

Even in the absence of an appreciable energy gain, reduction of CO_2 emissions are sometimes claimed as a useful result. Unfortunately, the process appears to be more a CO_2 source than a sink. Therefore, both energy and global warming reasons cannot be claimed to support this option. In addition,

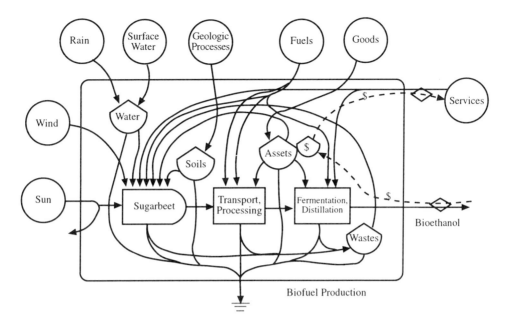

FIGURE 4.7 Diagram of biofuel production from sugarbeet.

other environmental costs are paid to provide the yield: evapotranspired water, soil erosion, release of agrochemicals into the environment, release of other pollutants from ethanol combustion, that are very often not accounted for. On the contrary, this can be done by means of the emergy approach, providing indices of efficiency, renewability, and environmental pressure (for example, land demand and excessive exploitation of local environmental resources), to be compared to other possible processes. Total contributions from earth cycle, topsoil, rain, and irrigation water, are, respectively, 33 and 17% of total emergy use in sugarbeet and ethanol production, showing a large direct support from the ecosystem, even if much lower than the indirect inputs under human control, including human labor (usually not accounted for). The global solar transformity of ethanol from sugarbeet is about 1.42 E5 sej/J, while fossil fuels transformities range 4 to 6 E4 sej/J. Sometimes, energy and carbon balances appear to be more positive, as in the case of ethanol from sugarcane (E.R. = 1.50 − 2.50; released/uptaken CO_2 = 0.5 − 0.9) or methanol from wood (E.R. = 1.10; released/uptaken CO_2 = 0.71), but even in these more favourable cases transformities are in the range 1 to 3 E5 sej/J, largely higher than fossil fuels (Figure 4.8). It appears that the environmental process of producing fossil fuels requires a lower convergence of solar emergy, and is more efficient and more "renewable,"* than ethanol production in an agro-industrial process. Other indices based on emergy allow a global comparison among different biofuel production processes, to find those that perform better (Table 4.8).

4.6.2 Emergy Evaluation of a National Economic System

An emergy analysis of the Italian economic system (Figure 4.8) has been performed comparing data of the years 1984, 1989, 1991, and 1995 (Himschoot, 1988, revised by the authors of this paper; Ulgiati et al., 1994b; Ulgiati et al., 1995b; Russi, 1999).

---

* This paradox of renewability deserves attention. Renewability is a concept involving the turnover time of a given resource. Production of biomass typically requires a short turnover time, under natural conditions, but yield per hectare is constrained by available resources. When large amounts of fossil resources are used in the form of fertilizers and fuels to boost the process and increase its yield per hectare, the turnover time is expanded up to the replacement time of fossil fuels. Biomass produced in large amounts by means of fossil sources is therefore nonrenewable biomass.

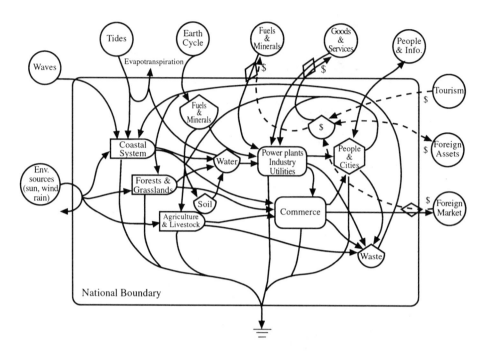

FIGURE 4.8 Emergy diagram of a national economy (Ulgiati, S. et al., 1994).

TABLE 4.8
Transformities and Other Emergy Indices for Selected Biomass Fuels

Fuel	Transformity (sej/J)	EYR	ELR	ESI	E.R.
Bioethanol from corn, Italy[a]	1.23E+05	1.61	5.08	0.32	0.96
Biodiesel from sunflower[a]	1.41E+05	1.70	3.96	0.43	0.97
Bioethanol from sugarbeet[b]	1.42E+05	1.18	11.93	0.10	0.69
Biodiesel from rapeseed[c]	1.46E+05	1.74	3.82	0.46	1.00
Methanol from wood[d]	1.51E+05	2.54	1.94	1.31	1.10
Bioethanol from sugarcane[e]	1.65E+05	2.19	6.41	0.34	1.71
Biodiesel from soybeans[a]	2.41E+05	1.68	3.44	0.49	0.58

[a] Ulgiati, unpublished manuscript.

[b] This chapter.

[c] Ulgiati and Brown, 1998.

[d] Our calculations from data in Ellington et al., 1993.

[e] Our calculations from data in De Carvalho Machedo 1., 1992.

Legend: E.R. = Output/input Energy Ratio; EYR = Emergy Yield Ratio; ELR = Environmental loading Ratio; ESI = Emergy Sustainability Index.

Table 4.9 shows a comparison among the main indices and ratios calculated from available data. A fruitful comparison can also be made among emergy indices and traditional energy and monetary ones: the integration of the different approaches may provide a comprehensive view of the system behaviour.

TABLE 4.9
Trend of Emergy and Energy Indicators for Italy in Selected Years

Flow/Index	1984[a]	1989[b]	1991[c]	1995[d]
	Italy			
1 Renewable sources used (sej/yr) (R)	1.21E+23	1.21E+23	1.21E+23	1.21E+23
2 Nonrenewable indigenous sources(sej/yr) (N)	3.00E+23	3.57E+23	5.02E+23	4.78E+23
2a Dispersed rural sources		1.33E+21	4.62E+22	4.01E+22
2b Concentrated use		3.54E+23	4.54E+23	4.35E+23
2c Exported without use		1.83E+21	1.48E+21	2.29E+21
3 Imported emergy (sej/yr) (F)	5.37E+23	7.89E+23	1.75E+24	1.01E+24
4 Total emergy available (sej/yr)	9.58E+23	1.27E+24	2.37E+24	1.61E+24
5 Emergy actually used within (sej/yr)	9.58E+23	1.26E+24	1.41E+24	1.54E+24
6 Exported emergy (sej/yr)	2.36E+23	3.12E+23	8.52E+23	4.56E+23
7 Fraction of use, derived from indigenous sources	0.44	0.38	0.44	0.39
8 Fraction of use that is free renewable (R/Y)	0.13	0.10	0.09	0.10
9 Fraction of use that is imported (F/Y)	0.56	0.62	0.56	0.61
10 Fraction of use that is electrical	14.3%	13.0%	12.7%	12.5%
11 Environmental loading ratio, ELR ((N + F)/R)	6.91	9.47	10.70	11.72
12 Emergy density (sej/m²/yr)	3.18E+12	4.20E+12	4.69E+12	5.12E+12
13 Emergy investment ratio, EIR	1.27	1.65	2.81	1.69
14 Emergy yield ratio, EYR	1.78	1.61	1.36	1.59
15 Gross national product ($)	3.90E+11	8.66E+11	1.15E+12	1.07E+12
15a Gross national product (£)	5.85E+14	1.19E+15	1.43E+15	1.75E+15
15b GNP in terms of £1984	5.85E+14	8.06E+14	8.77E+14	9.68E+14
15c GNP in terms of $1984	3.33E+11	4.59E+11	4.99E+11	5.51E+11
16 Inflation rate (£/£1984)[e]	1.00	0.48	0.63	0.80
17 Total energy use (J)	5.53E+18	6.81E+18	6.97E+18	5.32E+18
18 Emergy intensity (Y/GNP ratio, sej/£)	1.64E+09	1.06E+09	9.89E+08	8.82E+08
18a Emergy intensity (Y/GNP ratio, sej/£) in terms of £1984	1.64E+09	1.57E+09	1.61E+09	1.59E+09
18b Emergy intensity (Y/GNP ratio (sej/$))	2.46E+12	1.46E+12	1.23E+12	1.44E+12
18c Emergy intensity (Y/GNP ratio sej/$) in terms of $1984	2.87E+12	2.76E+12	2.83E+12	2.79E+12
19 Emergy intensity (= energy/GNP ratio, J/$)	1.42E+07	7.86E+06	6.06E+06	4.96E+06
19a Energy intensity (= energy/GNP ratio, J/$) in terms of $1984	1.66E+07	1.48E+07	1.40E+07	9.65E+06
20 Imports/exports (emergy basis)	2.27	2.50	2.05	2.22
21 Imports/exports (money basis)	n.a.	1.03	1.03	0.88
22 Emergy to energy ratio	1.73E+05	1.86E+05	2.03E+05	2.90E+05
23 Population	5.66E+07	5.67E+07	5.68E+07	5.73E+07
24 Energy use per person (J/person/yr)	9.77E+10	1.10E+11	1.11E+11	9.27E+10
25 Emergy use per person (sej/yr/person)	1.69E+16	2.23E+16	2.49E+16	2.69E+16
26 Emergy value of labor (sej/J)	5.61E+06	7.40E+06	8.26E+06	8.91E+06
27 Emergy sustainability index, ESI = EYR/ELR	0.26	0.17	0.13	0.14

Exchange ratio £/$1984: 1756 £/$, Bank of Italy Governor's report, 1990

[a] Himschoot, 1988, revised and partially modified by Ulgiati, 1996.

[b] Ulgiati et al., 1994.

[c] Ulgiati et al., 1995b.

[d] Daniela Russi, 1999, University of Siena, Class Report.

[e] Bank of Italy Governor's report, 1990 and 1995.

4.6.2.1 Monitoring the Performance of the Italian Economic System

The population of Italy showed no significant increase during the period from 1984 to 1995 (Table 4.9, Item 23), stabilizing around 57 million people (ISTAT, 1990a, 1993b, 1997).

The Gross National Product (GNP) in 1984 was $390 billion, $866 billion in 1989, $1150 billion in 1991, and $1070 billion in 1995, showing a sharp increase over time followed by an apparent stabilization (Table 4.9, Item 15). If 1984 is taken as a baseline and inflation is accounted for, the annual GNP rise was in the range 2 to 4%.

The average energy use, mostly from fossil fuels, can be calculated as 97.7 GJ per person in 1984, 110 GJ per person in 1989, 111 GJ per person in 1991, and 92.7 GJ per person in 1995 (Table 4.9, Item 23). For comparison, the average emergy use per person was 1.69E16 sej per person in 1984, 2.23E16 sej per person in 1989, 2.49E16 sej per person in 1991, and 2.61E16 sej per person in the year 1995 (Table 4.9, Item 25). It can be clearly seen that the rate of annual energy increase is largely lower than the rate of emergy increase. In addition, energy use is declining, while emergy use is not: this different behaviour suggests that substitution of high energy strategies with low energy strategies might have been supported by high emergy technologies and materials. As a consequence, the ratio of annual emergy to energy use (Table 4.9, Item 22) grew from 17.30E4 sej/J in the year 1984 up to 29.00E4 sej/J in 1995: it provides a sort of global transformity at the national level, indicating that each joule of energy processed within the national economy was supported by a huge and still growing amount of emergy. As emergy represents the convergence of different kinds of commodities and resources, not only in the form of energy carriers, this confirms that (1) the economic growth relies on a larger basis than simply actual energy, and (2) the environmental support to the system is a prerequisite of energy processing within the system itself.

An increase in the efficiency of the processing ability of the whole system may be suggested by the lowered ratio of electrical to total emergy use: it was 14.3% in 1984, 13.0% in 1989, down to 12.5% in 1991 (Table 4.9, Item 10). Electricity is a high quality flow and should not be misused. Inefficient electric uses like home heating appear to be declining in Italy, so that electricity is only used when comparable advantages can be expected in terms of feedback control and reinforcement. The ratio of imported emergy (F) to indigenous emergy resources (R + N) is called Emergy Investment Ratio (EIR). It increased from 1.27 in 1984 to 2.81 in 1991, then declined (Table 4.9, Item 13). Despite a higher efficiency in energy use, more emergy investment was needed in 1991 compared to 1984 and 1995 to exploit one unit of local resource: higher EIRs make the Italian economic system less competitive with other national economies performing at lower ratios.

The ratio of global emergy use to GNP (sej/currency) is also a crucial parameter. One dollar spent in countries with higher emergy/$ ratios (for example, oil-exporting countries) buys more emergy, than is embodied in the currency that is paid for (Ulgiati et al., 1994b). Countries whose economies do not rely upon an expanding monetary basis usually have a high emergy/currency ratio. As a consequence, Italy's emergy/$ ratio makes trade with less industrialized countries more advantageous to Italy. A developed country like Italy maintains a position advantage when importing from less developed areas.

A declining resource-use/GNP ratio (Table 4.9, Items 18 to 18c) might be explained by:

- Decrease of total emergy use (which is not our case here);
- Money circulation increasing faster than resource use;
- Increase of money circulation can be due to;
- Inflation (decrease of buying power of currency);
- Increased ability of the global economy to process primary resources for the market (more income for exported or sold goods);
- Increase of high value-added products (like services and information);
- Increase of the services component of the economy.

Finally, if sej/$ units are used, results can also be affected by geopolitical reasons, pushing upward or downward the exchange ratio of dollars, without any clear relationship with actual emergy used. Disaggregating these different effects might help to understand the dynamics of the relationship between the economic system and resource availability. In order to do so, we suggest:

1. The emergy buying power of the currency to be evaluated by means of the emergy/GNP ratio, in terms of the actual GNP for the year under consideration;
2. The efficiency of the global economy as a "machine" processing resources to GNP to be better evaluated by omitting inflation effects (i.e., by referring GNP to a reference year);
3. International market geopolitical problems to be avoided by using local currency;
4. After this evaluation is done, comparison to other national values given in sej/$ can also be performed. The dollar/local currency ratio should be corrected to account for the real buying power of both currencies.

Italian emergy/GNP ratio dropped from 1.64E9 sej/£ in 1984 to 8.82E8 sej/£ in 1995 (Table 4.9, Item 18). When inflation is accounted for, the ratio stabilizes around 1.60E9 sej/£ with negligible oscillations (Table 4.9, Item 18a). Therefore the "economy machine" was found to be acting at a quasi-constant efficiency over 10 years. Instead, calculating GNP in U.S. $ yields an emergy/GNP ratio declining from 2.46E12 sej/$ to 1.23E12 sej/$ in the period 1984–1991, then rising again to 1.44E12 sej/$ (Table 4.9, Item 18b), clearly showing how the different behaviour of the currencies in the international market affect the value of the index and, therefore, the comparison to other countries.

As a consequence of increased emergy use and stable population, emergy availability per person also increased by approximately 6% per year (Table 4.9, Item 25). A larger share of resources has been available to the Italian population, thus contributing to a better standard of living (more high quality goods available per person). This opportunity of a better material standard of living is counterbalanced by an annual 7% increase of the Environmental Loading Ratio (Table 4.9, Item 11): the larger pressure of the economic activity on the locally available natural resources should be considered as an alarming signal of excessive exploitation of the environment by the economy (land and natural environment becoming limiting factors). We already underlined that the ELR is linked to the fraction of renewable to total emergy used, ϕ_R: a 7% annual increase in the ELR means a parallel decrease in the fraction of renewable inputs to the system itself [fraction free renewable $= 1/(1 + \text{ELR})$]. The renewable emergy use was 13% of the total in 1984, dropping to about 10% in 1995 (Table 4.9, Item 8), thus raising many questions not just on Italy's sustainable development, but even on the possibility of maintaining the present welfare level.

Finally, the ESI declined from the already low value 0.26 to 0.14 in 1995 (Table 4.9, Item 27), highlighting the global fragility of an economy that is increasingly based on nonrenewables as well as on imports from outside. This finding may be contrasting with the above explained advantage of importing from countries with high emergy/$ ratio. Importing at a favourable exchange ratio makes the national economy richer among other nations, not necessarily stronger and environmentally sound. Brown and Ulgiati (1997) have calculated the same trend for the Taiwan economy from 1960 to 1990.

4.6.2.2 Emergy as a Basis for Policies

We are certainly aware that reliable analysis of economic and environmental trends of a country cannot be inferred by means of data from only four different years within an 11-year range. More regular and longer periods of observation may be needed to quantify and evaluate trends and suggest policies, whatever is the approach used for the analysis. Anyway, trends of emergy indicators give a clear and useful picture, in order to help economy and policy makers in understanding and planning the dynamics of their country (or better, the dynamical equilibrium of nature and the economy).

When indices and flows based on emergy are calculated, it clearly appears that monetary and energy indices only account for one side of the picture. For example, consider the difference between import/export ratios on a monetary basis and an emergy basis (Table 4.9, Items 20 and 21). When a monetary basis is considered, trade is quite balanced between Italy and foreign countries; on the contrary, when trade is evaluated on an emergy basis (i.e., quality of traded goods is accounted for), it appears that the import of real wealth in the form of imported solar emergy is much more favourable to Italy than to its trading partners. This is because most of Italy's imports come from less developed countries (higher emergy/$ ratio), where $1 buys more emergy than in Italy. A balanced exchange in dollars does not mean a real trade equity. In developed countries it often means exploitation of less developed countries and more emergy inflowing than going out. This kind of emergy unbalance has been indicated as the source of possible instability of world markets, due to overexploitation of natural resources of developing countries, even when it does not appear as such in the financial accounting (Repetto, 1992; Odum, 1994a).

Another clear example of how the emergy analysis can provide information not otherwise obtainable from traditional economic indicators is the Environmental Loading Ratio, ELR. It is about 10:1 for Italy (Table 4.9, Item 11). Such a larger amount of nonrenewable and imported resources in comparison to locally renewables (on the average, 10:1) indicates the huge work provided by the environment over larger space and time scales. Despite an increase in energy efficiency, this larger amount of resources per person results in a lower sustainability for the country. More environment is being "spent" to support the higher standard of living of Italy 1991 and 1995, in comparison to 1984 and 1989; this trend cannot be considered a sustainable one for future generations. If we give a deeper insight into the composition of the nonrenewable fraction of national emergy use, we find out that a large part of input is minerals other than fossil fuels. We usually consider fuels as the basic source, because we are persuaded that fuels are a scarce resource. We should start to consider many other resources as scarce, due to the huge environmental damages that are caused when they are exploited. We have already recalled in previous Section 4.5.2 the environmental consequences of excessive sand and gravel withdrawal from river beds. Also think of the enormous amounts of mineral ore that is dug to extract small percents of pure uranium or gold (Glasby, 1988), as well as to the huge water withdrawal for industrial and urban use, diverting this water from its basic task of stabilizing biosphere dynamics through evapotranspiration process and photosynthesis. Despite its importance, water (natural capital) is assigned small or no economic value in evaluating processes under human control (man-made capital). There is no doubt that water availability is going to be, and sometimes already is, the emergency of the next decade, due to the destabilization of water cycle induced by human activities. The environmental damages and the unsustainability of such uncontrolled withdrawals are well accounted for in emergy analyses, where the large emergy content and high transformity of these inputs clearly defines their low renewability at present exploitation rates and points out the need of assigning an environmental and economic value (and maybe constraints to use) to apparently abundant resources, in consideration of their role within the biosphere dynamics.

A third example of useful findings from emergy analysis can be inferred from the emergy evaluation of labor. Some energy analysts give small or no energy value to human labor, thus also contrasting with the huge problem of the economic cost of labor in modern societies. We evaluated the emergy support per joule of applied labor, as the ratio of solar emergy per person per year to total applied energy per person per year (Ulgiati et al., 1994b, 1995b). It was 5.61E6 sej/J in 1984, 7.40E6 sej/J in 1989, up to 8.91E6 sej/J in 1995 (Table 4.9, Item 26). This means that the emergy cost of labor (its true cost in the thermodynamic scale of the global ecosystem) is increasing, due to the demand for a higher standard of living of the average worker. When labor is accounted for as input to a given process, it may happen that its weight is much higher than expected (Ulgiati et al., 1994b): this may suggest (and it actually happens) to policymakers and business people to downsize labor-intensive processes or to lower the average standard of living, in periods of declining emergy availability.

4.7 THERMODYNAMICS OF OSCILLATING SYSTEMS

The optimum efficiency and thus the optimum self-organization suggested by the Maximum Empower Principle cannot be obtained once forever. As environmental conditions change, the response of the system will adapt, so that maximum power output can be maintained. The system should be able to tune its performance according to the changing environment.

A concept of "ecosystem integrity" has been introduced (Kay, 1991) to ascertain and describe "*the ability of an ecosystem to maintain its organization in the face of changing environmental conditions.*" Systems self-organizing pathways usually develop until a point in the state space is reached where "*the disorganizing forces of external environmental change and the organizing thermodynamic forces are balanced.*" (Kay, 1991). This point is referred to as the *optimum operating point.* Integrity is defined as the ability of the system to attain and maintain its optimum operating point.

Steady state and no-growth paradigms have been proposed in the last two decades: many scientists and policymakers believed that human societies could grow up to a special state, where resource supply and use are balanced. This should be considered a sustainable steady state (Goldsmith et al., 1972; Daly, 1977). After the balance point has been reached, only *development* of societies (better use of available resources) instead of *growth* (size and consumptions increase due to a larger supply of resources) should be possible (Daly, 1977). More recently, Daly (1990) has proposed a "*quasi-sustainable use of nonrenewables, by limiting their rate of depletion to the rate of creation of renewable substitutes.*"

The impossibility not just of growth, but also of a steady state economy have already been stated by Georgescu-Roegen (1976), who claimed that a finite amount of available resources could only support a population for a limited time, after which the human species would have to disappear. The only possible trend for humankind is a steady declining state.

We believe that the above statements do not highlight that the whole planet is a self-organizing system, where storages of resources are continuously depleted and replaced, at different rates, and matter is recycled and organized within a self-organization activity driven by solar, geothermal, and gravitational energy. "*The real world is observed to pulse and oscillate. There are oscillating steady states. In most systems, including those which people are part of, storages are observed to fill and discharge as part of oscillations.*" Odum (1994b).

There is an optimum loading for maximum power transformation in many kinds of processes (Odum and Pinkerton, 1955; Curzon and Ahlborn, 1975). It has been hypothesized (Odum, 1994b) that one of the reasons oscillating steady states displace steady ones may be that energy transformations have better loading ratios between inputs and outputs if there is an alternation in the growth of interconnected storages of producers and consumers. Simple models can be used to generate the essence of the observed alternation of production and pulsing consumption (Alexander, 1978). Richardson and Odum (1981) and Richardson (1988) studied the power-maximizing properties of this design.

Net primary production and storage of resources develop faster than consumers assets, until the system reaches a threshold where autocatalytic and higher order pathways are accelerated. Consumer assets show a sudden increase at the expenses of the environmental storage. When this latter is used up, consumer assets decrease, allowing a new cycle of environmental storage. In the case of the global economy, the storage are oil, minerals, topsoil, and other slowly renewable environmental resources, while the consumer assets are human economies and civilization. According to this so-called "pulsing paradigm," "*sustainability concerns managing, and adapting to the frequencies of oscillation of natural capital that perform best. Sustainability may not be the level 'steady-state' of the classical sigmoid curve but the process of adapting to oscillation. The human economic society may be constrained to track the thermodynamics that is appropriate for each stage of the glabal oscillation,*" (Odum, 1994b).

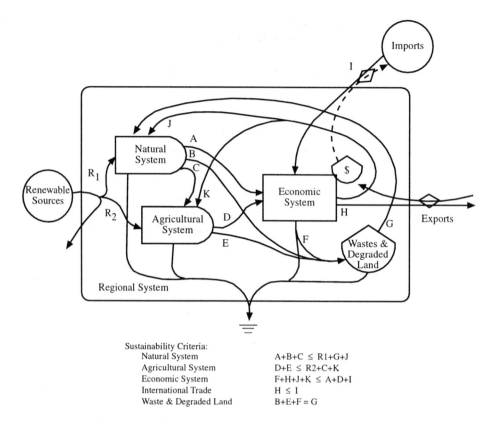

Sustainability Criteria:
Natural System	$A+B+C \leq R1+G+J$
Agricultural System	$D+E \leq R2+C+K$
Economic System	$F+H+J+K \leq A+D+I$
International Trade	$H \leq I$
Waste & Degraded Land	$B+E+F = G$

FIGURE 4.9 Systems diagram of an economy that is the interplay of renewable and nonrenewable energies. Shown are three scales for which sustainability should be determined: ecological scale, regional scale, and intraregional scale (Brown, M. T. and Lopez, S., 1995).

Finally, natural and man-made systems in the biosphere must perform according to a dynamic average equilibrium among components. The Maximum Empower Principle dictates that the emergy flow is maximized (optimized) at each level of the hierarchy according to the existing contraints, so that the global system performs at maximum power output. A sustainability criterion is that each component sector contributes to the global system an emergy flow consistent with the contribution it receives back from the system itself. According to the diagram of Figure 4.9 (Brown and Lopez, 1995), this criterion translates into an input–output relationship for each component sector. A storage of waste products is also shown in Figure 4.9, to be recycled by means of natural and man-made devices, so that matter is again made available to the global cycle of the biosphere.

4.8 CONCLUDING REMARKS

Discussion still continues between emergy analysts and critics. An example of such long lasting debate is given by Herendeen (1992), Herendeen (1994), and Brown and Herendeen (1996). The first two papers compare several efforts, all claiming to assess a total environmental cost. According to Herendeen (1994), "*no effort gets close to the goal of total environmental cost, except emergy analysis....However, this approach, which quantifies the environment's services in terms of the solar energy which has established and drives the biosphere, is now inadequately developed....*" In his previous papers, Herendeen stressed the emergy methodological approach, listing a number of comments and criticisms, on the basis of selected papers from Odum and his colleagues (Odum, 1988a; Huang and Odum, 1991; Odum and Arding, 1991). His conclusion was that emergy analysis *"still lacks a sufficiently clear methodological exposition for other analysts to apply it."* Finally, Brown and Herendeen (1996) have jointly performed a careful comparison between emergy analysis and embodied energy analysis, answering to some of the previously raised questions. They acknowledge that emergy analysis has developed *"...the conceptual and empirical basis that all energies are not of the same quality. Embodied Energy Analysis, and the units of embodied energy, do not recognize the qualities of energy across the energy spectrum of the biosphere, but instead account for only what has been termed 'cultural' energies. In so doing, about half of the total energy driving the economies of the biosphere is ignored."*

The emergy accounting approach, partially described in this chapter, provides a way to calculate the global environmental cost of a process, i.e., the amount of the environmental work supporting it. The total available emergy flow drives the system behaviour according to the Maximum Empower Principle, determining the size of the system itself and its growth rate.

Emergy flows and storages can be calculated for every system under study. After a suitable time, the system develops an optimum pattern of self-organization that maximizes power, so that optimum transformities can be accepted as measures of quality, i.e., a measure of resources convergence, hierarchical level, and feedback control ability (Ulgiati and Brown, 1998a). The behaviour of self-organizing, complex systems at every level of the natural hierarchy follows the availability of emergy flows and storages, with oscillating patterns of growth and decline. The same is true with thermodynamic optimum efficiency of resources use.

We have shown that emergy based indicators have the ability to account for the thermodynamic behaviour of a system as well as to assess the renewability of its products and its equilibrium with the surrounding environment. We believe that the emergy accounting has developed in the last years a sufficiently clear methodological approach for other analysts to apply it, offering suitable tools to allow the investigation of processes at the interface of human society and nature.

The emergy approach is flexible enough to be easily applied to the study of many, different systems, from natural to man-made ones. When it is applied to processes and systems including humans, public policies can be developed on the basis of the assessment of global environmental costs, thus integrating and enriching the usual market, monetary evaluations.

ACKNOWLEDGMENTS

Financial support for the research activity leading to this chapter has been provided by the Italian Ministry of University and Scientific Research, Grant 60%–Chemistry, 1996.

REFERENCES

Alexander, J. F., Energy Basis of Disasters and the Cycles of Order and Disorder. Ph.D. dissertation, Department of Environmental Engineering Sciences, University of Florida, Gainesville, FL, 232 pp., 1978.

Anderson, J. W., *Bioenergetics of Autotrophs and Heterotrophs*. Arnold, London, 59 pp., 1980.

Bartolelli, M., Adilardi, G., and Bartolelli, V., Analisi preliminare delle destinazioni energetiche alternative dei prodotti e sottoprodotti agricoli: il possibile contributo di alcune colture erbacee. *L'Informatore Agrario*, XXXVIII(31): pp. 22023–22038, 1982.

Bastianoni, S., Brown, M. T., Marchettini, N., and Ulgiati, S., Assessing energy quality, process efficiency and environmental loading in biofuels production from biomass. In *Biomass for Energy, Environment, Agriculture and Industry*. Ph. Chartier, A.A.C.M. Beenackers and G. Grassi (eds.) Pergamon-Elsevier Science, New York, pp. 2300–2312, 1995.

Biancardi, E., La barababietola da zucchero. *Le Scienze (Italian translation of Scientific American)*, No. 194: pp. 120–130, 1984.

Biondi, P., Panaro, V., and Pellizzi G. (Eds.), Le richieste di energia del sistema agricolo italiano. ENEA-CNR-PFE, LB-20, 389 pp., 1989.

Brown, M. T., Energy basis for hierarchies in urban and regional systems. Ph. D. dissertation. Department of Environmental Engineering Sciences. University of Florida, Gainesville, FL, 1980.

Brown, M. T. and Arding, J. E., Transformities Working Paper. Center for Wetlands, Environmental Engineering Sciences, University of Florida, Gainesville, FL., 1991.

Brown, M. T. and Herendeen, R. A., Embodied Energy Analysis and Emergy Analysis: A comparative view. *Ecological Economics* 19: 219–235, 1996.

Brown, M. T. and Ulgiati, S., Emergy-based indices and ratios to evaluate sustainability: monitoring economies and technology toward environmentally sound innovation. *Ecological Engineering,* 9:51–69, 1997.

Brown, M. T. and Ulgiati, S., Emergy Evaluation of the Biosphere and Natural Capital. Ambio, 2000, in press.

Brown, M. T. and Lopez, S., Emergy analysis of rebuilding options after hurricane Andrew. Chapter 5 of: Emergy Evaluation of Energy Policies for Florida, a Report to the Florida Energy Office, Brown, M. T., Odum, H. T., McGrane, G., Woithe, R. D., Lopez, S., and Bastianoni, S. (eds.), 1995.

CCPCS, Commission Consultative pour la Production de Carburant de Substitution, Rapport des Travaux du Groupe Numero 1. Paris, 1991.

Christaller, W., *Central Places in Southern Germany* (Trans, by G.C.W. Baskin), Prentice-Hall, Englewood Cliffs, NJ, 1996.

CNR-PFMA, Consiglio Nazionale delle Ricerche, Progetto Finalizzato Meccanizzazione Agricola, Possibilità produttive, tecniche di raccolta, utilizzazione attuale e alternativa dei sottoprodotti agricoli. E., Natalicchio and C., Semenza Editors, Quaderno n. 23, 1981.

Curzon, F. I. and Ahlborn, B., Efficiency of a Carnot engine at maximum power output. *American Journal of Physics,* 43:22–24, 1975.

Daly, H. E., *Steady State Economics: the Economics of Biophysical Equilibrium and Moral Growth.* Freeman W. H. and Company, San Francisco, 1977.

Daly, H. E., Toward some operational Principles of Sustainable Development. *Ecological Economics,* 2:1–6, 1990.

De Carvalho Macedo, I., The sugarcane agroindustry. Its contribution to reducing CO_2 emissions in Brazil. *Biomass and Bioenergy,* 3(2): 77–80, 1992.

Desmarquest, J. P., CO_2 emissions resulting from the use of fuels from biomass. Paper presented by the Institute Francais du Petrole at the IX International Symposium on Alcohol Fuels, Firenze (Italy), November 12–15; *Proceedings,* pp. 927–932, 1991.

Doherty, S. J., Emergy Evaluations of and Limits to Forest production. Ph.D. dissertation. Department of Environmental Engineering Sciences. University of Florida, Gainesville, FL. 214 pp., 1995.

Ellington, R. T., Meo, M., and El-Sayed, D. A., The net greenhouse warming forcing of methanol produced from biomass. *Biomass and Bioenergy,* 4(6): 405–418, 1993.

ENEA, National Renewable Energy Committee, Italy, Insolation in selected areas of Italy. CESEN-ENEA Database, Italy, 1989.

ERL, Environmental Resources Limited, Study of the environmental impacts of large scale bio-ethanol production in Europe: final report. London, 1990.

Georgescu-Roegen, N., *Energy and Economic Myths.* Pergamon Press, New York, 1976.

Giampietro, M., Using hierarchy theory to explore the concept of sustainable development. *Futures,* 26(6): 616–620, 1994.

Giampietro, M. and Pimentel, D., Energy analysis models to study the biophysical limits for human exploitation of natural processes. In *Ecological Physical Chemistry,* Rossi C. and Tiezzi E. (eds.), Elsevier Science Publishers B.V., Amsterdam, pp. 139–184, 1991.

Giampietro, M., Ulgiati, S., and Pimentel, D., Feasibility of large-scale biofuel production. Does an enlargement of scale change the picture? *BioScience,* 47(9): 587–600, 1997.

Giannantoni, C., Approccio integrato (energetico, exergetico ed emergetico) all'Analisi dei Processi Industriali. Applicazione all'impianto di cogenerazione di Moncalieri (Azienda Energetica Municipalizzata di Torino). ENEA, Roma, ERG-ING-EIAE 97009, 25/2/1997, 1997.

Giannantoni, Environment, Energy, economy, politics and rights: a global approach for evaluating the environmental impact of human activities. In Ulgiati, S., Brown, M. T., Giampietro, M., Herendeen, R. A., and Mayumi, K. (eds.), *Advances in Energy Studies. Energy Flows in Ecology and Economy.* Musis Publisher, Rome, Italy, pp. 541–558, 1998.

Giannantoni, Towards a mathematical formulation of the Maximum Em-Power Principle. *Proceedings of the First Biennial Emergy Analysis Research Conference: Energy Quality and Transformities.* Gainesville, FL, September 2–4, Department of Environmental Engineering Sciences, University of Florida, Gainesville, FL, 1999.

Glasby, G. P., Entropy, Pollution and environmental degradation. *Ambio,* 17(5): 330–335, 1988.

Goldsmith, E., Allen, R., Allaby, M., Davoll, J., and Lawrence, S., A blueprint for survival, *The Ecologist,* 2(1): 1–43, 1972.

Haukoos, D. S., An emergy analysis of various construction materials. Class report, ENV 6905, Environmental Engineering Sciences, University of Florida, Gainesville, FL, Dr. Mark T. Brown, supervisor, 1994.

Henning, D., *Atlas of the Surface Heat Balance of the Continents.* Gebruder Borntraeger, Berlin, 402 pp., 1989.

Herendeen, R., Total environmental cost: comparison of efforts and specific comments on emergy analysis. *Proceedings of the 36th annual Meeting of the International Society for the Systems Sciences,* Denver (Colorado), July 12–17, L. Peeno (ed.), pp. 173–182, 1992.

Herendeen, Needed: examples of applying ecological economics. *Ecol. Econ.,* 9:99–105, 1994.

Himschoot, A. R., Emergy analysis of 1984 Italy. Class report, EES 5306, performed under H.T. Odum supervision, at the Department of Environmental Engineering Sciences, University of Florida, Gainesville, FL, 1988.

Huang, S. L. and Odum, H. T., Ecology and economy: emergy synthesis and public policy in Taiwan. *Journal Environm. Management,* 92: 313–333, 1991.

Kay, J. J., A nonequilibrium thermodynamic framework for discussing ecosystem integrity. *Environm. Management,* 15(4): 483–495, 1991.

IEA, International Energy Agency, Biofuels. Energy and Environment, Policy Analysis Series, Paris, France, pp. 117, 1944.

IGDA, Istituto Geografico De Agostini, Calendario Atlante. Novara, Italy, 1975.

ISTAT, Istituto Nazionale di Statistica, Annuario Statistico Italiano. Roma, Italy, 1990a.

ISTAT, Istituto Nazionale di Statistica, Statistiche dell' Agricoltura, Zootecnia e Mezzi di Produzione. Roma, Italy, 1990b.

ISTAT, Istituto Nazionale di Statistica, Statistiche Ambientali. Roma, Italy, 1991.

ISTAT, Istituto Nazionale di Statistica, Statistiche Ambientali. Roma, Italy, pp. 261, 1993a.

ISTAT, Istituto Nazionale di Statistica, Annuario Statistico Italiano. Roma, Italy, 1993b.

ISTAT, Istituto Nazionale di Statistica, Annuario Statistico Italiano. Roma, Italy, 1997.

Jørgensen, S. E., *Integration of Ecosystem Theories: A Pattern.* Kluwer Academic Publishers, Dordrecht (The Netherlands), Second Edition, 388 pp., 1997.

Landsberg, P. T., *Thermodynamics with Quantum Statistical Illustrations.* Wiley Interscience, New York, 1961.

Lapp, C. W., Emergy analysis of the nuclear power system in the United States. Class report, EES 6916, Environmental Engineering Sciences, Prof. H. T. Odum, supervisor, 1991.

Loddo, M. and Mongelli, F., Heat Flow in Italy. Pageoph., Vol. 117, Birkhauser Verlag, Basel, pp. 135–149, 1978/79.

Losch, A., *The Economics of Location.* (Trans by U. Waglom and W. F. Stalpor), Yale University Press, New Haven, CT, 1954.

Lotka, A. J., Contribution to the energetics of evolution. *Proceedings of the National Academy of Sciences,* U.S.A., 8: pp. 147–150, 1922a.

Lotka, A. J., Natural selection as a physical principle. *Proceedings of the National Academy of Sciences,* 8: 151–155, 1922b.

Lotka, A. J., The law of evolution as a maximal principle. *Human Biology,* 17: 167, 1945.

Lyons, G. J., Lunny, F., and Pollock, H. P., A procedure for estimating the value of forest fuels. *Biomass,* 8: 283–300, 1985.

Magaldi, D., Bazzoffi, P., Bidini, D., Frascati, F., Gregori, E., Lorenzoni, P. Miclaus, N., and Zanchi, C., Studio interdisciplinare sulla classificazione e la valutazione del territorio: un esempio nel comune di Pescia (Pistoia). Istituto Sperimentale Studio e Difesa del Suolo, Firenze, Italy, Annali Vol. XII, pp. 31–114, 1981.

Marrow, J. E., Coombs, J., and Lees, E. W., An assessment of bio-ethanol as a transport fuel in the U.K. Vol. 1, ETSU-R-44. Department of Energy/HMSO, London, 1987.

Odum, E. P., *Basic Ecology.* CBS College Publishing, 1983.

Odum, H. T., *Systems Ecology. An Introduction.* John Wiley & Sons, New York, 644 pp., 1983a.

Odum, H. T., Maximum power and efficiency: a rebuttal. *Ecological Modelling,* 20: 71–82, 1983b.

Odum, H. T., Self organization, transformity and information. *Science,* 242: 1132–1139, 1988a.

Odum, H. T., Energy, environment and public policy. A guide to the analysis of systems. United Nations Environment Programme, UNEP Regional Seas Reports and Studies, No. 95, 109 pp., 1988b.

Odum, H. T., Emergy and Public Policy. Part I-II. Department of Environmental Engineering Sciences, University of Florida, Gainesville, FL, 1994a.

Odum, H. T., The emergy of natural capital. In *Investing in Natural Capital,* Jansson, A. M., Hammer, M., Folke, C., and Costanza, R., (eds.), Island Press, Covelo, CA, pp. 200–212, 1994b.

Odum, H. T., *Environmental Accounting. Emergy and Environmental Decision Making.* Wiley & Sons, New York, 1996.

Odum, H. T. and Arding, J. E., Emergy analysis of shrimp mariculture in Ecuador. Department of Environmental Engineering Sciences, University of Florida, Gainesville, FL. Working paper prepared for Coastal Resources Center, University of Rhode Island, Narragansett, 1991.

Odum, H. T. and Pinkerton, R. C., Time's speed regulator: the optimum efficiency for maximum power output in physical and biological systems. *American Scientist,* 43: 331–343, 1955.

Odum, H. T., Odum, E. C., and Brown, M. T., *Environment and Society in Florida.* Lewis Publishers, New York, 449 pp., 1998.

Odum, H. T., Brown, M. T., Whitfield, D. F., Lopez, S., Woithe, R., and Doherty, S., Zonal Organization of Cities and Environment. A study of Energy Systems Basis for Urban Society. Progress Report, Center for Environmental Policy, Environmental Engineering Sciences, University of Florida, Gainesville, 1996.

O'Neill, R. V., DeAngelis, D. L., Waide, J. B., and Allen, T. F. H., *A Hierarchical Concept of Ecosystems* Princeton University Press, Princeton, 1986.

OTA, Energy from Biological Processes. Office of Technology Assessment, U.S. Congress, Washington, D.C. U.S. Government Printing Office, 1980.

Pellizzi, G., A procedure to evaluate energy contribution of biomass. *Energy in Agriculture,* 5: 317–324, 1986.

Petela, R., Exergy of Heat Radiation. *J. Heat Transfer Ser. C,* 86: 187–192, 1964.

Pimentel, D., *Handbook of Energy Utilization in Agriculture.* CRC Press, Boca Raton, FL, 1980.

Pimentel, D., Warneke, A. F., Teel, W. S., Schwab, K. A., Simcox, N. J., Ebert, D. M., Baenisch, K. D., and Aaron, M. R., Food versus biomass fuels: socioeconomic and environmental impacts in the United States, Brazil, India and Kenya. *Advances in Food Research,* 32: 185–238, 1988.

Pimentel, D., Harvey, C., Resosudarmo, P., Sinclair, K., Kurz, D., McNair, M., Crist, S., Shpritz, L., Fitton, L., Saffouri, R., and Blair, R., Environmental and economic costs of soil erosion and conservation benefits. *Science,* 267: 117–1123, 1995.

Rees, W. E. and Wackernagel, M., Ecological footprints and appropriated carrying capacity: measuring the natural capital requirements of the human economy. In *Investing in Natural Capital.* A. M. Jansson, M. Hammer, C. Folke and R. Costanza (eds.), Island Press, Washington, D.C., pp. 362–390, 1994.

Repetto, R., Accounting for environmental assets. *Scientific American,* 266, June, 94, 1992.

Richardson, J. R., Spatial patterns and maximum power in ecosystems. Ph.D. dissertation, Department of Environmental Engineering Sciences. University of Florida, Gainesville, FL, 254 pp., 1988.

Richardson, J. R. and Odum, H. T., Power and a pulsing production model. In *Energy and Ecological Modeling,* W. J. Mitsch, R. W. Bosserman, and J. M. Klopatek (eds.), Elsevier Press, Amsterdam, 1981.

Russi, Daniela, Class report, Course in Fundamentals of Environmental Impact Assessment, Department of Chemistry, University of Siena, Italy, Dr. Sergio Ulgiati, supervisor, 1999.

SERI, Solar Energy Research Institute, The Production of Herbaceous Feedstock for Renewable Energy. SERI/SP-273-2302, 44 pp., 1986.

Sachs, R. M., Crop feedstocks for fuel alcohol production. *California Agriculture,* 34(6), 1980.

Scienceman, D. M., Energy and emergy, In *Environmental Economics. The Analysis of a Major Interface,* G. Pillet and, T. Murota, R. Leimgruber, Geneva, pp. 257–276, 1987.

Smil, V., *General Energetics. Energy in the Biosphere and Civilization.* Wiley Interscience, New York, p. 369, 1991.

Stuani, E., Iurcotta, E., and Genta, U., *Manuale Tecnico del Geometra e del Perito Agrario.* Signorelli, Milano, 1179 pp., 1986.

Szargut, J., Morris, D. R., and Steward, F. R., *Exergy Analysis of Thermal, Chemical and Metallurgical Processes.* Hemisphere Publishing Corporation, London, p. 332, 1988.

Triolo, L., *Energia Agricoltura Ambiente Libri di Base.* Editori Riuniti, Roma, Italy, 1989.

Triolo, L., Mariani, A., and Tomarchio, L., L'uso dell'energia nella produzione agricola vegetale in Italia: bilanci energetici e considerazioni metodologiche. ENEA, Italy, RT/FARE/84/12, 1984.

Ulgiati, S. and Bastianoni, S., Monitoring a System's Performance and Efficiency by means of Emergy Based Indices and Ratios. European Community, Environmental Research Programme. Area III, Economic and Social Aspects of the Environment. Final Report, Contract No. EV5V-CT92–0152, 1995a.

Ulgiati, S. and Bastianoni, S., A proposed bio-physical multicriteria approach (BPMA). The case of biofuels. European Community, Environmental Research Programme. Area III, Economic and Social Aspects of the Environment. Final Report, Contract No. EV5V-CT92-0152, 1995b.

Ulgiati, S. and Bianciardi, C., Describing states and dynamics in far-from-equilibrium systems. Needed a metric within a system state space. *Ecol. Modell.*, 96, 75–89, 1997.

Ulgiati, S. and Brown, M. T., Assessing Resource Quality by means of Emergy Accounting Techniques: Transformities as Quality Indicators. *International Journal of Global Energy Issues*, 1998a.

Ulgiati, S. and Brown, M. T., Monitoring patterns of sustainability in natural and man-made ecosystems. *Ecol. Modell.*, 108: 23–36, 1998b.

Ulgiati, S., Brown, M. T., Bastianoni, S., and Marchettini, N., Emergy based indices and ratios to evaluate the sustainable use of resources. *Ecol. Eng.*, 5(4): 519–531, 1995a.

Ulgiati, S., Brown, M. T., Giampietro, M., Herendeen, R. A., and Mayumi K. (eds.), *Advances in Energy Studies. Energy Flows in Ecology and Economy*. Musis Publisher, Roma, Italy, p. 642, 1998.

Ulgiati, S., Cassano, M., and Pavoletti, M., Bridging nature and the economy. The energy basis of growth and development in Italy. European Community, Environmental Research Programme. Area III, Economic and Social Aspects of the Environment. Final Report, Contract No. EV5V-CT92-0152, 1995b.

Ulgiati, S., Odum, H. T., and Bastianoni, S., Emergy analysis of Italian agricultural system. The role of energy quality and environmental inputs. In *Trends in Ecological Physical Chemistry*, L. Bonati, U. Cosentino, M. Lasagni, G. Moro, D. Pitea, and A. Schiraldi (eds.), Elsevier Science Publishers, Amsterdam, pp. 187–215, 1993.

Ulgiati, S., Odum, H. T., and Bastianoni, S., Emergy use, environmental loading and sustainability. An emergy analysis of Italy. *Ecol. Modell.*, 73: 215–268, 1994.

Wrighgt, P. G., *Proc. Roy. Soc., A*, 317, 477, 1970.

Wyman, C. E., Bain, R. L., Hinman, N. D., and Stevens, D. J., Ethanol and methanol from cellulosic biomass. In *Renewable Energy Sources for Fuels and Electricity*. T. B. Johansson, H. Kelly, A. K. N. Reddy, and R. H. Williams (eds.), Island Press, Washington, D.C., 1200 pp., 1993.

CHAPTER 5

This chapter presents a close connection between thermodynamics and mathematical theory of stability. Thermodynamic laws are applications of the direct Lyapunov methods to special dynamical systems which leads to a close relationship between the Second Law of Thermodynamics and mathematical stability. The stability of the equilibrium is according to the Second Law of Thermodynamics, when the entropy is maximal, while the stable structure of a biological community is the consequence of interactions between populations. As any biological system is an open system, the total entropy is changed in an arbitrary way. The author discusses in this context the "old" phrase that communities richer in comprising species are more stable. He is able to demonstrate clearly that it cannot be the case from a mathematical/thermodynamic stability viewpoint, which is also the general ecological understanding today.

He has furthermore disclosed that exergy is a Lyapunov function. This is an important result for some of the later chapters where exergy is widely used to approach ecological understanding (Chapters 9–10, and 12–14).

The statement brought forward by the author is that thermodynamics and the mathematical stability concepts are closely related. A comprehensive treatment of the biological concept of stability from a mathematical point of view is presented in the classical book by Svirezhev and Logofet, *Stability of Biological Communities*, (Mir, Moscow, 1983).

5 Thermodynamics and Theory of Stability

Yuri M. Svirezhev

CONTENTS

5.1   Introduction 117
5.2   A Few Words about the Stability Concept in Ecology...117
5.3   About One Class of Lyapunov Functions 119
5.4   Lyapunov Function for Volterra Equations 121
5.5   Extreme Properties of Volterra Equations for Competing Species:
      One More Lyapunov Function...121
5.6   Stability Concept in Thermodynamics 122
5.7   Exergy as a Lyapunov Function 123
5.8   Problem of Additivity and Some Thermodynamic Constraints 124
5.9   Thermodynamic Basis of Volterra's Equations for Competing Species............................126
5.10 Phenomenological Thermodynamics of Evolutionary Models 129
5.11 Some Generalisation of the Exergy Concept 131
References 132

5.1   INTRODUCTION

There is a very deep connection between thermodynamics as a physical theory and the mathematical theory of stability. One of the most important concepts in the theory of stability is the Lyapunov functions concept. Positive functions defined in a phase space of dynamical system possess the property of either monotonous increase or monotonous decrease along trajectories. They can be considered as some special class of goal functions. On the other hand, the main thermodynamic laws (the Second Law of Thermodynamics, the Prigogine theorem) state the similar properties of monotonocity for special functions called potentials, entropy, etc. These functions are Lyapunov functions and thermodynamic laws are applications of the direct Lyapunov method to special dynamical systems.

5.2   A FEW WORDS ABOUT THE STABILITY CONCEPT IN ECOLOGY

It is intuitively clear that both an ecosystem and a biological community that exists a sufficiently long time in a more or less invariant state (this property is often called *persistence*) should possess intrinsic abilities to resist perturbations coming from the environment. This ability is usually called "stability." Though the notion seems obvious, it is quite a problem to provide a precise and unambiguous definition for it. This heavily overloaded term has no established (stable) definition so far. For instance, the theory of stability, which can be considered as a branch of theoretical mechanics, uses about 30 different definitions of stability. So we can consider a definition of stability (and also entropy) as a "fuzzy" one. Paraphrasing von Neumann, we can say ". . . nobody knows what stability means in reality, that is why in the debate you will always have an advantage."

Among these definitions we can select two large classes which differ in respect to the requirements coming under the headline of "stability." The first group of requirements concerns preservation of the number of species in a community. A community is stable if the number of member-species remains constant over a sufficiently long time. This definition is the closest to various mathematical definitions of stability.

The second group refers to populations rather than to community, which is considered to be stable when numbers of component populations do not undergo sharp fluctuations. This definition is closer to the thermodynamic (or more correctly, to the statistical physics) notion of system stability. In thermodynamics (statistical physics) a system is believed to be stable when large fluctuations, which can take the system far from the equilibrium or even destory it, are unlikely (see, for instance, Landau and Lifshitz, 1964). Evidently, the general thermodynamic concepts (for instance, the stability principle associated in the case of closed systems with the Second Law and, in the case of open systems, with the Prigogine theorem) should be applicable to biological (and, in particular, ecological) systems. As an illustration of such a phenomenon, I would like to consider a very well known problem of the relationship between the biological diversity of a community and its stability.

Ecologists consider it almost as an axiom that communities which are more complex in structure and richer in comprising species are more stable. Any popular textbook on ecology (for instance, E. Odum's classic book, *Fundamentals of Ecology*, 1968) would convince you of this. This is explained in the following way: different species adapt differently to environmental variations. Therefore, a variety of species may respond with more success to different environmental variations than a community composed of a small number of species and, hence, the former will be more stable.

Perhaps, this motivates the fact that specific diversity indices (in particular, the Shannon information entropy called also the *information diversity index*),

$$D = -\sum_{i=1}^{n} p_i \ln p_i, \qquad p_i = N_i \Big/ \sum_{i=1}^{n} N_i, \qquad (1)$$

where n is the number of species (or some other groups) in a community and N_i is the population size of ith species) are used as a measure of stability (Margalef, 1951; MacArthur, 1955).

In accordance with this "logic," the community is the most stable if D is maximal. But, as could readily be shown in this case, the community structure is such that specimens of any species occur with the same frequency

$$(\max_{p_i} D \text{ is attained at } p_i^* = 1 / n).$$

In other words, the diversity of a community is maximal when the distribution of species is uniform, or when there are no abundant or rare species, and no structures. However, observations in real communities show that this is never the case, and that there is always a hierarchical structure with dominating species. What is the reason for such a contradiction? Probably, it is the formal application of models and concepts taken from physics and information theory to systems that do not suit the definition. Both the Boltzmann entropy in statistical physics and the Shannon entropy in information theory make sense only for populations of weakly interacting particles. A typical example of such system is the ideal gas, the macroscopic state of which is an additive function of the microscopic states of its molecules.

Let us remember the original formulation of the Boltzmann entropy: $S_B \sim \ln W$ where W is a probability of state of the system. In this general formulation the Boltzmann formula applies to any system, not only to systems with weak interactions. But, as soon as we use the standard formula

$$S = -k \sum_{i=1}^{n} p_i \ln p_i$$

we use implicitly the classic thermodynamic model of an ideal gas.

The introduction of an entropy measure to such sets is well founded. And, in addition, the stability of the equilibrium when the entropy is maximal is associated with the Second Law. On the other hand, the stable structure of a biological community is the consequence of interactions between populations rather than the function of characteristics of individual species, i.e., a biological community is a typical system of strongly interacting elements. But as soon as we become concerned with such systems the entropy measure is no longer appropriate. There is one more argument against the use of the diversity as a goal function relating to stability. The entropy increases (tending to a maximum) only in closed systems, but any biological system is an open system in a thermodynamic sense, so that its total entropy is changed in an arbitrary way. When in equilibrium (we speak of a dynamic equilibrium), the rate of the entropy production inside a system is positive and minimal. This is the Prigogine theorem. In this case, in relation to stability, the goal function is the rate of the entropy production, not the entropy.

Notice, however, that in numerous competitive communities in the initial stages of their successions, far from climax, an increase in diversity may be observed. It seems that in these cases diversity is a "good" goal function for stability. This is explained as the following: in the initial stages, far from equilibrium, the competition is still weak and the community may well be regarded as a system with weak interaction.

Summarising all the above mentioned reasoning we can say that the connection between diversity and stability is not evident and univalent as it seemed earlier.

5.3   ABOUT ONE CLASS OF LYAPUNOV FUNCTIONS

Let us consider the following dynamical system:

$$\frac{dC_i}{dt} = f_i(C_1,\dots,C_n), \quad i = 1,\dots,n \tag{2}$$

where the vector $\mathbf{C} = \{C_1, \dots, C_n\} \in P^n$, P^n is a positive orthant of Euclidean phase space E^n, i.e., the domain of the state space in which all the state variables C_i are non-negative. This is a typical form of equation describing ecological objects, since all ecological values are positive. In particular, in mathematical ecology the values C_i are interpreted as either biomasses or densities of corresponding species, age cohorts, and other ecological groups constituting a biological community. We shall assume that the system occupies a definite fixed spatial volume. Hence, the values C_i can consider both total amounts of biomass or some substance and biomass densities per volume unit or a volume concentration of some substances. Later we choose the first interpretation, namely, C_i is both the biomass of ith species and the amount of some ith substance in the system.

We also assume that the system (2) has a nontrivial equilibrium $\mathbf{C}^*$ which is situatied within P^n Let us check the following class of functions as a candidate for Lyapunov functions:

$$L = \sum_{i=1}^{n} C_i^* \varphi(C_i / C_i^*), \tag{3}$$

where $(\xi_i = C_i/C_i^*)$

$$\varphi(1) = \frac{d\varphi}{d\xi}(1) = 0; \qquad \frac{d^2\varphi}{d\xi^2} > 0 \quad \text{for any } \xi \geq 0. \tag{4}$$

In other words, the function $\varphi(\xi)$ must be convex for positive ξ. It is obvious that $L(C^*) = 0$. Because the first variation of L in a vicinity of C^* is equal to

$$\delta L = \sum_{i=1}^{n} \frac{\partial L}{\partial C_i} \delta C_i = \sum_{i=1}^{n} \frac{\partial \varphi}{\partial \xi_i} \delta C_i, \tag{5}$$

then $\delta L(C^*) = 0$ for any variations δC^*.

Calculating the second variation we get

$$\delta^2 L = \frac{1}{2} \sum_{i=1}^{n} \sum_{j=1}^{n} \frac{\partial^2 L}{\partial C_i \partial C_j} \delta C_i \delta C_j = \frac{1}{2} \sum_{i=1}^{n} \frac{d^2\varphi}{d\xi_i^2}(\xi_i) \frac{(\delta C_i)^2}{C^*_i} > 0 \tag{6}$$

for any nonzero variations δC. Thus, L is a convex function of C having an isolated minimum at the point C^* and increasing monotonously for any $C \in P^n$.

The equilibrium C^* is stable if the derivative dL/dt taken along the trajectory of (2)

$$\frac{dL}{dt} = \sum_{i=1}^{n} \frac{d\varphi}{d\xi}(\xi_i) f_i(C_1, \ldots, C_n) \leq 0. \tag{7}$$

for all trajectories (the Lyapunov stability theorem, see Malkin, 1967).

On the other hand, the equilibrium C^* is unstable if the derivative dL/dt^3 0 even for only one trajectory of (2) (the Chetaev instability theorem).

Let $\varphi(\xi) = a(1 - \xi)^2, a > 0$, then the function $L = \sum_{n=1}^{n} a_i C_i^* (1 - \xi_i)^2$ is a Lyapunov function. Setting $a_i = C_i^*/n$ we get

$$L = \frac{1}{n} \sum_{i=1}^{n} (C_i - C_i^*)^2, \tag{8}$$

i.e., the value L may be considered as a mean square measure of distance between a current state of the system and its equilibrium. If this distance decreases in time (i.e., the derivative $dL/dt < 0$ along the system trajectory) then, in accordance with the Lyapunov stability theorem, we hope that the system moves to a stable equilibrium. Note that I speak of "hope," since we must check all the trajectories (or a statistically "sufficient" large number of them) in order to be able to speak correctly about a stability of equilibrium. If, on the other hand, this measure increases along some trajectory, then (in accordance with the Chetaev instability theorem) the equilibrium is unstable and this trajectory resides from it.

All these results can be interpreted easily from a thermodynamic viewpoint. Indeed, in thermodynamics the value L is a mean square of fluctuations around an equilibrium or a *power of fluctuations*. Therefore, we can state that if a power of fluctuations decreases in time then the system moves to a stable equilibrium. Note that if the movement to a stable equilibrium can be called an *evolution of the system* then the foregoing statement can be reformulated in the following way: the power of fluctuations decreases during the process of the system's evolution. Since the state $\mathbf{C}^*$ is a *goal of evolution* in this sense, it is natural to consider the function L as a *goal function*.

5.4   LYAPUNOV FUNCTION FOR VOLTERRA EQUATIONS

Let the function φ be $\varphi(\xi) = \xi - \ln \xi - 1$; then the corresponding Lyapunov function will be

$$L = \sum_{i=1}^{n} (C_i - C_i^*) - C_i^* \ln(C_i / C_i^*). \tag{9}$$

If our dynamical system is the system of Volterra equations

$$\frac{dC_i}{dt} = C_i \left(\varepsilon_i - \sum_{j=1}^{n} \gamma_{ij} C_j \right), \qquad i = 1, \ldots, n \tag{10}$$

where $\varepsilon_i > 0$ is the Malthusian coefficients (i.e., the difference between intrinsic birthrate and mortality for "prey" species and $\varepsilon_j < 0$ is mortality coefficients for "predator" species, with a community matrix $1 = |\gamma_{ij}|$, we immediately get

$$\frac{dL}{dt} = -\sum_{i=1}^{n} \sum_{i=1}^{n} \gamma_{ij} \delta C_i \delta C_j, \tag{11}$$

and the nontrivial equilibrium $\mathbf{C}^*$ is always stable if the matrix $\mathcal{G}$ is positive definite. This is a well-known result in the general theory of Volterra systems (see Svirezhev and Logofet, 1978).

5.5   EXTREME PROPERTIES OF VOLTERRA EQUATIONS FOR COMPETING SPECIES: ONE MORE LYAPUNOV FUNCTION

We consider again the Volterra system (10) assuming a symmetry for $\mathcal{G}$, i.e., $\gamma_{ij} = \gamma_{ji}$, and $\varepsilon_i > 0$. Such a system is a very popular model for a community of competing species. The transformation

$$\eta_i = \pm \sqrt{C_i}, \qquad i = 1, \ldots, n$$

transfers the positive orthant P^n into the complete coordinate space R_η^n in which the trajectories of the system (10) are trajectories of the steepest ascent for the function

$$W = \frac{1}{4} \sum_{i=1}^{n} \varepsilon_i \eta_i^2 - \frac{1}{32} \sum_{i=1}^{n} \sum_{j=1}^{n} \gamma_{ij} \eta_i^2 \eta_j^2 = \sum_{i=1}^{n} \varepsilon_i C_i - \frac{1}{2} \sum_{i=1}^{n} \sum_{j=1}^{n} \gamma_{ij} C_i C_j. \tag{12}$$

Then the system (10) can be rewritten in the gradient form

$$\frac{d\eta_i}{dt} = \frac{\partial W}{\partial \eta_i}, i = 1, \ldots, n; \qquad \frac{dW}{dt} = \sum_{i=1}^{n} \frac{\partial W}{\partial \eta_i} \frac{d\eta_i}{dt} = \sum_{i=1}^{n} \left(\frac{\partial W}{\partial \eta_i} \right)^2 \geq 0. \tag{13}$$

From (13) it follows that the value W decreases in the process of the system's evolution attaining a maximum at the equilibrium; i.e., W may be considered as a goal function for the competitive community (in detail see Svirezhev and Logofet, 1978). Then the function $L = W(\mathbf{C}) - W(\mathbf{C}^*) \leq 0$ is a Lyapunov function for (10). Note that this Lyapunov function does not belong to the introduced class and, as opposed to those functions, the equilibrium $\mathbf{C}^*$ must not necessarily be nontrivial; it may be situated on the appropriate coordinate hyperplanes. This implies that in the process of the community evolution one or several species have to be eliminated.

All these results also permit a sensible interpretation. The value $R(\mathbf{C}) = \Sigma_{i=1}^{n} \varepsilon_i C_i$ in essence accounts for the rate of biomass gain in the case that competition and any kind of limitation by resources are absent, and the growth is only determined by the physiological fertility and natural mortality of the organisms. Therefore it is natural to define the value R as the *reproductive potential of the community*. The value

$$D = \frac{1}{2} \sum_{i=1}^{n} \sum_{i=1}^{n} \gamma_{ij} C_i C_j$$

may be used to measure the *rate of energy dissipation* resulting from inter- and intraspecific competition. Therefore we shall refer to D as the *total expenses of competition*. Hence the increase in D in the process of evolution may be interpreted as the goal of the community to maximise the difference between its reproductive potential and the total expenses of competition. This goal can be achieved in several ways: either the reproductive potential is maximised at fixed expenses of competition (*r-strategy*) or the competition expenses are minimised for a limited reproductive potential (*K-strategy*). There may also be some intermediate cases.

5.6   STABILITY CONCEPT IN THERMODYNAMICS

Let us consider the problem from a thermodynamic viewpoint. In the vicinity of an equilibrium the entropy of an open system is presented in the form

$$S = S^{eq} + (\delta S) + \frac{1}{2} (\delta^2 S). \tag{14}$$

Differentiating (14) in respect to time, we get ($\Delta S = S - S^{eq}$)

$$\frac{d\Delta S}{dt} = \frac{dS}{dt} = \frac{d(\delta S)}{dt} + \frac{d(\delta^2 S)}{dt}. \tag{15}$$

As usual we present the difference dS in the form $dS = d_e S + d_i S$ where $d_i S$ is the entropy change that takes place by means of internal processes and $d_e S$ is the entropy change caused by independent changes in the interaction between the system and its environment. If to assume that the difference $d_i S$ has a higher order of smallness than $d_e S$, then

$$\frac{d_e S}{dt} = \frac{d(\delta S)}{dt} \quad \text{and} \quad \frac{d_i S}{dt} = \frac{1}{2} \frac{d(\delta^2 S)}{dt}. \tag{16}$$

In accordance with the Second Law, the entropy within any system must increase (nondecrease). Then

$$\frac{d_iS}{dt} = \frac{1}{2}\frac{d(\delta^2 S)}{dt} \geq 0. \tag{17}$$

In thermodynamics the second variation $\delta^2 S$ is presented as a quadratic form of $\delta C_i = C_i - C_i^*$ and, if it is negative definite, then the corresponding equilibrium is stable (in a thermodynamic sense). Returning to our model we assume that the equilibrium $\mathbf{C}^*$ is nontrivial. Then, remembering that $\sum_{i=1}^{n} \gamma_{ij} C_j^* = \varepsilon_i$; $i = 1,\ldots, n$, we can rewrite the expression for $L = W(\mathbf{C}) - W(\mathbf{C}^*)$ in the form

$$L = -\frac{1}{2}\sum_{i=1}^{n}\sum_{j=1}^{n}\gamma_{ij}(C_i - C_i^*)(C_j - C_j^*) = -\frac{1}{2}\sum_{i=1}^{n}\sum_{j=1}^{n}\gamma_{ij}\delta C_i \delta C_j. \tag{18}$$

Setting $\delta^2 S = 2L$ we can consider $\delta^2 S$ as a Lyapunov function. In accordance with the Lyapunov stability theorem, if $\delta^2 S < 0$ (this inequality is a condition for thermodynamic stability) and $d(\delta^2 S)/dt \geq 0$ (this is the Second Law) then the equilibrium is stable (in a Lyapunov sense). It is obvious if the matrix $\mathcal{G}$ is positive definite then $d(\delta^2 S)/dt \geq 0$ (one of the sufficient conditions for the stability of $\mathbf{C}^*$). An interesting analogy between the theory of competing communities and the thermodynamics of chemical systems follows from the expression for $\delta^2 S$ in the previous formula:

$$\delta^2 S = -\frac{1}{T}\sum_{i=1}^{n}\sum_{i=1}^{n}\frac{\partial \mu_i}{\partial C_j}\delta C_i \delta C_j, \quad \text{where} \quad \mu_i = \mu_i(C_1,\ldots,C_n)$$

is the chemical potential of ith component depending on all the concentrations in the general case (see, for instance, Rubin, 1967). Comparing the expressions for $\delta^2 S$ we can see that the coefficients of competition are analogous to the partial derivatives of chemical potentials so that

$$\gamma_{ij} \sim \frac{1}{T}\frac{\partial \mu_i}{\partial C_j}.$$

I think that these and other analogies will be helpful if we intend to construct some form of phenomenological thermodynamics of biological communities.

Since L (and also correspondingly, $\delta^2 S$) increases when going further from the stable equilibrium, then this may be regarded as a peculiar form of the *Le Chatelier principle*: any displacement from a stable equilibrium increases the competition expenses within the community.

5.7 EXERGY AS A LYAPUNOV FUNCTION

It is obvious that our class of Lyapunov functions is very large; therefore, I would like to reduce it using some additional concepts, for instance, thermodynamic ones.

Before starting these thermodynamic speculations we consider the following function $\varphi(\xi) = \xi \ln \xi - \xi + 1$ (see (3)). It is obvious that

$$\varphi(1) = (d\varphi/d\xi)_{\xi=1} = 0; \qquad (d^2\varphi/d\xi^2)_{\xi=1} > 0$$

and the corresponding Lyapunov function will be

$$L = \sum_{i=1}^{n} \{C_i \ln(C_i/C_i^*) - (C_i - C_i^*)\}. \tag{19}$$

Comparing (19) and the expression for exergy suggested by Jørgensen (see Chapter 7) we see that Jørgensen's *exergy* is equal to the formally defined Lyapunov function for the system (2), i.e., $L = Ex$. Jørgensen's interprets the equilibrium $\mathbf{C}^*$ (a reference state) by identifying it with a thermodynamic equilibrium when the life is absent. Logically, we may say that the origin of life can be considered as the loss of stability for thermodynamic equilibrium and the movement of the system away from it along one trajectory. In this case (in accordance with the Chetaev instability theorem), if the exergy increases along this trajectory, i.e., the inequality $dL/dt = dEx/dt > 0$ takes place then the thermodynamic equilibrium is unstable. It is easy to see that this is the other formulation for Jørgensen's maximal principle. (Note here we implicitly assumed that the reference state, i.e., the thermodynamic equilibrium, is also one of possible equilibria of the considered dynamical system.)

5.8   PROBLEM OF ADDITIVITY AND SOME THERMODYNAMIC CONSTRAINTS

I would like to say a few words about specific thermodynamic constraints. Since all the C_i are nonnegative we can consider the values as some "masses" of ith species and the sum

$$A = \sum_{i=1}^{n} C_i$$

as a total mass of the system. In thermodynamics the role of A plays such a value as a total number of particles.

Let us look in the Landau and Lifshitz textbook *Statistical Physics* (1964). One can see that such thermodynamic values as entropy, energy, and all four thermodynamic potentials possess the so-called property of additivity. This means that if the amount (mass) of matter of the system changes then a corresponding value is changed by the same times. For instance, let the system energy, E, be a function of C_i (this is quite a reasonable assumption), $E = f(C_1, \ldots, C_n)$. The energy E is an extensive variable since the system energy is the sum of energies of its parts. If we remember that these variables represent the amounts of different substances in the system then we may speak about them as additive values. Indeed, the mass of joined systems is equal to the sum of masses of its component. Introducing the frequencies $p_i = C_i/A$ we define the vector $\mathbf{p} = \{p_1, \ldots, p_n)$ as a *structure* vector of the system. It is obvious that $\mathbf{p}$ is not the sum of partial structures. Since the energy possesses a property of additivity, i.e., the function $f(C_1, \ldots, C_n)$ must be a homogeneous function of the first order with respect to additive variables $C_1, \ldots, C_n$, then the following relation takes place:

$$E = f(C_1, \ldots, C_n) = Af(p_1, \ldots, p_n). \tag{20}$$

If the exergy of the system can be presented in the same form, $Ex = f_{Ex}(C_1,\ldots, C_n) = Af_{Ex}(p_1,\ldots, p_n)$ then in this case such a value as a *specific exergy* (exergy per mass unit, $ex = Ex/A$) makes sense. However, if we consider the formulas (10.15), (10.16) Chapter 10, we can see that $Ex(\mathbf{C}) = AEx(\mathbf{p}) + Ex(A)$, i.e., the *exergy does not possess a property of additivity*. In other words, the exergy does not have a thermodynamic limit, and the introducing of the concept of density of the exergy (or the concept of specific exergy) using a standard thermodynamic method is not correct. It is not surprising since the exergy is not a function of state (like entropy or energy), but the exergy is some degenerated functional since its value depends on two states: current and reference. On the other hand, it is intuitively clear that the exergy can be used as a characteristic of one mass unit for living matter (see Jørgensen, 1995). How to solve this contradiction?

We get the expression for specific exergy, $ex = Ex/A$, in the form

$$ex = Ex(\mathbf{p}) + Ex(A)/A = \sum_{i=1}^{n} p_i \ln \frac{p_i}{p_i^*} + \ln \frac{A}{A^*} - \left(1 - \frac{A^*}{A}\right), \qquad (21)$$

i.e., the specific exergy (the density of exergy) depends both on the total amount of matter in the system, A, and on its structure, $\mathbf{p}$. (It is necessary to remember that by omitting the factor RT before the expression for the exergy we measure it in dimensionless units.) It is easy to see that if we assume that the total mass of the system does not change in the course of evolution,

$$\text{i.e., } A = A^* = \text{const}$$

then $\quad Ex(A) = A \ln \dfrac{A}{A^*} - (A - A^*) = 0 \quad$ and $\quad ex(C_1, \ldots, C_n) = Ex(\mathbf{p}) = K$

where $\quad K = \sum_{i=1}^{n} p_i \ln(p_1/p_i^*)$

is so-called Kullback's measure for excess of information (see, for instance, Chapter 10). In other words we can consider that Kullback measure, K, as a specific exergy of one unit of mass (or biomass, if we talk about a living matter). Remembering an interpretation of K we can say that the specific exergy is equal to the *amount of information* for one biomass unit, which has been accumulated in the process of transition from some *reference state* corresponding to a thermodynamic equilibrium, i.e., some "*prevital*" state, to the current state of living matter. In other words, the specific exergy is equal to the information contents of one unit of biomass.

On the other hand, we can interpret two different items in the expression for the specific exergy from the thermodynamic point of view. As in thermodynamics of open systems, we divide the change of total entropy (dS) into two parts: the first is tied up to internal processes (d_iS) and the second results from exchange processes between the system and its environment (d_eS). In a similar way we may divide the total system exergy (Ex) into two parts:

$$Ex = Ex_i + Ex_e \text{ where } Ex_i = Ex_{\text{inf}} = AK, \text{ and } Ex_e = Ex_{\text{mat}} = A \ln(A/A^*) - (A - A^*).$$

We identify the value Ex_i with internal processes changing only the system structure and the value Ex_e with external processes of exchange between the system and its environment. This leads to the change only in the total biomass. In this case, the value Ex_i can be considered as a thermodynamic value possessing additivity. If the system structure does not change, then this value changes proportionally to the increase or decrease of a total mass (biomass) of the system. The second item increases non-proportionally with the growth of a mass.

Analogously, the total specific exergy can be presented as a sum of two items:

$$ex = ex_i + ex_e = K + \left[\ln \xi + \frac{1}{\xi} - 1 \right] \qquad \text{where } \xi = A / A^*. \qquad (22)$$

If the first item, $ex_i = K$, is determined by structural changes then the second item,

$$ex_e = \ln \xi + \frac{1}{\xi} - 1,$$

is determined by a change of a total mass of the system. It is nonmonotone in respect to ξ which is a relative change of mass of the system in relation to a "reference mass" of some "prevital" substance. For instance, if we presuppose that the Earth biosphere has originated from the Earth crust, then the biosphere and crust masses can be considered as A and A^*, correspondingly. On one hand, $A \ll A^*$ and

$$\xi \ll 1 \text{ (in accordance to Vinogradov, 1959, } \xi \approx 10^{-4})$$

$$\text{so that} \qquad \frac{1}{\xi} \gg |\ln \xi - 1| \approx 10^0 \div 10^1.$$

On the other hand, the chemical composition of living matter does not principally differ from the composition of the crust. So it should also be that $ex_i \approx 10^0 \div 10^1$. Therefore, the main contribution into the specific exergy of living matter is provided by processes that maintain the separation of total living matter or, on the whole, from nonliving matter, which had been a material basis for the origin of life.

Note that there is a singularity here, since $ex_e \to \infty$ when $\xi \to 0$. It formally means that the specific exergy of living matter, i.e., an ability to perform some useful work (for instance, to turn the global biogeochemical cycles), must be very high in order to "push" the *biosphera machina*. This is a very interesting problem of the "Primary Push" in the origin of biosphere.

5.9 THERMODYNAMIC BASIS OF VOLTERRA'S EQUATIONS FOR COMPETING SPECIES

Let us consider a biological community of species with concentrations C_i, $i = 1,..., n$, competing for one resource with concentration R. The resource compartment is considered as an environment for the biological community (the system). We assume (a simple hypothesis) that the rate of resource uptake is equal to

$$\sum_{i=1}^{n} \alpha_i R C_i.$$

Since the resource flow supports the biological system then we can identify this flow with a flow of negentropy from environment into the system. In other words, the change of entropy caused by exchange processes between the system and its environment is equal to

$$\frac{d_e S}{dt} = \beta_e = - \sum_{i=1}^{n} \alpha_i R C_i. \qquad (23)$$

It is obvious that the thermodynamic equilibrium is not interpreted as a life, i.e., all $C_i^0 \equiv 0$. Then the values grad $C_i \cong C_i - C_i^0 = C_i$ can be considered as thermodynamic forces, X_i. We assume that thermodynamic fluxes can be presented in the form

$$J_i = \sum_{j=1}^{n} \gamma_{ij} X_j = \sum_{j=1}^{n} \gamma_{ij} C_j \qquad \text{where} \qquad \gamma_{ij} = \gamma_{ji} \geq 0.$$

These are well known as Onsager's reciprocal relations. In accordance to one of the main formulas of thermodynamics of open systems (De Groot and Mazur, 1963) the change of entropy caused by irreversible processes within the system is equal to

$$\frac{d_i S}{dt} = \beta_i = \sum_{i=1}^{n} J_i X_i = \sum_{i=1}^{n} \sum_{j=1}^{n} \gamma_{ij} C_i C_j. \tag{24}$$

In accordance to thermodynamics of open systems, the following two conditions must be fulfilled in a stable dynamic (not thermodynamic!) equilibrium.

$$\mathbf{C}^* = \{C_i^*, \ldots, C_n^*\}, \mathbf{C}^* > 0$$

1.
$$\beta_i^* = -\beta_e^*, \quad \text{i.e.,} \quad \sum_{i=1}^{n} \sum_{j=1}^{n} \gamma_{ij} C_i^* C_j^* = R^* \sum_{i=1}^{n} \alpha_i C_i^*, \tag{25}$$

2.
$$\beta_i^* = \min_{\mathbf{C}} \beta_i > 0, \quad \text{i.e.,} \quad \sum_{i=1}^{n} \sum_{j=1}^{n} \gamma_{ij} C_i^* C_j^* = \min_{\mathbf{C}} \left\{ \sum_{i=1}^{n} \sum_{j=1}^{n} \gamma_{ij} C_i C_j \right\} > 0 \tag{26}$$

(Prigogine's theorem).

The relations (25), (26) determine the coordinates of equilibrium. For this we must find the minimum of quadratic form β_i (condition 2) situated on the plane $\beta_e = -\beta_i^*$ (condition 1), i.e., the minimum of β_i under the constraint $\beta_e = -\beta_i^*$. Using the method of Lagrange multipliers we get

$$2 \sum_{j=1}^{n} \gamma_{ij} C_j^* + \lambda \alpha_i R^* = 0, \qquad i = 1, \ldots, n. \tag{27}$$

Multiplying both sides of (27) by C_i^* and adding together we get the formula for the determination of λ,

$$2\beta_i^* - \lambda \beta_e^* = 0, \qquad \lambda = \frac{2\beta_i^*}{\beta_e^*} = -2. \tag{28}$$

And, finally, the equations for equilibrium coordinates will be

$$\alpha_i R^* - \sum_{j=1}^{n} \gamma_{ij} C_j^* = 0, \qquad i = 1, \ldots, n. \tag{29}$$

Since Equation (27) is a necessary condition for minimum, then a sufficient condition will be a positive definiteness for the quadratic form β_i. In fact, in this case, it is positive in the equilibrium. (Note that we did not use any differential equations, like Volterra's.)

Let us return to Volterra's system (10) in which $\varepsilon_i = \alpha_i R^*$ and $\gamma_{ij} = \gamma_{ij} \geq 0$:

$$\frac{dC_i}{dt} = C_i \left(\alpha_i R^* - \sum_{j=1}^{n} \gamma_{ij} C_j \right), \qquad i = 1, \ldots, n \tag{30}$$

We see that the coordinates of nontrivial (i.e., all $C_i^* > 0$) equilibrium points are determined by (29), and the equilibrium is stable if the quadratic form

$$\sum_{i=1}^{n} \sum_{j=1}^{n} \gamma_{ij} C_i C_j$$

is positive definite (it was proved in Section 5.4). It is interesting that we proved here the same statement using only thermodynamic methods and approaches. Certainly, another verbal model was used to deduce Volterra's equation by a standard way. For instance (we mentioned it above), $\varepsilon_i > 0$ is the difference between intrinsic birthrate and natural mortality. In the "thermodynamic" interpretation, a mortality is only a result of competition, no specimen survives up to its own age limit, and any individual death is a consequence of competition. This is very close to the reality of Nature.

Reciprocal Onsager's coefficients in thermodynamic formalism correspond to coefficients of competition in the Volterra equations. The latter is certain since a competition is a typical irreversible process leading to dissipation of energy and biomass.

Note that in all our arguments we assume the constancy of environment, which appears in the constancy of resource concentration. It is possible, if an external reservoir containing a resource is very large in comparison with the biological system (the latter is a standard assumption for different thermodynamic considerations) and an inverse influence of the system on its environment can be neglected. However, it is interesting to consider a situation when we cannot neglect this. Then we must supplement the system (30) with one more equation for R.

If we assume that the resource is restored by a constant inflow Q, then the equation will be

$$\frac{dR}{dt} = Q - \sum_{i=1}^{n} \alpha_i R C_i. \tag{31}$$

A constancy of resource concentration, R^*, will be supported if the constraint

$$\sum_{i=1}^{n} \alpha_i C_i = a = \text{const}$$

is fulfilled. It is easy to show that the minimum of positive definite quadratic form

$$\beta_i = \sum_{i=1}^{n} \sum_{j=1}^{n} \gamma_{ij} C_i C_j$$

under this constraint is reached in point $\mathbf{C}^*$ determined by (29). In accordance with the *duality principle* in the optimisation theory (Gill et al., 1981) this is equivalent to the problem of maximisation for the linear form

$$a = \sum_{i=1}^{n} \alpha_i C_i$$

under the constraint $\beta_i = $ const. Since $R^* = Q/a$ then the maximisation of a means the minimisation of R^*. These results may be interpreted in the following, somewhat speculative manner.

If a system such as a biological community of competing populations in a process of interaction with its environment tends to a stable equilibrium with nonzero values of population sizes, then at the equilibrium,

- A system tends to minimise the production of entropy within a system (*the Prigogine theorem*);
- A system tends to arrange its structure and interaction with environment in such a way that the negentropy flux out of a system into an environment will be maximal (*the principle of maximum for the negentropy production*);
- With this, the concentration of resource tends to minimal (*the principle of maximal utilisation*).

5.10 PHENOMENOLOGICAL THERMODYNAMICS OF EVOLUTIONARY MODELS

We know that the concept of extensive and intensive variables plays a very important role in thermodynamics. For instance, energy, entropy, mass, and volume are extensive variables, but temperature and pressure are intensive ones. It is obvious that for extensive values a concept of density or a specific value can be introduced, while this cannot be done for intensive values. Let us try to introduce an analogous concept to our formalism. For this we return to the system (2).

It is obvious both C_i and A can be considered as some extensive variables, but the vector of structure $\mathbf{p} = \{p_1, \ldots, p_n\}$ is an intensive variable. (Let us remember that the structure of joined system is not the sum of partial structures; in order to get this structure we must use a special averaging procedure.)

We assume that the system (2) is presented in the form

$$\frac{dC_i}{dt} = \mu_i(C_1, \ldots, C_n)C_i, \quad i = 1, \ldots, n \tag{32}$$

where the so-called Malthusian functions μ_i are analytical with respect to variables C_i. When passing from C-variables to $\mathbf{p}$ and A ones we get

$$\frac{dp_i}{dt} = p_i(\mu_i - \hat{\mu}),$$

$$\frac{dA}{dt} = \hat{\mu}A, \quad \hat{\mu} = \sum_{i=1}^{n} \mu_i p_i. \tag{33}$$

It is obvious that the value $\hat{\mu}$ is the *mean Malthusian parameter*. If, in addition, the Malthusian functions depend only on p_i then the system dynamics does not depend on its "total mass," i.e., we can increase (or decrease) the total biomass A by several times, but the equation for A is not changed. In other words, in this case two systems with the same structures but with different masses will be dynamically equivalent. (We used this fact to create the so-called "phenomenological thermodynamics of interacting populations," see Svirezhev, 1991). From this point of view, such systems are very "thermodynamic" since their dynamics does not depend on their total "sizes" defined by their total masses.

The Onsager reciprocal relations lie at the basis of any phenomenological thermodynamics. In this case these relations are given the linear relation between μ and $\mathbf{p}$,

$$\mu_i = \sum_{j=1}^{n} \alpha_{ij} p_j, \, \alpha_{ij} = \alpha_{ji}; \quad i, j = 1, \ldots, n. \tag{34}$$

If we consider the system as one evolving population with n alleles A_i and interpret the coefficients α_{ij} as a fitness of genotypes (A_i, A_j), then Equations (33) become the classic Fisher–Haldane–Wright equations of evolutionary genetics (see, for instance, Svirezhev and Passekov, 1990). In this theory the value

$$\hat{\mu} = \sum_{i=1}^{n} \sum_{j=1}^{n} \alpha_{ij} p_i p_j$$

is called the *mean population fitness*, and the statement,

$$\frac{d\hat{\mu}}{dt} = 2 \sum p_i (\mu_i - \hat{\mu})^2 \geq 0,$$

i.e., the mean population fitness tends to maximum in the process of evolution is well known as the Fisher fundamental theorem of natural selection. How can the result be interpreted from thermodynamic point of view?

Let us represent the fitness coefficients in the form $\alpha_{ij} = F(1 - a_{ij})$ where F is the mean number of offspring per one individual, a_{ij} is the probability to live until a reproductive age for genotype (A_i, A_j) and to produce F offsprings.

$$\text{Then } FN \sum_{i=1}^{n} \sum_{j=1}^{n} a_{ij} p_i p_j = N(F - \hat{\mu})$$

offspring must perish in the course of one generation for the process of genetic evolution to continue. This is the cost of evolution! The loss of individuals is a typical irreversible process with a biomass dissipation, and the value $\beta_i = N(F - \hat{\mu})$ can be entirely considered as the dissipative function or the entropy production within the system. In accordance to Prigogine's theorem

$$\beta_i \rightarrow \min_{\mathbf{p}} (\beta_i) > 0$$

and the gene structure of population, $\mathbf{p}$, tends to the stable equilibrium, $\mathbf{p}^*$. Since the evolutionary dynamics do not depend on N, and F can be always chosen sufficiently big so that $(F > \hat{\mu}^*)$, then the statement about $\min_{\mathbf{p}} \beta_i$ will be equivalent to one about $\max_{\mathbf{p}} \hat{\mu}$. This means that in the thermodynamic theory of evolution Prigogine's theorem of the entropy production minimum in a dynamic equilibrium is equivalent to Fisher's fundamental theorem of natural selection in the classic evolutionary genetics.

5.11   SOME GENERALISATION OF THE EXERGY CONCEPT

In Section 5.7 it was shown that the exergy is a Lyapunov function, so that

$$Ex = L = \sum_{i=1}^{n} \{C_i \ln(C_i / C_i^*) - (C_i - C_i^*)\}. \tag{35}$$

The equilibrium $\mathbf{C^*} = \mathbf{C^0}$ (the reference state) is interpreted as a thermodynamic equilibrium corresponding to some "prebiological" state. Since the exergy increases along a trajectory going from the reference state then the equilibrium must be unstable. The chain of these arguments is logically faultless, except for one fact: all the concentrations concerning living matter must be equal to zero in the thermodynamic equilibrium when no life exists. Then the corresponding items in the expression (35) for exergy tend to infinity by the same token, "blocking up" an influence of other components. In order to bypass the difficulty S.E: Jørgensen (1992) suggested the concept of "inorganic soup," when these concentrations are very close to zero, but, nevertheless, differ from zero. Certainly, in this case the corresponding items are finite; however, their influence remains prevalent. Maybe there is a deep biological sense in this, but I would like to suggest another bypass based on a formal application of Lyapunov's theory.

Let some coordinates of equilibrium $\mathbf{C^*} = \mathbf{C^0}$ be equal to zero, so that $C_k^* \neq 0, k \in \omega, C_s^* = 0$, $s \notin \omega$ where ω is some subset of $[1,2,...,n]$. Instead of Ex determined by (35), we consider the function

$$\hat{E}x(\mathbf{C}) = \sum_{k \in \omega} \left\{ C_k \ln \frac{C_k}{C_k^*} - (C_k - C_k^*) \right\} + \sum_{s \notin \omega} C_s. \tag{36}$$

Since only the positive variations, $\delta C_s^*, s \notin \omega$, are admissible, then the function $\widehat{B}(\mathbf{C})$ is positive,

$$\hat{E}x(\mathbf{C^*}) = 0,$$

and it is a Lyapunov function. If the exergy, defined by this way, increases along even one trajectory then the reference state is unstable. Let us illustrate this by one simple example. Suppose there is a system consisting of a living biomass, C, and an inorganic resource, R. Then the exergy will be

$$\hat{E}x = R \ln \frac{R}{R^0} - (R - R^0) + C.$$

Being given the dynamic equations for C and R in the form describing a simplest biological cycle, $dR/dt = -\alpha RC + mC; dC/dt = \alpha RC - mC$; so that $R + C = R^0$, we get:

$$d\hat{E}x/dt = \ln(R/R^0)(dR/dt) + dC/dt = (dC/dt) \cdot (1 - \ln(R/R^0)) > 0.$$

Since $R = R^0 - C < R^0$ for any $C > 0$ then the second multiplier is always positive. For the exergy to increase, it is necessary that

$$dC/dt = C(\alpha R - m) = C[(\alpha R^0 - m) - C] > 0.$$

In the vicinity of $C = 0$ this is fulfilled if $R^0 > m/\alpha$. This means:

- For a biological cycle to start "turning," an initial value of "turned" matter will be more than the ratio (m/α). With a small amount of matter, when the exergy would not increase, the cycle would not turn!

Of course, the example is trivial, and the same result could be achieved in a different way, without the use of the exergy concept. But in my opinion in this case, the general idea is more important than a concrete result.

REFERENCES

De Groot, S. R. and Mazur, P., *Non-equilibrium Thermodynamics.* North-Holland Press, New York, 510 pp., 1963.

Gill, P. E., Murray, W., and Wright, M. H., *Practical Optimisation.* Academic Press, London, 1981.

Jørgensen, S. E., *Integration of Ecosystem Theories: A Pattern.* Kluwer Academic Publishers, Dordrecht, 383 pp., (2nd Edition 1997), 1992.

Kullback, S. *Information Theory and Statistics.* Wiley, New York, 395 pp., 1959.

Landau, L. and Lifshitz, E., *Statistical Physics,* Nauka, Moscow, 568 pp., 1964.

MacArthur, R. H., Fluctuations of animal population and a measure of community stability. *Ecology,* 36:533–536, 1955.

Malkin, I. G., *Theory of Motion Stability.* Nauka, Moscow, 533 pp., 1967.

Margalef, R. A., A practical proposal to stability. *Publ. de Inst. de Biol. Apl. Univ. de Barselona,* 6:5–19, 1951.

Rubin, A. B., *Thermodynamics of Biological Processes.* Moscow State Univ. Press, Moscow, 240 pp., 1967.

Svirezhev, Yu. M., Vito Volterra and the modern mathematical ecology. In: Volterra, V., (ed.), *Mathematical Theory of Struggle for Existence.* Nauka, Moscow (the postscript to the Russian translation of this book), 1976.

Svirezhev, Yu. M., Phenomenological thermodynamics of interacting population. *J. Theor. Biol.,* 52, No. 6, 883–899, 1991.

Svirezhev, Yu. M., Thermodynamic orientors: how to use thermodynamics concepts in ecology. In *Eco Targets, Goal Functions, and Orientors,* Müller, F. and M. Leupelt (eds.), Springer, Berlin, 102–122, 1998.

Svirezhev, Yu. M. and Logofet, D. O., *Stability of Biological Communities.* Nauka, Moscow, 324 pp., 1978. (English version: 1983 Mir, Moscow).

Svirezhev, Yu. M. and Passekov, V. P., *Fundamental of Mathematical Evolutionary Genetics.* Kluwer Academic Publishers, Dordrecht, 1990.

Vinogradov, A. P., *Chemical Evolution of the Earth.* The U.S.S.R. Academic Scientific Publisher, Moscow, 1959.

CHAPTER 6

This chapter discusses thermodynamics on agricultural systems (man-made and controlled ecosystems). It is the so-called eco-energetic approach which looks into all the inputs and outputs of energy to an agroecosystem, but most often not including the solar radiation. The energetic efficiency is found as the ratio energy outputs to energy inputs. Two additional concepts are used in the thermodynamic analysis and comparison of agroecosystems: (1) energy flow density, defined as the energy flows per unit of area and (2) redundancy index which can be calculated as $1 - H/\ln N$, where H is Shannon's index (also sometimes called information entropy) and N is the total number of flows. The redundancy index is smaller the more homogeneous the energy allocation is among the flows and the more aligned the hierarchy of the flow.

The analysis has been expanded compared with the more classical eco-energetic approach with determination of the input and output of information (sometimes called information entropy referring to that "not having the information" would correspond to entropy). If we do not count the solar radiation, the input of information for a natural ecosystem is 0, and we have therefore a gain in information (and in exergy, compare with the previous chapter where this thermodynamic concept is introduced), corresponding to the diversity of output flows. This is in contrast to industrial agriculture where the input of information (fertilisers, fossil fuel, machinery, electricity, and so on) is more than the output of information. The industrial agrosystem can only be maintained due to this input of information (we may also express it as input of exergy).

Five specific farms have been examined by these methods of thermodynamic analysis: (1) a Dutch farm, year 1800, (2) a Dutch farm, year 1965 (representing very intensive modern farming), (3) A Lithuanian farm, 1985, (4) a Kursk farm, 1985, and (5) a Hungarian farm, 1980s. It can be shown that there is a good relationship between redundancy index and energy flow density. The higher the energy flow density, the lower the redundancy index, because it means increased branching of the artificial energy. The energy flow density increases in the sequence $1 \to 3 \to 4 \to 5 \to 2$. However, the energy load per annum, expressed in GJ/ha, is increasing in the sequence $1 \to 2 \to 3 \to 4$ and 5, and the gain in information is following approximately the same trends. The gain in information is actually negative corresponding to a loss for farms 3, 4, and 5. A gain in information equal to 0, corresponds approximately to 18 GJ/ha.

Energy analyses of agroecosystems are very important and can give important information about the allocation of the energy used, but as demonstrated in this chapter a second law–based examination, looking into the change in information is necessary, if we want to assess whether or not an agricultural system is sustainable. Solar energy input is not included in these calculations, but it is completely acceptable in association with a sustainability analysis.

6 Application of Thermodynamic Concepts to Real Ecosystems: Anthropogenic Impact and Agriculture

Yuri M. Svirezhev

CONTENTS

6.1 Description of Agroecosystems Using Energy Flows, Eco-Energetics Analysis 135
6.2 Ulanowicz's "Ascendancy" and Flow Diversity of Agroecosystem 138
6.3 Different Information Measures Defined According to the Typical Structure
of an Agroecosystem...139
6.4 Change of Information Entropy in Agroecosystems: Artificial Energy
Load and Stability...142
6.5 Farms: Energy and Information Characteristics...143
6.6 Change of Information Entropy and Energy Load 145
6.7 Limits of Agriculture Intensification and Its Entropy Cost 147
6.8 Case Study: The Hungarian Agriculture 148
6.9 Partially Closed Agrosystem..150
References 152

6.1 DESCRIPTION OF AGROECOSYSTEMS USING ENERGY FLOWS, ECO-ENERGETICS ANALYSIS

Following a classic "ecological" tradition going from Lindeman, an agroecosystem is considered a transformer of the input flow of "artificial" energy into the output flow of agricultural production (Pimentel et al., 1973; Deleage et al., 1979). This approach, called "eco-energetics analysis," is based on:

1. Categorising of material and energy flows that are the most significant for agro-ecosystem;
2. Determination of energy equivalents for these flows.

Thereby one can define the intensity levels for all the flows considered and estimate the flow intensities in general energy units. This allows comparison of the different flows and calculation of the ratio of outputs to inputs, i.e., the efficiency η of the transformer, in this case of the agroecosystem.

As a rule, η exceeds one for crop production systems, because the approach does not take into account the "natural" energy of the sun as an input flow into the system. Why is it so? The flows of different kinds of artificial energy are approximately of the same order of magnitude, but the solar energy flow and that of artificial energy are not comparable as the difference is two orders in favour of solar energy. So they cannot be compared in calculating efficiency. Certainly, the "solar energy" item in the total energy flow might be taken as consisting of photosynthetic active energy only (~1% of the total solar energy), but in such a case uncertainty arises as to the exact measurement of photosynthesis efficiency. Therefore, eco-energetics analysis disregards "natural" solar energy.

Traditionally, the utilised energy of anthropogenic origin is divided into two types: direct and indirect ("grey") energy. Direct energy input implies the flows of resources directly related to *energetics*: oil, coal, peat, electricity, etc. (Deleage et al., 1979). By indirect energy we denote the flows of resources which are not actually energetic but take part in the operation of the system. These flows involve mineral fertilisers, pesticides, machinery, agrosystems infrastructure, and some other resources.

The energy content of the flows is estimated by taking into account the *total primary energy* expenses needed for their formation. Thus, for example, the energy content of electric energy is the heat equivalent of the fuel burnt at power stations to produce this amount of electricity. The most questionable elements in such an approach relate to the quantification of multiple processes participating in the production flow. For example, when estimating the flow "human labour," one can account for not only the calories of the nutrition necessary for maintaining the physical activities required, but one can also account for all the other "energetic" expenses to provide an adequate living standard (e.g., using gasoline for private motor vehicle transport (Smile et at., 1983). In our paper the methodology and recalculation factors from Pimentel's book (1980) were used as it is still the most complete summary of the results including the most significant production processes in agriculture.

The method, having been described in detail by Svirezhev et al. (1986), will be illustrated by the analysis of a farm from the Central Chernozem Region of Russia.

The following input flows were taken into account:

- Direct energy — fuel (diesel, petroleum, oil), electric energy, human labour;
- Indirect energy — fertilisers (N,P,K), agro-chemicals, agricultural machinery, infrastructure, combifodders.

Energy flows for the agroecosystem are shown in Figure 6.1. *The total input* of artificial energy is estimated at 268 TJ (TJ = 10^{12} joules), indirect energy input being equal to 184.5 TJ, or 69% of total energy input. The input energy for crop production is equal to 217.4 TJ (an additional internal system's flow of manure from husbandry equals to 107 TJ). This energy is transformed into 442.3 TJ of crop production, subtracting the production used for seeds. Thus, crop production energy efficiency defined as outputs/inputs ratio in energy units equals 2.0; i.e., 2 unit energy of crop production per unit energy input. A part of crop production is used for feeds.

It should be noted that in livestock production analysis the energy content is calculated as the quantity of energy necessary for producing the forage, rather than that contained in the forage. In this example, the energy expenses in forage production amount to 94 TJ. In Figure 6.1 the value of 232 TJ corresponds to the forage production in caloric equivalent. The major output flows of the agroecosystem include: crop production — 209.9 TJ, and livestock production — 19 TJ.

Livestock receives 50.4 TJ of energy from outside the system and 94 TJ — as forage energy, internal system's flow of straw equals to 107 TJ. This energy is transformed into 19 TJ of energy in livestock products. Let us introduce the definition of energetic efficiency.

$$\eta = \frac{E_{\text{out}}}{E_{\text{in}}} \tag{1}$$

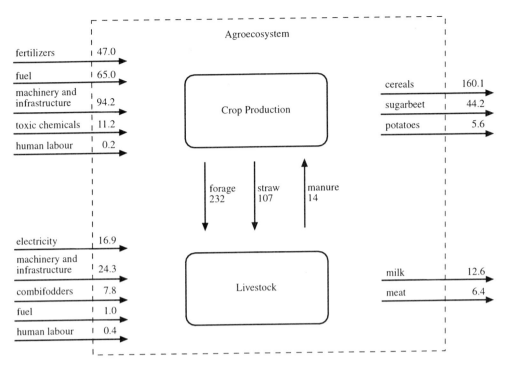

FIGURE 6.1 The structure of energy flows in an agroecosystem (farm from Central Chernozem region, 1985, units: TJ).

TABLE 6.1
Energy Efficiency of Different Agroecosystems

Agroecosystem	η
Hungary	0.87
France	0.69
Kursk farm	0.64
Farm from Lithuania	0.61
Israel	0.40
U.K.	0.34

where E_{out} is the total output flow, E_{in} is the input flow; all the flows involved in the process of energy transformation are considered here. In this case $E_{in} = 144.3$ TJ, $E_{out} = 19$ TJ, so the livestock production energy efficiency is $\eta = 0.13$.

The total production output equals 228.9 TJ, and total energy efficiency of the system equals to 0.64.

It is clear that the values of energy efficiency may be considered as comparative characteristics of different agricultural systems, the possible levels of system comparison varying from a single unit (a farm, collective farm) up to a whole country. In Table 6.1 values are given for the farm units of the Central Chernozem Region, Lithuania, as well as for agricultural systems of Hungary (Svirezhev et al., 1986, 1995), France, Israel, and the U.K. (Deleage et al., 1979).

6.2 ULANOWICZ'S "ASCENDANCY" AND FLOW DIVERSITY OF AGROECOSYSTEM

Obviously, the agroecosystem description given above, as a system of energy flows, provides its natural structure, and therefore the corresponding information measure can be defined for this structure. However, before formulating this definition, let us dwell on a new concept of "ascendancy" introduced by Ulanowicz (1986).

Ulanowicz postulates that a specific indicator, termed "ascendancy," can be considered as a measure of the ecosystem "organisation level," proceeding, by intuition, from the postulate: the more a system is "organized," the more it is stable, resistant, and able to compensate disturbing effects of the environment. It should be noted that this postulate is also an empirical generalisation.

According to Ulanowicz, the expression for "ascendancy" can be written in the form

$$As = x \sum_{i=0}^{n+2} \sum_{j=0}^{n+2} \frac{x_{ij}}{x} \cdot \ln\left(\frac{x_{ij}}{x_i} \bigg/ \sum_{s=0}^{n+2} \frac{x_{sj}}{x} \right) \tag{2}$$

where x_{ij} is the flow from compartment i to compartment j, $x_i = \sum_{j=0}^{n+2} x_{ij}$ is the total flow through the ith compartment, $x = \sum_{i=0}^{n+2} x_i = \sum_{i=0}^{n+2} \sum_{j=0}^{n+2} x_{ij}$ is the total flow through the system, and n is the number of compartments in the system.

The ratio $p_{ij} = x_{ij}/x_i$ can be considered as the probability for the movement of energy or substance from the ith compartment to the jth one. Index $i = 0$ corresponds to the input flow, $i = n + 1$ corresponds to the dissipation flow (for example, respiration), and $i = n + 2$ corresponds to the output flow. It is obvious from the definition that the Ulanowicz "ascendancy" represents an information entropy measure defined for the system, which is described by flows, with the structure of the latter determining the structure of the system. It is well known that such sort of measures are connected very closely with thermodynamic concepts.

Here for the first time we encounter a different type of system description where instead of describing its state through the values of its components we achieve it by measuring the values of flows between compartments. This can be illustrated by the following example.

Let the system consist of n components, each described by the value of z (for example, the amount of some substance in the ith compartment). Then the system state is defined by vector $z = (z_1, \ldots, z_n)$. We can calculate the quantity of information contained in the system for each of its states.

$$I_z = - \sum_{i=1}^{n} p_i \cdot \ln p_i, \tag{3}$$

where $p_i = z_i / \sum_{j=0}^{n} z_j$. However, this measure does not reveal the relationship between the compartments, i.e., the system structure. It is clear that two different structures resulting in the same value of z vector will have the same value of information measure I_z. To avoid this ambiguity, we can supplement the description with a dynamic model (giving additional n links by differential equations), e.g.,

$$\frac{dz_i}{dt} = q_{0i} - q_{i0} + \sum_{j=1}^{n} q_{ji} + \sum_{j=1}^{n} q_{ij}, \tag{4}$$

where q_{ij} is the matter flow from the ith compartment to the jth compartment, "0" indicates the position outside the system, i.e., q_{0i} the input and q_{i0} the output flows. Consequently, we can determine the system state, using the *flows matrix*.

$$Q = \|q_{ij}\|, \quad i, j = 0, 1, \ldots, n.$$

This matrix gives information about the system structure, both qualitative and quantitative. The information measure can be determined for this matrix in the following form:

$$I_q = -\sum_{i=0}^{n} \sum_{j=0}^{n} p_{ij} \cdot \ln p_{ij}, \qquad \text{where} \quad p_{ij} = q_{ij} \bigg/ \sum_{i=0}^{n} \sum_{j=0}^{n} q_{ij}. \tag{5}$$

Comparing (4) and (5) we can see that "ascendancy" according to Ulanowicz and the information measure determined on the set of all the system flows do not differ from each other. Let us show that if we describe the system in terms of flows, the information content is approximately twice that of a description in terms of states of the compartments. In fact, in case of a large n and more or less equal flows, we get

$$I_z = -\sum_{i=1}^{n} p_i \cdot \ln p_i \approx \ln n. \qquad \text{But since } I_q = -\sum_{i=1}^{n} \sum_{j=1}^{n} p_{ij} \cdot \ln p_{ij} \approx \ln n^2 = 2 \cdot \ln n,$$

$$\text{then } I_q = 2 \cdot I_z.$$

In the previous section we presented the method of describing an agroecosystem in terms of energy flows (both external and internal input-output). However, in contrast to Ulanowicz's formalism described above, this one has its specific feature: the input and output flows of energy are subdivided. If the flows are aggregated (considering total input and output flows) the information about the structure of agricultural system will be inevitably lost. By the way, it is such loss of information that takes place in calculating the system's energy efficiency when the relationship of total flows is calculated. Therefore, the next section will be devoted to considering different information measures defined for special structures, such as the one shown in Figure 6.1.

6.3   DIFFERENT INFORMATION MEASURES DEFINED ACCORDING TO THE TYPICAL STRUCTURE OF AN AGROECOSYSTEM

To represent a standard flow structure of an agroecosystem, a generalisation of Figure 6.1 (see Figure 6.2) will be used. An agroecosystem consists of two compartments: crop production (1) and livestock (2). Exchange of energy by flows takes place between these compartments: flows $X_1, \ldots, X_p$ from compartment 1 to 2; and $Y_1, \ldots, Y_q$ from compartment 2 to 1. The input flows: into compartment 1 are $U_1, \ldots, U_n$; into compartment 2 are $V_1, \ldots, V_m$. The output flows are $u_1, \ldots, u_e$ and $v_1, \ldots, v_s$, correspondingly. In addition to the flows of artificial energy, solar energy, flow E_s is delivered to the input of compartment 1. Its flow is not controllable, and $E_s \gg \sum_{i=1}^{n} U_i + \sum_{i=1}^{m} V_i$, so that in the calculation of the information entropy the fraction corresponding to solar energy (from the total energy delivered to the system), is close to zero ($p_s \cong 1$, hence, $p_s \cdot \ln p_s \cong 0$).

Therefore, this flow will not be taken into account in analysing the structure. Though this can result in formal violation of the energy conservation law, such violation will not influence the

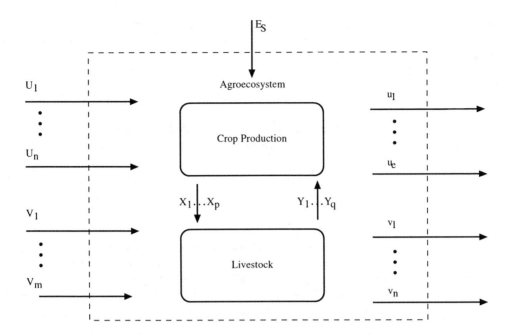

FIGURE 6.2 Typical flow structure of agroecosystem.

qualitative conclusions. As a result (as we have already mentioned) the coefficient of energy efficiency in crop production can exceed 1.

Let us consider all of the agroecosystems in general. These can be described by vectors of energy flows:

$$U\{U_1,\ldots, U_n\}, \qquad V\{V_1,\ldots, V_m\}, \qquad X\{X_1,\ldots, X_p\},$$
$$Y\{Y_1,\ldots, Y_q\}, \qquad u\{u_1,\ldots, u_e\}, \qquad v\{v_1,\ldots, v_s\}.$$

All these flows contribute to the formation of the agroecosystem structure. So we shall use the sum of all the energy flows as its integral characteristics:

$$E = \sum_{i=1}^{n} U_i + \sum_{i=1}^{m} V_i + \sum_{i=1}^{p} X_i + \sum_{i=1}^{q} Y_i + \sum_{i=1}^{e} u_i + \sum_{i=1}^{s} v_i,$$

and their frequency distribution $\mathbf{P} = (P_1,\ldots, P_n)$ with the components

$$P_i = \begin{cases} U_i/E; & i = 1,\ldots, n \\ V_i/E; & i = n + 1,\ldots, n + m \\ X_i/E; & i = n + m + 1,\ldots, n + m + p \\ Y_i/E; & i = n + m + p + 1,\ldots, n + m + p + q \\ u_i/E; & i = n + m + p + q + 1,\ldots, n + m + p + q + e \\ v_i/E; & i = n + m + p + q + e + 1,\ldots, n + m + p + q + e + s \end{cases}$$

The system state will be described by the pair {E, **P**}, where E is a scalar and **P** is a vector. Of course, when calculating the total energy characteristic E, we do not observe the energy conservation law since the input and output flows are simply summed, but in our case we are interested in the system's structure (branching of flows, their different orientations, etc.), rather than in its efficiency. Furthermore, in order to compare different systems, we shall use the energy density per unit of land, i.e., E/S, where S is the area of the system. Note that functions *defined on* **P** do not depend on S. Let us define on **P** on the function.

$$H = -\sum_{i=1}^{N} p_i \cdot \ln p_i,$$

which can be considered as either the information entropy of the system or its flow diversity, i.e., a function analogous to standard definition but defined on the energy flows, not on frequencies of particles or species. As a matter of fact, function H is equivalent to Ulanowicz's ascendancy, but it is defined in a more formal way.

Since the maximum value of H equals $\ln N$, where N is the total number of flows, then, by implication, the value of H depends on this number. However, it is desirable to characterise the system's structure using a function independent of N. Information theory as well as its applications in biology and geography uses the so-called *redundancy index* (or informational index of connection ability by Pielou (1969) or by Puzachenko and Moshkin (1981), which equals

$$h = \frac{H_{max} - H}{H_{max}} = 1 - \frac{H}{H_{max}}, \qquad 0 \le h \le 1. \tag{6}$$

Using the value of $\ln N$ as H_{max}, i.e., considering entropy to be maximal in systems with homogenous distribution, we obtain

$$h = 1 + \frac{\sum_{i=1}^{n} p_i \cdot \ln p_i}{\ln N}, \qquad 0 \le h \le 1. \tag{7}$$

Formula (7) shows that the more homogenous energy allocation is among the flows and the more aligned the hierarchy of the flow, the smaller is the redundancy index. We observe that the redundancy index h reduces along with a reduction of the system's hierarchical organisation. Therefore, the value of h may be considered as an index for the description of the degree of system hierarchy. Introducing some additional flows always results in increased h, and making the system more homogeneous reduces it.

Considering the above, we shall characterise the state of a system, say the kth system by a pair of values $\{\varepsilon_k = E_k/S_k, h_k\}$ representing its state by a point on plane: $\{\varepsilon, h\}$. Our hypothesis is the following: *on plane $\{\varepsilon_k, h_k\}$ there exists the curve $h = f(\varepsilon)$ along which the agroecosystem evolution is realised*, i.e., the values of h and ε, which are the information and energetic indices occurring in reality, are connected. This curve can be called *the curve of structural evolution for agroecosystems*.

The existence of this connection is our speculative hypothesis, however, based on general thermodynamic concepts. Entropy and energy are known to be connected in thermodynamics, this connection being none other than reflection of some general laws of nature. We shall try to empirically determine the form of the curve.

6.4 CHANGE OF INFORMATION ENTROPY IN AGROECOSYSTEMS: ARTIFICIAL ENERGY LOAD AND STABILITY

Let us consider a system characterised by input and output flows (see Figure 6.3). We assume that all flows are measured by the same units (for instance, either in calories or in joules). The input and output of the system can be described by the total energy input, $L = \sum_{i=1}^{m} U_i$, and output, $l = \sum_{i=0}^{e} u_i$, and also the input and output information entropies, H_{in} and H_{out}, correspondingly,

$$H_{in} = - \sum_{i=1}^{n} \frac{U_i}{L} \cdot \ln \frac{U_i}{L}, \qquad H_{out} = - \sum_{i=1}^{e} \frac{u_i}{l} \cdot \ln \frac{u_i}{l} \qquad (8)$$

Changes of information entropy in passing of energy flows through the system will be equal to $\Delta H = H_{out} - H_{in}$. We consider two different cases.

1. Let $H_{out} > H_{in}$ so that $\Delta H > 0$. The system transforms its input, making its output more homogenous as a number of different components in the output flow increase. By the same token the system is an *organising* one. In this case the energy input is usually low, and most input energy is dissipated supporting its self-organisation. The typical example of such systems is any natural ecosystem, the energy input of which is only the flow solar energy, and a result of its transformation within a system is a very high biological diversity. In this sense we can identify a system *stability*.

The similar situation is for a primitive agroecosystem using as energy input only the flow of solar energy (human labour and livestock exploitation are considered flows inside the system). As a rule, the diversity of output flows in these ecosystems is very large (which is inherent for subsistence economy). In fact, in an extreme case $e \gg 1$ and $n = 1$ (see Figure 6.4a) then $\Delta H = H_{out} - (H_{in} = 0) > 0$.

Since the entropy increment is positive, the system is stable (in the sense defined above). As already mentioned, its stability is provided by solar energy dissipated by different components of a primitive agroecosystem — thus it is provided by its species diversity.

2. Let $H_{out} < H_{in}$ so that $\Delta H < 0$. The system transforms its input, making its output more heterogeneous, and a number of different components in the output flow decreases.

In this case the system is a *disorganising* one. The typical example of such systems is any industrial agroecosystem. Now let us consider an extreme case of such a system with different types of artificial energy being supplied as input and with only a single energy type as output (single crop) (Figure 6.4b). An industrial agroecosystem producing only a single crop is, of course, a theoretical abstraction, but as we said above, it is an extreme case that we consider. Then $e = 1$, $n \gg 1$, and $\Delta H = . = (H_{out} = 0) - H_{in} < 0$. (Note since the flow of artificial energy is much less than that of solar energy, the term corresponding to the latter must be excluded from the calculation of H_{in} — see Section 6.3).

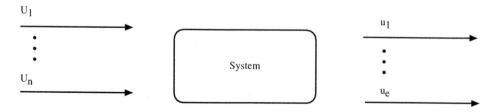

FIGURE 6.3 System described by energy flows.

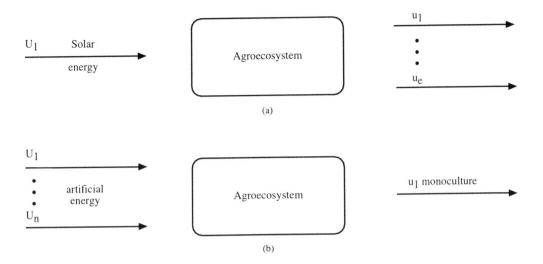

FIGURE 6.4 Structure of input and output energy flows for primitive (a) and industrial (b) agroecosystems.

From the point of view of common sense, an industrial agroecosystem is unstable, since if the flows of artificial energy are stopped, a succession to a "wild" type of ecosystem begins. As a result, an agroecosystem will disappear. The only way to make it stable is to dissipate some definite quantity of artificial energy inside the system so that total entropy production inside it (taking into account a dissipation of artificial energy) would be positive.

From the information point of view, an industrial agroecosystem is a "black hole" for information, since information "disappears" inside a system. On the other hand, information cannot disappear, principally. For such a system to exist, "disappeared" information must be compensated. The compensating information is the information about organisation of industrial agroecosystem, i.e., the information contained in technologies, new varieties, etc. From this point of view the role of diversity of artificial energy flows as the main factor for their stabilisation becomes clear.

Now let the agroecosystem state be described by two parameters: the value of information entropy increment, ΔH, i.e., that of information entropy produced by the agroecosystem, and the density of artficial energy input flow, $W = E_{in}/S$ where S is the agroecosystem area. The latter is an energy load on agroecosystem determining a dissipation of this energy inside. Its state can be described by a point on the plane $\{W, \Delta H\}$. The general concepts clearly show that if W increases then ΔH is reduced and even becomes negative. As in the previous section, we assume that a definite relationship exists between ΔH and W that is described by curve $\Delta H = f(W)$ on the plane $\{W, \Delta H\}$ along which agroecosystem evolution proceeds from primitive to industrial ones. To estimate the form of this curve we shall use experimental data on energy flows of different farm types.

6.5 FARMS: ENERGY AND INFORMATION CHARACTERISTICS

Above we have considered the structure of the energy flows of a contemporary farm from Central Chernozem Region (near Kursk city, Russia, so we call it Kursk farm). In addition, data on energy structure will be given for four other farms: a farm in Holland in 1800 (primitive agroecosystem), a farm in Holland in 1965 (see Figures 6.5 and 6.6), and collective farms in Lithuania in 1985 and in Hungary in 1980s (industrial agroecosystems). The data about Dutch farms in 1800 and 1965 were taken from Tooming (1984); those on the Lithuanian and Hungarian farms from Svirezhev et al. (1986, 1995), and from Semyonov et al. (1987).

The values of the redundancy index h and the density of total energy flow $\varepsilon = E/S$ were estimated using the data from the observed farms. As can be seen in Figure 6.7, the redundancy

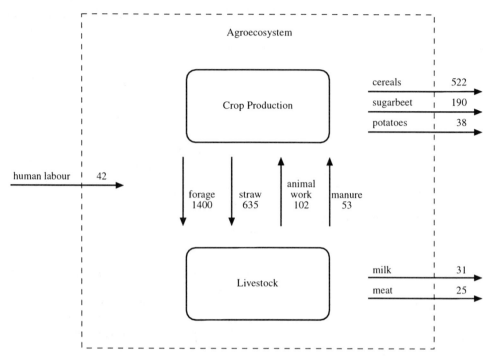

FIGURE 6.5 The structure of energy flows of agroecosystem (farm from Holland, 1800, units: GJ).

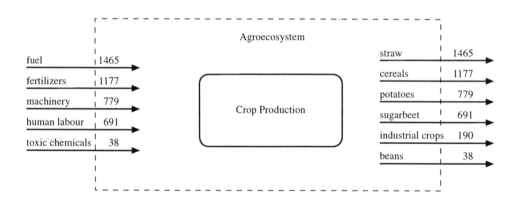

FIGURE 6.6 The structure of energy flows of agroecosystem (farm from Holland, 1965, units: GJ).

index h decreases with increased density of total energy flow ε. It is interesting that h is smallest for the most industrialised agroecosystem (Dutch farm, 1965). It is approximately the same for the two farms of the former USSR (from Kursk and Lithuania), though they are situated in different geographical regions. The value of h for Hungary is intermediate. For the primitive agroecosystem (Holland, 1800) this index is largest. The energy index behaviour characterising total quantity of energy flowing through the system and circulating inside it is vice versa, it increases when going over from a primitive agroecosystem to an industrial one. Then the trajectory describing the connection between h and ε can be plotted by these points (see Figure 6.7). This curve can be considered as the trajectory of evolution from primitive to industrial agroecosystems.

This evolution inherently involves an increase in artificial energy flowing through the system and its inside circulation. But it is interesting that the redundancy index concomitantly reduces.

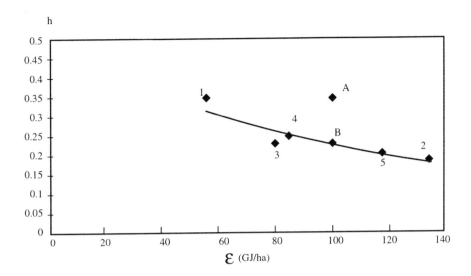

FIGURE 6.7 Dependence of redundancy index h on the energy flow density ε. 1 — Dutch farm, 1800; 2 — Dutch farm, 1965; 3 — Lithuanian farm, 1985; 4 — Kursk farm, 1985; 5 — Hungarian farm, 1980s.

This reduction indicates that the evolution is accompanied by increased branching of artificial energy flows and an increase of homogenisation of energy flows. The degree of *hierarchical structuring* reduces, all the flows become equally important, and not one can be excluded in a modern farm without causing damage to the agroecosystem functioning. In other words, the more industrialised the system, the more homogeneous its energetic structure becomes.

Kursk and Lithuanian farms occupy very close locations on the plane $\{\varepsilon, h\}$ although they are situated in geographically different regions. This means that the integral indices used characterise farm structure and type of agroecosystem organisation rather than geographical location.

The last point: our hypothesis of the existence of an agroecosystem evolution curve is an empirical generalisation, rather than a strong logical conclusion. If the hypothesis is correct, we can use the curve in Figure 6.7 for prognosis of future agroecosystems organisation from an evolutionary point of view. For example, if a future agroecosystem will be at point A, we can conclude that it is inadequately organised from an evolutionary viewpoint. Therefore, the structure of its energy flows should be changed, so that the new location on the plane is somewhere near the evolutionary curve (e.g., at point B).

6.6   CHANGE OF INFORMATION ENTROPY AND ENERGY LOAD

Using these data we can calculate the information entropy increment ΔH, as well as the energy load W (Table 6.2) for different farms.

On the plane $\{W, \Delta H\}$ the corresponding points are plotted and the linear regression is drawn (see Figure 6.8).

It is interesting to note that Kursk, Lithuanian and Hungarian farms differed by the total input energy, technology, and geographical location, but have close locations on the plane. It can be explained by some general regularities defined by internal organisation of the system rather than by natural conditions. Their closeness in terms of the criteria W and ΔH indicates their similar organisation. The thermodynamic origin of these criteria and the fact that thermodynamic laws are the general laws of Nature allow us to believe that these criteria can determine the degree of organisation and energy efficiency of agroecosystems. We can assume that in the course of agroecosystem

TABLE 6.2

Information Entropy Indices and Energy Load for Different Agroecosystems

Agroecosystem	H_{in}	H_{out}	ΔH	W GJ/ha
Dutch farm, 1800	0	1.10	1.10	0.8
Dutch farm, 1965	1.18	1.64	0.46	16.0
Lithuanian farm, 1985	2.06	1.22	−0.84	25.7
Kursk farm, 1985	1.68	0.92	−0.76	27.0
Hungarian farm, 1980s	1.32	0.80	−0.52	27.0

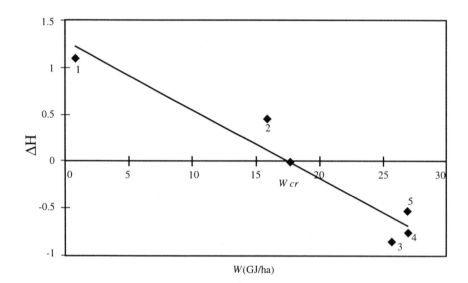

FIGURE 6.8 Dependence of information entropy increment ΔH on the energy load W. 1 — Dutch farm, 1800; 2 — Dutch farm, 1965; 3 — Lithuanian farm, 1985; 4 — Kursk farm, 1985; 5 — Hungarian farm, 1980s.

industrialisation they evolve along the curve shown in Figure 6.8. The increase of energy load results in decreased entropy production by the system at the expense of its environmental degradation. But this growth is limited since the influence of a degrading environment can cause the degradation of the agroecosystem. As a maximum load assessment, the value of W_{cr} can be determined by equation $\Delta H = 0$. The boundary $\Delta H = 0$ can be considered as that of system stability. From Figure 6.8 it follows that $W_{cr} \cong 18$ GJ/ha.

If we compare this value with other estimates obtained from different considerations (see, for instance, Bulatkin [1982] where $W_{cr} = 14$ GJ/ha; Novikov [1984] and M. Simmons [Zhuchenko, 1988] where $W_{cr} = 15$ GJ/ha) we shall see that they are very close.

The positive value of ΔH for an industrial Dutch farm in 1965 can be explained by the following: at this farm the optional energy load has been chosen ($W = 16$ GJ/ha $< W_{cr} = 18$ GJ/ha).

Considering all the above we conclude that a continued increase of the energy density of an agroecosystem must eventually result in a corresponding increase of negative entropy production; this causes increased risk of system degradation. To avoid such a development, the energy load must be reduced to $W < W_{cr}$ simultaneously changing the structure of input and output flows so that information entropy production by this system remains positive, i.e., that the system is maintained.

6.7   LIMITS OF AGRICULTURE INTENSIFICATION AND ITS ENTROPY COST

It is known that intensification of agriculture (the increase of crop production) correlates with increase of artificial energy flow in the ecosystem. Indeed, the increase of fertiliser input and usage of complex infrastructure, pesticides, herbicides, etc., i.e., all that is called a "modern agriculture technology," results in greater crop production. This is a typical pattern of development agriculture in industrial countries (industrial agroecosystems).

We have seen above that energy and information characteristics are not independent and they are connected with each other. From the first view, it seems that following these relations and increasing an artificial energy input, an agriculture production can also increase. However, this is not true and there are limits, determined by physical laws. In other words, we pay the cost for increasing the productivity of agriculture, which is a degradation of the environment, in particular, soil degradation (for details, see Svirezhev and Svirejeva-Hopkins, 1998).

Let us remember Section 5.6 of Chapter 5; namely, formula (14) describing the "degradative" part of the entropy in a disturbed ecosystem.

$$\sigma T = W_f + W_{ch} + P_1 - P_0. \tag{9}$$

If the first term (W_f) can be associated to a direct flow of such type of artificial energy as electricity, fossil fuels, etc. (energy load) then the second term (W_{ch}) is associated to an inflow of chemical elements that maintain molar concentrations C_i ($i = 1,\dots, n$) inside the system (chemical load). We assume (as we have already assumed above) that the inflow W_f is dissipated inside the system and transformed into heat. How to calculate W_{ch}? There are two ways for this.

The first way (and this was mentioned above) is a calculation of some energy equivalent of W_{ch} using the eco-energetics analysis by Pimentel et al. (1973). Applying the special algorithm, fertilisers are represented in energy units. As a result we can join the values W_f and W_{ch} into one value $W = W_f + W_{ch}$, which we called the *artificial energy*. Then if we know the crop production of an agroecosystem (y) we can calculate the *coefficients of energy efficiency* ($\eta = y/W$). The coefficient η gives us an empirical linear relation between y and W so that $y = \eta W$.

The coefficient η is some modification of the efficiency coefficient well known in thermodynamics, which is a consequence of the First Law. In thermodynamics it is usually less than one; on the contrary, the coefficient η may be more than one. The point is that we must really take into account the solar energy (E_s) and the correct form for efficiency will be $\eta' = y/(W + E_s)$. What does Pimentel's coefficient η mean in this case? It is obvious that $y = y(W, E_s)$. If this value will be linearised at $W = 0$, we get

$$y(W, E_s) \approx y(0, E_s) + \left(\frac{\partial y}{\partial W}\right)_0 W. \tag{10}$$

If we assume now that the first term is negligibly small in comparison with the second one and the drivative $(\partial y/\partial W)_0 = \eta = \text{const}$ for considered values W we immediately get Pimentel's relation $\eta = y/W$. Of course, there is some incorrectness in this approach, but, on the other hand, why not use these existing results?

There is a second method to use when the chemical load is described directly (in detail, see Section 5.6, Chapter 5). Let C_i and C_i^0 be the current concentrations of chemical elements and their basic concentrations in a "successionally closed" ecosystem, respectively. If q_i are the flows of these elements into the system then

$$W_{ch} = RT \sum_{i=1}^{n} \ln \left(\frac{C_i}{C_i^0}\right) q_i. \tag{11}$$

If the annual total (gross) agroecosystem production is equal to P_1, the net production is equal to $(1 - r)P_1$ where r is the respiration coefficient, then the term $r P_1$ describes the respiration losses. The kth fraction of the net production is being extracted from the system with the yield, so the crop yield equals

$$y = k(1 - r)P_1. \qquad (12)$$

The remaining fraction $(1 - k)(1 - r)P_1$ is transferred to the litter and soil. If we accept the stationary hypothesis then we must assume that the corresponding amount of litter and soil organic matter must be decomposed. Concerning the extracted fraction of production we assume that this fraction does not take part in the local entropy production, and, finally, the entropy balance of this system is

$$\sigma T = W + (1 - k)(1 - r)P_1 + rP_1 - P_0 \qquad (13)$$

$$\underbrace{\qquad}_{\text{decomposition}} \qquad \underbrace{\qquad}_{\text{respiration}} \quad \text{"entropy pump"}$$

where P_0 is the gross production of a successionally closed ecosystem. (In our case this is commonly called "old field succession.")

Note that for the first ("robust") estimation we joined the energy flow (fuels, machinery, etc.) and chemical flow (fertilisers and pesticides) in one (artificial energy) inflow, W, i.e., we used the first way.

Since really we know only the values y and W (and their ratio η), it is better to rewrite (13) in the form.

$$\sigma T = W + \frac{1 - s}{s} y - P_0 \qquad \text{where} \quad s = k(1 - r). \qquad (14)$$

If we use the relation $y = \eta W$ then

$$\sigma T = W\left(1 - \eta + \frac{\eta}{s}\right) - P_0 = y\left(\frac{1}{\eta} + \frac{1}{s} - 1\right) - P_0. \qquad (15)$$

6.8   CASE STUDY: THE HUNGARIAN AGRICULTURE

This approach has been applied to the analysis of several agricultural systems, in particular, to the analysis of crop production (maize) in Hungary in the 1980s. The necessary data about this system has been taken from Semyonov et al. (1987).

The average yield of maize was 4.9 t/ha (in dry matter) which equals to $0.735 * 10^{11}$ J/ha. For maize production in Hungary $\eta = 2.7$ and $k = 0.5$. The steppe community (Hungarian "puszta") is a successionally closed ecosystem for a corn field after cultivation is stopped (grassland of the temperate zone). The gross production of the steppe is equal to $P_0 = 2800$ kcal/m^2 = $1.18 * 10^{11}$ J/ha (Lang, 1990). For a maize crop in the temperate zone $r \cong 0.4$ (Lang, 1990) then $s = k(1 - r) = 0.3$. Substituting these values into (14), we get $\sigma T = 0.8 * 10^{11}$ J/ha. On the other hand, the artificial energy input is equal to $W = 0.27 * 10^{11}$ J/ha. Therefore, *compensation for environmental degradation requires a 300% increment in energy input with all the additional energy spent only for soil reclamation, pollution control, etc. with no increase in the crop production.* Note that all the values are calculated *per year*, i.e., all the values are *annual rates* of entropy production, of energy production (power), etc.

Let us remember the condition of dynamic equilibrium for open system: the change of its total entropy in the course of characteristic time of the system must be equal to zero. A system that accumulates an entropy cannot exist for a long time and it will inevitably self-destroy. (There is a reason the Second Law is the scientific basis of any "scientific" eschatological concept.) Certainly, we implicitly assume that the considered agroecosystem would be a long living system.

Let us assume that the characteristic time for agriculture is equal to 1 year. (Note if an agroecosystem with crop rotation is considered equal to a rotation period, the entropy balance must be calculated for total rotation.)

The system will be in a dynamic equilibrium and exist for an infinitely long time (an entropy does not grow) if $\sigma = 0$. Using the part of (15) containing only W, under the condition $\sigma = 0$ we get

$$W_{cr} = \frac{P_0}{1 - \eta + \frac{\eta}{s}} = 16 \text{ GJ/ha.} \tag{16}$$

Let us compare this value with the values of "the limit energy load" obtained above (18 GJ/ha), by Bulatkin and Novikov (14 GJ/ha), by Simmons (15 GJ/ha) from different concepts and considerations, and we shall see that they are very close. Is this not a very curious coincidence?

Note the "limit energy load" concept is a typical empirical one and it means the maximal value of the total anthropogenic impact (including tillage, fertilisation, irrigation, pest control, harvesting, grain transportation and drying, etc.) on one hectare of agricultural land. All these values are evaluated in energy units (in accordance to Pimentel's method). It is assumed that when the anthropogenic impact exceeds this limit then the agroecosystem begins to destroy (soil acidification and erosion, chemical contamination, etc.).

Using the part of (15) containing only y under the same condition, we can calculate

$$y_{cr} = \frac{P_0}{\frac{1}{s} + \frac{1}{\eta} - 1} = 2.9 \text{ t/ha.} \tag{17}$$

This is the evaluation of the "sustainable" crop, i.e., the maximal crop production (in dry matter) for "sustainable" or "ecological" agriculture.

Let us suppose that the leading, main degradation process, which comprehends all the degradation processes, is soil erosion. If we have the thermodynamic model of soil erosion, we can estimate the annual erosion losses resulting from intensive agriculture. How does one estimate the annual entropy production, corresponding to erosive loss of 1 ton of soil? Consider the soil erosion as a combination of two processes: the burning of organic matter contained in the soil (~4% of carbon in mass units), which results in destruction of the chemical structure, and the mechanical destruction of soil particles (from 10^{-1} to 10^{-4} cm in size. The latter is the size of dust particles, which can be weathered by wind and be washed out by water. Then

$$\sigma_s T = 0.17 * 10^{10} + 0.14 * 10^{10} = 0.31 * 10^{10} \text{ J/tons*ha*year}$$

where the first addendum corresponds to a chemical destruction and the second corresponds to a mechanical destruction (Svirezhev, 1990, 1999). The annual total erosion loss of soil from one hectare is equal to $\sigma/\sigma_s = 26$ tons.

Consequently, high crop production will cost the loss of 26 tons of soil annually. According to U.S. standards, no more than 10 t of soil may be lost from a hectare. Obviously, 26 t per hectare is the extreme value: the actual losses are less, since there are other degradation processes, like environmental pollution, soil acidification (this factor is very important for Hungary), etc.

There is increasing talk about how chemical "no-till" agriculture actually allows topsoil to accrete. It is being touted as a "sustainable" agriculture. Let us consider this concept from a thermodynamic point of view. If you look at the formula for $\sigma_s T$ you see that the second addendum corresponds to a mechanical destruction. The main reason of such a destruction is a tilling. No-till agriculture means that the value $\sigma_s T$ is reduced approximately two times. The first impression is that we achieve our goal and reduce the soil erosion. However, it is impossible to get something free of charge and, if we want to save the same crop production, we must increase an energy input (W) up to the former value replacing the mechanical component with some chemical one. As a result we get approximately the same value of the entropy overproduction (σ) but the result of compensation would be another. For instance, instead of soil erosion we would get an increase of chemical contamination.

Within the framework of the thermodynamic approach we can calculate the entropy of these processes as well. For example, the entropy contribution to the acidification of soil can be calculated in terms of appropriate chemical potentials. However (and this is the principal constraint of thermo-dynamic approach), we cannot predict the way in which the degradation of the environment will be realised: by the strong mechanical degradation of soil and weak chemical pollution, the high acidification of soil, the strong chemical contamination by pesticides and fertilisers, or some inter-mediate variants and their various combinations. For the solution of this problem some additional information is needed.

On the other hand, this approach gives possibilities to estimate the "entropy fee," which mankind pays to get high crop yield, and for intensification of agriculture. Overproduction of entropy can be compensated by processes of environmental degradation, in particular, by soil degradation. It is known that the loss of about 40% of soil results in a decrease of crop yield of 5 to 7 times (Dobrovolsky, 1974). This is a typical agricultural disaster. But it is a disaster from an anthropo-centric or physics laws point of view. From these points of view, a fall of crop yield by reason of soil degradation is a natural reaction of the physical system, tending to decrease an internal production of entropy and minimise its overproduction. It is the consequences of the Prigogine theorem.

We suspect that this statement summons very serious objections, especially among biologists. Their essence may be the following. *Of course, the Prigogine theorem is applying to the physical world. The same theorem does NOT, however, apply to the biological realm, and it is the juxtapo-sition of the countervailing tendencies of the physical and biological realms that lends tragic overtones to the overwhelming biological trend over that of the physical.* In general, we agree with the above sentence but, as to the self-justification, we can say that here we speak as strong followers of reductionism. We understand that this point of view is a biological heresy but we consider an agriculture system only as a physical one. In the framework of our approach we reduce a variety of all processes inside an agrosystem to one process of heat producing and its dissipation. In other words, we measure the entropy production by a total thermal effect of different physical and chemical processes taking place in the system. We understand that it is a very serious simplification of the biological realm but (it seems to us) this approach gives some practical results. Namely, in this sense, we explain the applicability of Prigogine's theorem.

The corresponding estimations for Hungary show that if the intensive production of maize were continued, it would end in an agricultural disaster in 30 to 40 years.

6.9 PARTIALLY CLOSED AGROSYSTEM

Let us return to agriculture history and consider a more traditional agrosystem in which only $(1 - m)$th fraction of crop is exported out of the system; the balance is used as a cattle forage. We assume that αth fraction of the latter is returned into the system as an organic fertiliser (manure). Then the total energy input can be presented in the form $W = W_{in} + W_{org} = W_{in} + \alpha m y$, where W_{in} is the inflow of artificial energy (fuels, chemical fertilisers, pesticides, etc.) and $W_{org} = \alpha m y$ is the energy inflow corresponding to an organic fertiliser produced inside the system. As before, $y = k(1 - r)P_1 = sP_1$.

Formula (13) for the entropy balance is rewritten in the form

$$\sigma T = W_{in} + \alpha m y + (1 - k)(1 - r)P_1 + rP_1 - P_0 = W_{in} + \left(\frac{1}{s} + \alpha m - 1\right)y - P_0. \quad (18)$$

Certainly, a return of manure in the form of organic fertilisers takes place with some time delay, but, under a steady-state assumption, this time delay plays no role.

In accordance to Pimentel's concept we have either $y = \eta_1 W_{in}$ (if the flow of organic fertilisers is not considered as external and not taken into account), or $y = \eta_2 W = \eta_2(W_{in} + \alpha m y)$ (if this flow is considered as external and it is added to W_{in}). The choice between these two alternatives is defined by the title of this section: "Partially closed agrosystem." The returned flow of manure is produced within the system and, by the same token, this flow must not be taken into account in the calculation of the efficiency coefficient for agroecosystem producing a culture with crop y. Then $y = \eta_1 W_{in}$. As we mentioned above, this system is more traditional.

On the other hand, we may imagine that the same amount of organic fertilisers, $\alpha m y$, is exported from another region that is not connected with our system. Then, in accordance to Pimentel, $y = \eta_2(W_{in} + \alpha m y)$. This system can be considered as more industrialised. It is obvious that from the latter we get

$$y = \frac{\eta_2}{1 - \alpha m \eta_2} W_{in}. \quad (19)$$

If we assume that the crops are equal for both the systems, then

$$\eta_1 = \frac{\eta_2}{1 - \alpha m \eta_2}. \quad (20)$$

From (20) follows that the efficiency coefficient for a partially closed agroecosystem (which is more traditional and closer to a natural one) is higher than the coefficient for an industrialised agroecosystem. The returned fraction m is more (i.e., the system is more closed); the efficiency coefficient η_1 is higher.

It is easy to see, in the limiting case we get: if $\eta_2 \to 1/\alpha m$ then $\eta_1 \to \infty$, but since the crop y must be finite then, at the same time, $W_{in} \to 0$. Formally, it means that this limiting system is fully closed, and any external artificial energy is not necessary to maintain its functioning. Certainly, this is a theoretical abstraction, but it can be used for some real estimation. For instance, if we remember that for maize production in Hungary $\eta_2 = 2.7$, then for full closing the value αm must be equal to 0.37. Even if we assume that $m = 1$, i.e., all the crop is used as a cattle forage inside the system, then, in this limiting case, the coefficient of transformation of the forage into organic fertiliser (manure), α, must be equal to 0.37. On the other hand, we know (see, for instance, Bulatkin 1982 or Pimentel 1980) that this coefficient cannot be higher than 0.15 to 0.20. This simple calculation shows that it is impossible to transform an agrosystem into a fully closed ecosystem like a natural one.

And finally, we calculate y_{cr}, i.e., a "sustainable" crop. From (18) we get for the "sustainability" ($\sigma = 0$)

$$y_{ct} = \frac{P_0 - W_{in}}{\frac{1}{s} + \alpha m - 1}. \quad (21)$$

Since the values W_{in} and m are connected with each other (increasing m, we may decrease W_{in}), then we cannot answer uniquely the influence of the closing degree m on the value of "sustainable" crop y_{cr}. Nevertheless, if formula (19) is used for definition of W_{in} we get

$$y_{ct} = \frac{P_0}{\dfrac{1}{\eta_2} + \dfrac{1}{s} - 1}. \tag{22}$$

If we compare (17) and (22) and remember that η_2 is the efficiency coefficient for industrialised (nonclosed) agrosystems, we can be convinced of their full identity. Therefore, the value y_{cr} does not depend on the characteristics of closing m and the coefficient of transformation α. The closing of the agrosystem does *not change* its "sustainable" crop.

REFERENCES

Bulatking, G. A., *Energy Efficiency of Fertilisers in Agroecosystems*. The USSR Academy of Sciences. Moscow–Puschino (in Russian), 1982.

Deleage, J. P., Juliene, J. M., Sauget-Naudin, N., and Souchon, C., Eco-energetics analysis of an agricultural system: the French case in 1970. *Agro-Ecosystems,* 5: 345–365, 1979.

Dobrovolsky, G. V., *Soil Geography*. Moscow University Press, Moscow (in Russian), 1974.

Lang, I., Personal communication, 1990.

Novikov, Yu. F., Energy balance of agro-industrialised systems and bioenergetics of agroecosystem. The USSR Academy of Agriculture Sciences Reports: 5: 4–9 (in Russian), 1984.

Pielou, E. C., *An Introduction to Mathematical Ecology*. John Wiley & Sons, New York, 1969.

Pimentel, D., Hurd, L. E., and Bellotti, A. C., Food production and energetic crisis. *Science,* 182: 443–449, 1973.

Pimentel, D., (Ed.), *Handbook of Energy Utilization in Agriculture*. CRC Press, Boca Raton, FL, 475, 1980.

Puzachenko, Y. G. and Moshkin, A. V., Information and logical analysis in medical-geographical studies. In: *Summa Sciences: Geographical Series*. All-Union Institute of Scientific Information, Moscow. 3: 6–13 (in Russian), 1981.

Smile, V., Nachman, P., and Long II, Th. V., *Energy Analysis and Agriculture*. Westview, CO, 1983.

Semyonov, M. A., Brovkin, V. A., and Polenok, S. P., *Agroecosystems Modelling: Flows Model*. The USSR Acad. Comp. Centre, Moscow (in Russian), 1987.

Svirezhev, Yu. M., Entropy as a measure of environmental degradadation. *Proc. Int. Conf. on Contaminated, Soils, add. volume:* 26–27, Karlsruhe, Germany, 1990.

Svirezhev, Yu. M., Thermodynamics and ecology. *Ecol. Model.* 2000, in press.

Svirezhev, Yu. M., and Svirejeva-Hopkins, A., Sustainable biosphere: critical overview of basic concept of sustainability. *Ecol. Model.* 106: 47–61, 1998.

Svirezhev, Yu. M., Brovkin, V. A., and Denisenko, E. A., Agroecosystem Analysis Approach Based on the Flows of Artificial Energy and Information. IIASA Working Paper, WR-95–27, 1995.

Svirezhev, Yu. M., Brovkin, V. A., Denisenko, E. A., Semenov, M. A., and Polenok, S. P., Eco-energetics analysis of agrosystems. In: *Geosystems Monitoring. Structure and Functions of Geosystems*. The USSR Academy of Sciences Institute of Geography, Moscow: 205–227 (in Russian), 1986.

Tooming, H. G., *Ecological Principles of Maximal Crop Production*. Gidrometeoizdat, Leningrad (in Russian), 1984.

Ulanowicz, R., *Growth and Development Ecosystems Phenomenology*. Springer-Verlag, New York, 1986.

Zhuchenko, A., Personal communication, 1988.

Disorder is the information that
you don't have
Order is the information
you have

CHAPTER 7

Exergy has been used since the mid-1950s, introduced by Szargut, as the best concept to express the efficiency of any industrial process. The energy efficiency is always 100%, because energy is conserved, but our interest in engineering is on how much work do we utilise of the available amount of work? The answer to this question has required the introduction of exergy.

Exergy work capacity is a question of suitable gradients. Gradients in temperature, in pressure, in chemical potential, and so on, compared with the environment, can be utilised to do work. Exergy becomes, therefore, a measure of the available gradients and a measure of the distance from a thermodynamic equilibrium with the environment.

From this definition of exergy and the entailed characteristics of this concept, it is almost obvious that exergy has a potential as a good descriptor of ecosystem development. Development is associated with increasing gradients and increasing structure which is a contrast to the environment.

Boltzmann related entropy to probability (information) and when we translate his idea to exergy and information theory today, it is clear that exergy measures may also encompass information. In this context, we have information about one state that is valid among a given (high) number of possible states. As ecosystems have a very specific organisation and order selected from many possibilities, it is understandable that it is very important to include these perspectives of exergy application in system ecology. These perspectives are further developed in Chapter 13, where a hypothetical Fourth Law of Thermodynamics is formulated.

This chapter gives a short introduction to the application of exergy in an ecological context. It does not present new hypotheses or theories but is solely based on the use of exergy during the last two to three decades. The coming chapters all touch on the concept of exergy and will present new ideas, hypotheses, and theoretical considerations. It was therefore found appropriate in this volume to include a chapter on the basic idea behind the introduction of the concept, exergy. The presentation focuses mainly on the scientific idea behind the introduction of this concept, its relevance in ecology, and its possibility to measure information.

7 The Thermodynamic Concept: Exergy

Sven E. Jørgensen

CONTENTS

7.1 What is Exergy?..155
7.2 Exergy Balances..158
7.3 Exergy and Information 160
7.4 Dissipative Structure 161
References 162

7.1 WHAT IS EXERGY?

Exergy is defined as the amount of work (= entropy-free energy) a system can perform when it is brought into thermodynamic equilibrium with its environment. Figure 13.1 illustrates the definition. The considered system is characterised by the extensive state variables S, U, V, $N1$, $N2$, $N3$..., where S is the entropy, U is the energy, V is the volume, and $N1$, $N2$, $N3$... are moles of various chemical compounds, and by the intensive state variables, T, p, $\mu c1$, $\mu c2$, $\mu c3$... The system is coupled to a reservoir, a reference state, by a shaft. The system and the reservoir form a closed system. The reservoir (the environment) is characterised by the intensive state variables To, po, $\mu c1o$, $\mu c2o$, $\mu c3o$, ... and as the system is small compared with the reservoir, the intensive state variables of the reservoir will not be changed by interactions between the system and the reservoir. The system develops toward thermodynamic equilibrium with the reservoir and is simultaneously able to release entropy-free energy to the reservoir. During this process the volume of the system is constant as the entropy-free energy must be transferred only through the shaft. The entropy is also constant as the process is an entropy-free energy transfer from the system to the reservoir, but the intensive state variables of the system become equal to the values for the reservoir. The total transfer of entropy-free energy in this case is the exergy of the system. It is seen from this definition that exergy is dependent on the state of the total system (= system + reservoir) and not dependent entirely on the state of the system. Exergy is therefore not a state variable. In accordance with the First Law of Thermodynamics, the increase of energy in the reservoir, ΔU, is

$$\Delta U = U - U_O \tag{1}$$

Where U_o is the energy content of the system after the transfer of work to the reservoir has taken place. According to the definition of exergy, Ex, we have:

$$Ex = \Delta U = U - U_O$$
$$\text{As} \quad U = TS - pV + \Sigma \mu_c N_i \tag{2}$$

(see any textbook in thermodynamics), and

$$U_o = T_o S - p_o V + \sum_c \mu_{co} N_i, \qquad (3)$$

we get the following expression for exergy:

$$Ex = S(T - T_o) - V(p - p_o) + \sum_c (\mu_c - \mu_{co}) N_i \qquad (4)$$

As reservoir, reference state, we can, for instance, select the same system but at thermodynamic equilibrium, i.e., that all components are inorganic and at the highest oxidation state, if sufficient oxygen is present (nitrogen as nitrate, sulphur as sulphate and so on). The reference state will, in this case, correspond to the ecosystem without life forms and with all chemical energy utilised or as an "inorganic soup." Usually, it implies that we consider $T = T_o$ and $p = p_o$, which means that the exergy becomes equal to the Gibbs free energy of the system, or the chemical energy content + the thermodynamic information (see below) of the system. Notice that the equation shown above also emphasises that exergy is dependent on the state of the environment (the reservoir = the reference state), as the exergy of the system is dependent on the intensive state variables of the reservoir.

Notice that exergy is not conserved — only if entropy-free energy is transferred, which implies that the transfer is reversible. All processes, in reality, are, however, irreversible, which means that exergy is lost (and entropy is produced). Loss of exergy and production of entropy are two different descriptions of the same reality, namely, that all processes are irreversible, and we unfortunately always have some loss of energy forms that can do work to energy forms that cannot do work. So, the formulation of the Second Law of Thermodynamics by use of exergy is: all real processes are irreversible, which implies that exergy is lost. Energy is of course conserved by all processes according to the First Law of Thermodynamics. It is therefore wrong to discuss an energy efficiency of an energy transfer, because it will always be 100%, while the exergy efficiency is of interest, because it will express the ratio of useful energy to total energy, which always is less than 100% for real processes. It is therefore of interest to set up an energy balance in addition to an exergy balance for all environmental systems. Our concern is loss of exergy, because that means that "first class energy" that can do work is lost as "second class energy" that cannot do work.

Exergy seems more useful to apply than entropy to describe the irreversibility of real processes, as it has the same unit as energy and is an energy form, while the definition of entropy is more difficult to relate to concepts associated to our usual description of the reality. In addition, entropy is not clearly defined for systems far from equilibrium, particularly for living systems, according to Kay (1984) and Kay and Schneider (1992). Finally it should mentioned that the self-organising abilities of systems are strongly dependent on the temperature, as discussed in Jørgensen et al. (1997). Exergy takes the temperature into consideration as the definition shows, while entropy does not.

Notice also that information contains exergy. Boltzmann (1905) showed that the free energy of the information that we actually possess (in contrast to the information we need to describe the system) is $k * T * \ln I$, where I is the information we have about the state of the system, for instance, that the configuration is 1 out of W possible (i.e., that $W = I$) and k is Boltzmann's constant = $1.3803 * 10^{-23}$ (J/molecules*deg). It implies that one bit of information has the exergy equal to k $T \ln 2$. Transformation of information from one system to another is often almost an entropy-free energy transfer. If the two systems have different temperatures, the entropy lost by one system is not equal to the entropy gained by the other system, while the exergy lost by the first system is equal to the exergy transferred and equal to the exergy gained by the other system, provided that the transformation is not accompanied by any loss of exergy. In this case it is obviously more convenient to apply exergy than entropy.

Exergy is seen as closely related to information theory. A high local concentration of a chemical compound, for instance, with a biochemical function that is rare elsewhere, carries exergy *and* information.

On the more complex levels, information may still be strongly related to exergy but in more indirect ways. Information is also a convenient measure of physical structure. A certain structure is chosen out of all possible structures and defined within certain tolerance margins.

Biological structures maintain and reproduce themselves by transforming energy and information, from one form to another. Thus, the exergy of the radiation from the sun is used to build the highly ordered organic compounds. The information laid down in the genetic material is developed and transferred from one generation to the next.

The chromosomes of one human cell have an information storage capacity corresponding to 2 billion K-bytes! This would require 1000 km of standard magnetic tape to store on a macro computer! When biological materials are used to the benefit of mankind it is in fact the organic structures and the information contained therein that are of advantage, for instance, when using wood.

Information is of course of utmost importance for production systems, where the right management can be crucial for the efficiencies of the processes utilised in the system. Green accounting or environmental audit is a new method of environmental relevance to provide more information about the system.

It can be shown (Evans, 1969) that exergy differences can be reduced to differences of other, better known, thermodynamic potentials, see Jørgensen (1997), which may facilitate the computations of exergy in some relevant cases.

As seen, the exergy of the system measures the contrast — it is the difference in free energy if there is no difference in pressure, as may be assumed for an ecosystem or an environmental system and its environment — against the surrounding environment. If the system is in equilibrium with the surrounding environment, the exergy is, of course, zero.

Since the only way to move systems away from equilibrium is to perform work on them, and since the available work in a system is a measure of the ability, we have to distinguish between the system and its environment or thermodynamic equilibrium alias, for instance, an inorganic soup. Therefore it is reasonable to use the available work, i.e., the exergy, as a measure of the distance from thermodynamic equilibrium.

As we know that ecosystems, due to the through-flow of energy, have the tendency to develop away from thermodynamic equilibrium losing entropy or gaining exergy and information, we can put forward the following proposition of relevance for ecosystems: **Ecosystems attempt to develop toward a higher level of exergy.**

This description of exergy development in ecosystems makes it pertinent to assess the exergy of ecosystems. It is not possible to measure exergy directly — but it is possible to compute it, if the composition of the ecosystem is known.

If we presume a reference environment that represents the system (ecosystem) at thermodynamic equilibrium, which means that all the components are inorganic at the highest possible oxidation state if sufficient oxygen is present (as much free energy as possible is utilised to do work) and homogeneously distributed in the system (no gradients), the situation illustrated in Figure 13.2 is valid. As the chemical energy embodied in the organic components and the biological structure contributes far more to the exergy content of the system, there seems to be no reason to assume a (minor) temperature and pressure difference between the system and the reference environment. Under these circumstances we can calculate the exergy content of the system as coming entirely from the chemical energy:

$$\sum_c (\mu_c - \mu_{co}) N_i.$$

This represents the nonflow chemical exergy. The difference in chemical potential $(\mu_c - \mu_{co})$ between the ecosystem and the same system at thermodynamic equilibrium is determined. This difference is determined by the concentrations of the considered components in the system and in the reference state (thermodynamic equilibrium), as it is the case for all chemical processes. We can measure the concentrations in the ecosystem, but the concentrations in the reference state (thermo-dynamic equilibrium) can be based on the usual use of chemical equilibrium constants. If we have the process:

$$\text{Component A} \leftrightarrow \text{inorganic decomposition products} \tag{5}$$

It has a chemical equilibrium constant, K:

$$K = [\text{inorganic decomposition products}]/[\text{Component A}] \tag{6}$$

The concentration of component A is difficult to find, but is possible based upon the composition of A, to find the concentration of component A at thermodynamic equilibrium from the probability of forming A from the inorganic components.

We find, by these calculations, the exergy of the system compared with the same system at the same temperature and pressure but in form of an inorganic soup without any life, biological structure, information, or organic molecules. As $(\mu_c - \mu_{co})$ can be found from the definition of the chemical potential replacing activities by concentrations, we get the following expressions for the exergy:

$$Ex = RT \sum_{i=0}^{i=n} C_i(\ln C_i/C_{i,o}) \tag{7}$$

where R is the gas constant, T is the temperature of the environment (and the system; see Figure 13.2), while C_i is the concentration of the ith component expressed in a suitable unit, e.g., for phytoplankton in a lake, C_i could be expressed as mg/l or as mg/l of a focal nutrient. $C_{i,o}$ is the concentration of the ith component at thermodynamic equilibrium and n is the number of components. $C_{i,o}$ is, of course, a very small concentration (except for $i = 0$, which is considered to cover the inorganic compounds), corresponding to a very low probability of forming complex organic compounds spontaneously in an inorganic soup at thermodynamic equilibrium. $C_{i,o}$ is even lower for the various organisms, because the probability of forming the organisms is very low with their embodied information, which implies that the genetic code should be correct.

The total exergy of an ecosystem *cannot* be calculated exactly, as we cannot measure the concentrations of all the components or determine all possible contributions to exergy in an ecosystem. If we calculate the exergy of a fox, for instance, the above shown calculations will only give the contributions coming from the biomass and the information embodied in the genes, but not the contribution from the blood pressure, the sexual hormones, and so on. These properties are at least partially covered by the genes, but is that the entire story? We can calculate the contributions from the dominant components, for instance, by the use of a model or measurements that cover the most essential components for a focal problem. The *difference* in exergy by *comparison* of two different possible structures (species composition) is here decisive. Moreover, exergy computations always give only relative values, as the exergy is calculated relative to the reference system. These problems will be treated in further detail in Chapter 13.

7.2   EXERGY BALANCES

It would be very useful to set up an exergy balance for a production system. It is hardly possible to give general equations to be applied for all production systems, similar to the set of equations given above for ecosystems. However, some general considerations on the development of exergy balances for a production system could be made. Usually it will not involve the exergy of information that is included in the analysis, but that the input of first class energy in form of fossil fuel, electricity, and other energy forms are computed and compared with the exergy of the outputs and/or with the loss of exergy in form of nonuseful heat from the entire production systems. Useful heat is heat associated with a higher temperature than the environment. It has work capacity, as also heat input for maintenance of a suitable room temperature contains exergy, while the waste heat transferred to the temperature of the environment has no exergy content. It could be considered whether nonuseful heat could be made useful, but it would require, of course, that the heat still has a temperature which is higher than the environment.

In principle can (and should) the exergy balance of each process in a production system be analysed with the aim to find if the exergy input and/or the generation of nonuseful heat could be reduced? The exergy efficiencies of many of our production processes are surprisingly low. There has been an increasing interest in reducing the exergy consumption with the increasing exergy (energy) cost and decreasing access to low cost energy. The exergy consumption of a top, modern refrigerator is, for instance, only 30 to 40% of a 20-year-old refrigerator with the same capacity. By more comprehensive use of exergy balances, it will be possible to indicate how to very significantly reduce the overall exergy consumption in developed countries. If, on average, we could increase the exergy efficiencies from 40 to 60%, it would imply a reduction of our energy consumption and the associated emissions of pollutants by $33\frac{1}{3}\%$.

To distinguish between exergy and energy that cannot do work, we may call the latter anergy. We have that

$$\text{Energy} = \text{exergy} + \text{anergy} \tag{8}$$

In accordance with the Second Law of Thermodynamics, energy is always > 0 for any process.

Sama (1992) has proposed a separate name for the exergy lost, namely entrogy, which is synonymous with the more general energy. He expresses the Second Law of Thermodynamics as follows:

$$\Delta B \text{ (Exergy)} = \Delta H - T\Delta S \tag{9}$$

where H is enthalpi, T is the environmental absolute temperature, and $T\Delta S$ is the entrogy, which, by any realistic process, is > 0. Equation (9) is, by this interpretation, used for the total system = the considered system + the environment. Usually ΔS in this equation is > 0 for a considered production system, as most production processes in a system imply loss of heat to the environment, but in principle we could have a negative ΔS if heat is transferred from the environment to the system. In this case, we can get more exergy than corresponding to ΔH. This is, for instance, the case when we burn carbon. The exergy gained is about 4.3% higher than the combustion value for pure carbon due to the input of heat from the environment that can be partly transformed to exergy (work). This is of course not a violation of the Second Law of Thermodynamics as more than the 4.3% of the exergy is lost in the environment.

If we consider ΔS for the system only to cover the entropy produced by irreversibility, Equation (7) is, according to the general formulation, the loss of exergy calculated as the environmental absolute temperature times the entropy produced due to the irreversibility.

The use of equations (1) to (5) requires that the utilisation of exergy does not involve displacement work = $p_o(V - V_o)$, which is not the case by application of the (relative) exergy definition of an ecosystem according to Figure 13.2. In the application of exergy in practical engineering, the displacement work cannot be utilised and must of course be deducted. The same is valid for the entropy production (loss of exergy) due to utilisation of a difference in temperature. As there is no temperature difference between the environment and the system in Figure 13.2, this loss of exergy is not actual in our above-mentioned calculations of exergy for ecosystem, but it may be very actual for an engineering system. A simple example will illustrate these considerations. If a substance in the solid or liquid phase is considered, the exergy can be calculated by use of the following equation:

$$\text{Exergy (available shaft work)} = U - U_o + p_o (V - V_o) - T_o (S - S_o) \qquad (10)$$

The displacement work is negligible. For m kg of a substance at temperature T and with the specific heat capacity of c, we get

$$\text{Exergy} = mc(T - T_o) - mcT_o \ln T/T_o \qquad (11)$$

Notice that the index is applied for the environment. For ideal gases the equation is the same, provided that the specific heat capacity at constant pressure is applied. If the specific heat capacity at constant volume is applied, the displacement work must be considered.

If potential energy and kinetic energy are included in the calculations of exergy, they can be equated with exergy at the same magnitude, as they are mechanical energy forms.

Szargut (1978) calculates the exergy efficiency of technological processes by adding the exergy loss due to the emission of waste to the environment. This approach is indeed recommended as a very exergy effective process should generally not be used, if it produces waste products that can harm the environment. As the harm can be expressed as a loss in exergy, the overall exergy efficiency, including considering this loss of exergy, should be used for selection of technologies.

7.3   EXERGY AND INFORMATION

In statistical mechanics, entropy is related to probability. A system can be characterised by averaging ensembles of microscopic states to yield the macrostate. If W is the number of microstates that will yield one particular macrostate, the probability P that this particular macrostate will occur as opposed to all other possible macrostates is proportional to W. It can further be shown that

$$S = k^* \ln W \qquad (12)$$

where k is Boltzmann's constant, $1.3803 * 10^{-23}$ J/(molecules*deg). The entropy is a logarithmic function of W and thus measures the total number of ways that a particular macrostate can be microscopically constituted.

S may be called thermodynamic information, meaning the amount of information *needed* to describe the system, which must *not* be interpreted as the information that we actually possess. The more microstates there are and the more disordered they are, the more information is required and the more difficult it will be to describe the system.

Shannon and Weaver (1949) introduced a measure of information that is widely used as a diversity index by ecologists under the name of Shannon's index

$$H = -\pi \sum_{i=1}^{n} \log_2 \pi \qquad (13)$$

where pi is the probability distribution of species.

Shannon's index of diversity (Shannon and Weaver, 1949) is sometimes called entropy, but should not be confused with thermodynamic information. The symbol H is used to avoid confusion. The use of Shannon's index should be limited to a measure of diversity and communication, although as mentioned above, the two concepts to a certain extent are parallel, as both S and H increase with an increasing number of possible (micro)states.

If an ecosystem is in thermodynamic equilibrium, the entropy, S_{eq}, is higher than in nonequilibrium. The excess entropy may be denoted by the thermodynamic information and is sometimes also defined as Schrödinger's negentropy NE — an expression which should be omitted after introduction of the easier understandable concept, exergy,

$$I = Seq - S = NE \tag{14}$$

In other words, a decrease in entropy will imply an increase in information and loss of information implies increase of entropy. Further, the principle of the Second Law of Thermodynamics corresponds to a progressive decrease of the information content. An isolated system can evolve only by degrading its information.

I also equals Kullback's measure of information (Brillouin, 1962),

$$I = k* \sum_j pj* \ln(pj*/pj) \tag{15}$$

where $pj*$ and pj are probability distributions, *a posteriori* and *a priori* to an observation of the molecular detail of the system, and k is Boltzmann's constant. It means that I expresses the amount of information that is gained as a result of the observations. If we observe a system, which consists of two connected chambers, we expect the molecules to be equally distributed in the two chambers, i.e., $p1 = p2$ is equal to 1/2. If, on the other hand, we observe that all the molecules are in one chamber, we get $p1* = 1$ and $p2 = 0$. $R = k*A$, where A is Avogadro's number.

It is interesting in this context to draw a parallel with the discussion of the development of entropy for the entire universe. The classical thermodynamic interpretations of the Second Law of Thermodynamics predict that the Universe will develop toward "the heat death," where the entire Universe will have the same temperature, no changes will take place, and a final overall thermodynamic equilibrium will be the result. This prediction is based upon the steady increase of the entropy according to the Second Law of Thermodynamics: the thermodynamic equilibrium is the attractor. It can, however, be shown (see Frautschi, 1988), that the thermodynamic equilibrium is moving away at a high rate due to the expansion of the universe.

Due to the incoming energy of solar radiation, an ecosystem is able to move away from the thermodynamic equilibrium, i.e., the system evolves and obtains more information and organisation. The ecosystem must produce entropy for maintenance, but the low-entropy energy flowing through the system may be able to more than cover this production of disorder, resulting in an increased order or information of the ecosystem.

Information contains exergy. Boltzmann (1905) showed that the free energy of the information that we actually possess (in contrast to the information we need to describe the system) is $k * T * \ln I$, where I represents the pieces of information we have about the state of the system and k is Boltzmann's constant $= 1.3803 * 10^{-23}$ (J/molecules*deg). It implies that one bit of information has the exergy equal to $k T \ln 2$. Transformation of information from one system to another is often almost an entropy-free energy transfer. If the two systems have different temperatures, the entropy lost by one system is not equal to the entropy gained by the other system, while the exergy lost by the first system is equal to the exergy transferred and equal to the exergy gained by the other system. Therefore, it is obviously more convenient to apply exergy than entropy.

7.4 DISSIPATIVE STRUCTURE

As an ecosystem is nonisolated, the entropy changes during a time interval, dt, can be decomposed into the entropy flux due to exchanges with the environment, d_eS, and the entropy production due to the irreversible processes inside the system such as diffusion, heat conduction, and chemical reactions, d_iS. It can be expressed as follows (Glansdorf and Prigogine, 1971),

$$dS/dt = d_eS/dt + d_iS/dt \qquad (16)$$

or by use of exergy:

$$Ex/dt = d_eEx/dt - d_iEx/dt,$$

where d_eEx/dt represents the exergy input to the system and d_iEx/dt is the exergy consumed by the system for maintenance, etc.

For an isolated system, $d_eS = 0$, and the Second Law of Thermodynamics yields,

$$dS = d_iS \geq 0 \qquad (17)$$

In other words, d_iS, the internal entropy production (increase) can never be negative, while d_eS does not have a definite sign.

Equation (17) — among other things — shows that systems can only maintain a nonequilibrium steady state ($dS/dt = 0$) by compensating the internal entropy production ($d_iS/dt > 0$) with a negative entropy or better expressed with a positive exergy influx ($d_eS/dt < 0$, or $d_eEx/dt > 0$). Such an influx induces order into the system. In ecosystems, the ultimate exergy influx comes from solar radiation, and the order induced is, e.g., biochemical molecular order.

A special case of nonequilibrium systems is the steady state, where the state variable does not evolve in time. This condition implies that

$$d_eS = -d_iS < 0 \qquad (18)$$

Thus, to maintain a steady nonequilibrium state, it is necessary to pump a flow of exergy of the same magnitude as the internal exergy consumption into the system continuously.

If $d_eEx > -d_iEx$ (the exergy consumption in the system), the system has surplus exergy, which may be utilised to construct further order in the system, or as Prigogine (1980) calls it: dissipative structure. The system will thereby move further away from the thermodynamic equilibrium. The evolution shows that this situation has been valid for the ecosphere on a long-term basis. In spring and summer, ecosystems are in the typical situation that d_eEx exceeds $-d_iEx$. If $d_eEx < -d_iEx$, the system cannot maintain the order already achieved, but will move closer to the thermodynamic equilibrium, i.e., it will lose order. This may be the situation for ecosystems during fall and winter or due to environmental disturbances.

REFERENCES

Boltzmann, L., *The Second Law of Thermodynamics*. (Populare Schriften Essay no. 3 (address to Imperial Academy of Science in 1886). Reprinted in English in: *Theoretical Physics and Philosophical Problems, Selected Writings of L. Boltzmann*. D. Reidel, Dordrecht, 1905.

Brillouin, L., *Science and Information Theory,* Second edition. Academic Press, New York, 351 pp., 1962.

Evans, R. B., A proof that essergy is the only consistent measure of potential work for chemical systems (thesis). Darmouth College, Hanover, N.H., 1969.

Frautschi, S., Entropy in an expanding universe. Chapter 1 in Weber, B. H., Depew, D. J., and Smith, J. D. (Editors): *Entropy, Information and Evolution*, Bradford, MIT, Cambridge, 376 pp., 1988.

Glansdorff, P. and Prigogine, I., *Thermodynamic Theory of Structure, Stability, and Fluctuations*. Wiley-Interscience, New York, 1971.

Jørgensen, S. E., *Integration of Ecosystem Theories: A Pattern,* Second Edition. Kluwer Academic Publishers, Dordrecht, 400 pp., 1997.

Kay, J., Self Organisation in Living Systems [Thesis]. Systems Design Engineering. University of Waterloo, Ontario, Canada, 1984.

Kay, J. and Schneider, E. D., Thermodynamics and measures of ecological integrity. In: *Proc. "Ecological Indicators,"* pp. 159–182. Elsevier, Amsterdam, 1992.

Sama, D., A common sense 2nd law of heat exchanger network design. *Proc. ECOS'92,* Zaragoza, 1992.

Shannon, C. and Weaver, W., *The Mathematical Theory of Communication.* University of Illinois Press, Chicago, 125 pp., 1949.

Szargut, J., Minimisation of the consumption of natural resources. *Bull. Pol. Acad. Tech. Sci.,* No 6, 1978.

CHAPTER 8

This chapter takes off from the Second Law of Thermodynamics and discusses the possibility of applying this law and the minimum entropy production principle by Nicolis and Prigogine to living systems. The question is whether or not entropy is defined for far-from-equilibrium-systems. In accordance with Landsberg (1972), the answer is yes, a statement which he calls the Fourth Law of Thermodynamics. While the author of this chapter expresses his doubt about the measurement of entropy, he proposes to adapt (calculate) entropy productions (flows).

These calculations of entropy flows are demonstrated for several concrete cases, for instance for a deciduous plant leaf under sunlight, for a 50 kg white-tailed deer, for a lizard, and for the human body. The method is furthermore expanded to entire ecosystems, in the first to lake ecosystems then to the entire earth. In all cases the entropy flows are negative. The plants, the animals, and the ecosystems adsorb "negative entropy" from the environment in the sense of Schrödinger (1944).

A comparison of the entropy production of lakes at different eutrophication levels reveals an entropy principle of eutrophication: entropy production increases parallel to the eutrophication. Further interpretation of the results leads to what the author denotes a hypothetical entropy principle: the entropy production of a living system which is open and far from equilibrium consists of two (or more) phases, an early increasing stage and a later decreasing stage. This is (almost) consistent with the hypothesis in Chapter 14, where exergy destruction (which is necessary for maintenance) and exergy storage increase in the early phases, while exergy storage continues to increase in the mature stage due mainly to increase of information, while at least the specific exergy destruction decreases. Exergy is also used in this chapter, but the proposed principle is formulated with other words than the principles proposed in Chapters 12 and 14. However, the basic ideas are the same. It looks like thermodynamic principles of living systems are emerging.

8 Entropy and Exergy Principles in Living Systems

Ichiro Aoki

CONTENTS

8.1 Introduction 167
8.2 Entropy and Entropy Production: Quantifiability in Living Systems.................168
8.3 Entropy Flow and Entropy Production in Nature 169
    8.3.1 Plant..169
    8.3.2 Animal...170
    8.3.3 Human Body 171
    8.3.4 Lake Ecosystem 173
    8.3.5 The Earth...175
    8.3.6 Remarks Regarding the Calculations 175
8.4 Comparative Study of Entropy Production in Lake Ecosystems.......................177
    8.4.1 Lake Biwa and Lake Mendota...177
    8.4.2 Entropy Production in Ecological Systems 180
8.5 Entropy Principle in Living Systems..181
8.6 Exergy Principle in Living Systems 182
Appendix 183
Sample Methods for Calculation of Entropy Flow and Entropy Production 183
    A1 White-tailed Deer..183
    A2 Lake Ecosystem 184
        A2.1 Direct Solar Radiation..184
        A2.2 Diffuse Solar Radiation..185
        A2.3 Diffusely Reflected Solar Radiation 187
        A2.4 Infrared Radiation Emitted by Lake Surface....................................188
        A2.5 Infrared Radiation Incident on Lake Surface 188
        A2.6 Evaporation of Water and Heat Conduction-Convection 188
        A2.7 Entropy Production 189
References 189

8.1 INTRODUCTION

The energy concept, originated in physics, has been extensively employed in natural and even social sciences. In biological sciences, we can mention bioenergetics, biocalorimetry, and ecological energetics (or more specifically energy-flow analysis in ecosystems) as examples of use of the energy concept. However, little has been known about implications of entropy in nature, although entropy is as significant as energy from the thermodynamical viewpoint: the First Law of Thermodynamics is concerned with the concept of energy and the Second Law with entropy. Hence, the importance of the study from entropy viewpoints should be emphasized.

The Second Law of Thermodynamics is the law of the entropy concept. It states for an isolated system that the change of entropy content of the system in irreversible processes [ΔS (irrev)] is always larger than that in reversible processes [ΔS (rev)], and the latter is zero (as stated in usual textbooks):

$$\Delta S \text{ (irrev)} > \Delta S \text{ (rev)} = 0. \tag{1}$$

Since biological objects are not isolated systems, Equation (1) cannot be applied to biology.

Biological systems are open systems which exchange energy and matter with their surroundings. For open systems, the change of entropy content of a system (ΔS) is the sum of two terms: entropy flow ($\Delta_e S$) and entropy production ($\Delta_i S$). The entropy flow is the entropy that is brought into or out of the system associated with flows of energy and matter, and the entropy production is the entropy that is produced by irreversible processes occurring within the system. The Second Law for open systems asserts that the entropy production in irreversible processes [$\Delta_i S$ (irrev)] is always larger than that in reversible processes [$\Delta_i S$ (rev)], and the latter is zero (e.g., Nicolis and Prigogine 1977),

$$\Delta_i S \text{ (irrev)} > \Delta_i S \text{ (rev)} = 0. \tag{2}$$

Thus, the Second Law for open systems is formulated in the term of entropy production.

In open systems near to thermodynamic equilibrium, the entropy production always decreases with time and approaches a minimum stationary level according to the minimum entropy production principle (Nicolis and Prigogine 1977),

$$\frac{d}{dt} \Delta_i S \text{ (irrev)} < 0. \tag{3}$$

However, biological systems are not near to thermodynamic equilibrium, hence Equation (3) can not be applied to biological systems.

What is the entropy principle, if any, applicable to living systems which are open and far from equilibrium? In Section 8.5, a hypothetical entropy principle (Aoki 1989a, 1990a, 1995, 1998) is proposed for the development of living systems, based on the calculation and observation of entropic properties of lake ecosystems shown in Section 8.4.

By use of the Gouy–Stodola theorem (Bejan 1982), a corresponding exergy principle in living systems (Aoki 1991, 1996, 1998) is naturally derived from the entropy principle, and is presented in Section 8.6

8.2 ENTROPY AND ENTROPY PRODUCTION: QUANTIFIABILITY IN LIVING SYSTEMS

Thermodynamical variables are divided into two classes: state variables and process variables. With regard to the entropy concept, the state variable is entropy content and the process variables are entropy flow and entropy production.

As to the state variable, entropy content, it is possible to imagine entropy content of biological systems *in a purely conceptual context*. Biological organisms are in states far from equilibrium; however, even in such conditions, entropy content of organisms can be considered to exist. Prigogine argued that entropy of a nonequilibrium state can be defined if materials are dense and variations of densities with space and time are small; these conditions are fulfilled in biological systems (Nicolis and Prigogine 1977). Also, Landsberg (1972) asserted that, for a class of nonequilibrium states, extensive variables such as entropy content exist, and called this statement the "Fourth Law" of Thermodynamics. It will be evident that, according to Prigogine, biological systems are in the states for which the extensive variable entropy exists and the Fourth Law holds.

However, it is quite questionable whether entropy content can be measured for living systems. The measurement of the absolute value of entropy content of an organism needs measurements of heat capacity from the absolute temperature 0 [K] to ordinary temperatures and of latent heat at phase transition, if any, for the organism. In the measurement at an extremely low temperature, the organism is in a "dead" state: its internal organized structure and function are completely destroyed. The entropy content of organisms thus measured will have no significant meanings for organisms that must have the essential features of life: specific organization and biotic function. An alternative method to measure entropy content is to adopt entropy content relative to higher reference temperature (not 0 [K], and near to ordinary temperature), in which organisms can exist in living state. However, the choice of an adequate reference temperature causes a problem: reference temperature should differ for different kinds of organisms, because the temperature dependence of biochemical and physiological conditions differs for different organisms. Hence, the relative entropy content thus measured may not be useful for the comparative study of various kinds of organisms. (It should be noted, in biological science, that the comparative study is one of the most powerful tools for revealing fundamental characteristics of biological systems and has been recognized from the old age of C. Darwin to the present modern molecular biology and biochemistry.)

Also, the calculation of entropy content of living systems is an extremely formidable task (it is really impossible), even for the simplest organisms, i.e., bacteria. The situation is easily understood by recognizing that even a single bacterium contains various kinds of organized structure composed of tremendous variety and amounts of molecules undergoing intricate and specific biochemical reactions. (It differs extremely in complexity from, e.g., homogeneous ideal gas confined in a cylinder under constant temperature and pressure, for which the entropy content can easily be calculated as in usual textbooks.)

Hence, it is impossible to develop quantitative thermodynamical considerations based on measured or calculated entropy content of living systems.

However, entropy flow and entropy production — process variables — can quantitatively be estimated by use of some physical modelling and calculation from *observed* energetic data of biological objects. Sample methods of calculation are simply described in the Appendix. Thus, we can develop entropic considerations based on numerical values of entropy flow and entropy production. Numerical examples of entropy flow and entropy production in nature are illustrated in the next section.

Thus, entropy production is a significant quantity due to the property of Equation (2) and due to quantifiability for living systems. Entropy is produced anywhere at any time when processes are irreversible. The higher the irreversibility of process, the more entropy is produced. Hence, entropy production is a measure of the extent of irreversibility of processes. Since almost all motions and reactions actually occurring in nature are irreversible (except some carefully controlled laboratory experiments), entropy production is also a measure of the magnitude of activity of natural processes, which consists of physical activity (the strength of process of energy flow and transportation of matter), chemical activity (the strength of chemical reaction), and biological (including human) activity (the strength of biological interaction).

8.3   ENTROPY FLOW AND ENTROPY PRODUCTION IN NATURE

8.3.1   PLANT

Entropy flow and production for a deciduous plant leaf under sunlight (the energy flux of solar radiation is 1.20 [cal cm^{-2} min^{-1}]) are calculated from observed energetic data and results are shown in Figure 8.1 (Aoki 1987a). Units are [J m^{-2} s^{-1} K^{-1}]. The entropy inflows into the leaf

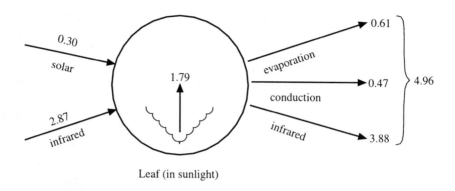

Leaf (in sunlight)

FIGURE 8.1 Entropy flow and entropy production for a deciduous plant leaf under sunlight. Units are [Jm^{-2} s^{-1} K^{-1}].

due to absorption of solar radiation and absorption of infrared radiation are 0.30 units and 2.87 units, respectively. The entropy production in the leaf is 1.79 units. The entropy outflow from the leaf is 4.96 units, which consists of 3.88 units by emission of infrared radiation, 0.47 units by heat conduction, and 0.61 units by evaporation of water. On the other hand, the entropy production in the leaf is nearly zero at night (Aoki 1987a, 1987b). Entropy flow and entropy production in conifer branches are also calculated (Aoki 1989b). It is shown that the entropy production in leaves is proportional to the solar radiation energy absorbed by leaves. Hence, the entropy production in leaves changes, keeping pace with solar radiation.

The most entropy production in plant leaves will be due to the physical process of scattering and absorption of incident solar radiation by various components and particles in leaves, followed by subsequent conversion of solar radiation energy to heat energy; this process causes entropy production. The contribution of photosynthesis to the entropy production is shown to be very small (Aoki 1987a). This is partly because photosynthesis uses only about 1% or less of solar radiation energy incident on plant leaves.

8.3.2 ANIMAL

Figure 8.2 shows the results of calculation of entropy flow and entropy production for a 50 [kg] white-tailed deer on a maintenance diet in a standing posture during a winter night (Aoki 1987c). The values are in units of [J s^{-1} K^{-1}]. The infrared radiation from the sky and from the ground is incident upon and absorbed by a white-tailed deer. The entropy inflow into the deer due to the infrared radiation is 1.66 units. The entropy of 0.46 units is produced by irreversible processes within the body of the white-tailed deer. The entropy outflow from the deer is 2.12 units, which consists of 1.82 units by emission of infrared radiation from the deer, 0.21 units by convection in the surrounding air, 0.07 units by evaporation of water from the skin and the lung of the deer, and 0.02 units by heat conduction to ingested food. The entropy production per effective surface area becomes 0.32 [J m^{-2} s^{-1} K^{-1}], and that per body weight is 0.0092 [J kg^{-1} s^{-1} K^{-1}]. The entropy production in the deer is not zero at night, contrasting with plant leaves (Section 8.3.1). The reason is that the biochemical and physiological activity in deer is naturally kept at a high level even at night. Thus, the difference between a plant and animal is also reflected in the values of entropy production. The methods of calculation of entropy flow and entropy production for the deer are simply given as examples in Appendix A1.

Entropy flow and entropy production in a lizard of 18.4 [g] located vertically on a tree trunk under sunlight in summer are also investigated (Aoki 1988a). The entropy production within the lizard is shown to be 34.9 × 10^{-4} [J s^{-1} K^{-1}] per individual, 0.46 [J m^{-2} s^{-1} K^{-1}] per surface area, and 0.19 [J kg^{-1} s^{-1} K^{-1}] per body weight. The net entropy flow due to mass exchange associated

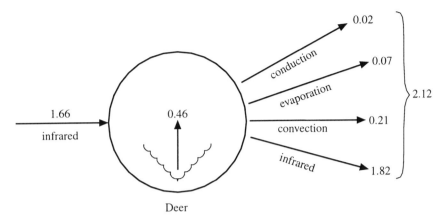

FIGURE 8.2 Entropy flow and entropy production for a white-tailed deer during a winter night. Units are $[Js^{-1}K^{-1}]$.

with respiration between the lizard and its environment is shown to be four orders of magnitude smaller than the net entropy flow associated with energy flow, and hence it can be neglected. The small effect of mass exchange, compared with energy exchange, is also observed in a human body (Section 8.3.3), and it may be a general law for organisms. This point will further be discussed in Section 8.3.6.

Entropy flow and entropy production of a swine in a calorimeter chamber are calculated from the results of measurement of energy flow into and out of a swine (Aoki 1992). The entropy production per surface area in swine shows some characteristic tendency: it increases from birth to about 8 months of age and decreases gradually thereafter. This "two-stage" trend of entropy production (early increase and later decrease) is in accord with the case of a human body (Aoki 1991, 1994), and is discussed further in the next section.

8.3.3 HUMAN BODY

Energy inflow into a naked human body in a respiration calorimeter is due to infrared radiation emitted by the inside walls of the calorimeter, and energy outflow from the human body is due to infrared radiation emitted by the body, convection in its surrounding air, and evaporation of water from the lung and the body surface. Energy exchange for the human body through radiation, convection, and evaporation have been measured by use of the radiometer and the respiration calorimeter in which a nude subject is laid in basal conditions. Also, mass-flows into and out of the human body have been measured. From these results, it is possible to calculate corresponding entropy inflows and outflows and to estimate the entropy production occurring within the human body. Entropy flow and entropy production thus obtained are shown in Figure 8.3, in units of $[J\ s^{-1}\ K^{-1}]$ per individual (Aoki 1989c). The incoming entropy consists of 3.16 units by absorption of infrared radiation and 0.04 units by O_2 uptake. The entropy production within the body is 0.26 units. The entropy in the body decreases by 0.06 units during the experiment of one hour (it will naturally become zero in the long term, e.g., one day). The outgoing entropy consists of 3.34 units by emission of infrared radiation, 0.04 units by convection, 0.08 units by evaporation of water, and 0.06 units by CO_2 liberation and by liquid H_2O supplied to the outside.

The entropy production per surface area in a human body in a basal condition are shown to be kept nearly constant irrespective of environmental factors of the body: surrounding temperature, wind, and clothing. This property may be called "entropy homeostasis" of human body in basal condition (Aoki 1990b). On the other hand, when naked subjects in the respiration calorimeter are in nonbasal

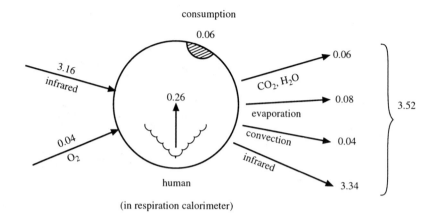

FIGURE 8.3 Entropy flow and entropy production for a human body in respiration calorimeter. Units are $[Js^{-1}K^{-1}]$.

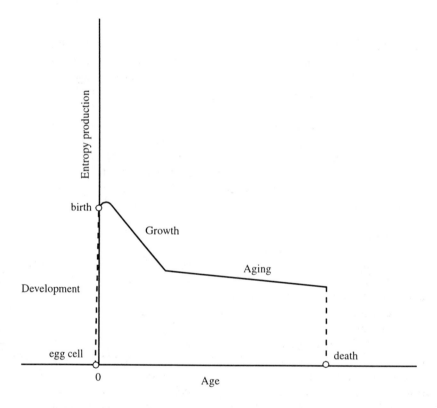

FIGURE 8.4 Schematic profile of entropy production for a human body life span. Scales are arbitary.

conditions (exercise and chill), entropy productions per surface area become 2'8 times as great as that in basal condition (Aoki 1990b). These increases in entropy production are due to the increase in heat production within the body.

The time-course of entropy production accompanying human development, growth, and aging is also investigated (Aoki, 1991, 1994). The entropy production per surface area rises from 0 to 2 years of age, decreases rapidly from 2 to 25 years of age, and then decreases gradually to 85 years of age (Aoki, 1991). Schematic profile of entropy production per surface area is illustrated in Figure 8.4.

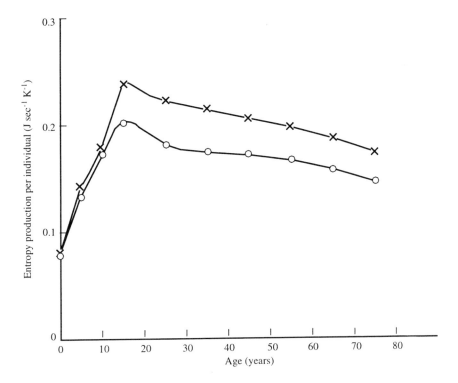

FIGURE 8.5 Entropy production per individual vs. 0–75 years of age for humans male and female. (x), male; (0), female.

Entropy productions per individual vs. 0–75 years of age and 0–19 years of age for male and female are shown in Figures 8.5 and 8.6, respectively (Aoki 1994). Thus, the entropy production per individual increases from birth to about 16 years of age for males and to about 14 years of age for females, and then gradually decreases afterward.

That is to say, the entropy production per surface area and per individual manifests two phases in the human life span: the early increasing stage and the later decreasing stage. This "two-stage principle" will reflect the essential and universal trend of development, growth, and aging for human life span from fertilized egg to death, and will also be applied to other organisms (Aoki 1991, 1994).

Sections 8.4 and 8.5 show that the two-stage tendency (early rise and later fall) of entropy production is also applied to the time-course of ecosystems, and will be a universal law for the development of living systems which have the opposing phases: growth and senescence (Aoki 1989a, 1995, 1998).

8.3.4 Lake Ecosystem

As pointed out by Hutchinson (1964), the study of large and complex ecosystems, such as lakes, consists of two different approaches: holological (holos = whole) and merological (meros = part). In the holological approach, an ecosystem is treated as a black box without scrutinising the internal contents of the system, and the attention is focussed on inputs and outputs to and from the ecosystem. On the other hand, in the merological approach, components or parts of a system are studied in detail and then integrated into a whole system, if possible.

The pioneering work for the holological approach to lakes was done by Birge (1915) and Hutchinson (1957) and others extended this line of research. These studies are all from an energy viewpoint.

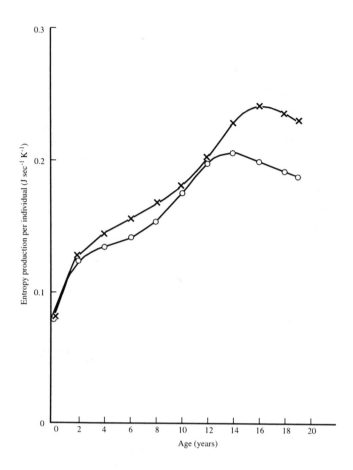

FIGURE 8.6 Entropy production per individual vs. 0–19 years of age for humans male and female. (x), male; (0), female.

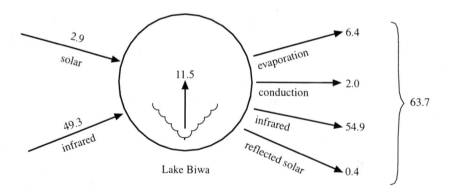

FIGURE 8.7 Entropy flow and entropy production for the northern basin of Lake Biwa. Units are [MJ m^{-2} year^{-1} K^{-1}].

Holological and entropic studies of lake ecosystems are made by Aoki (1987d, 1989a, 1990a, 1995, 1998). Figure 8.7 shows, as an example, the annual values of entropy flow and entropy production per surface area of lake water in the northern basin (the main part) of Lake Biwa, which is located at the central part of Japan (near Kyoto) (Aoki 1987d). The values are in units of

[MJ m^{-2} year^{-1} K^{-1}]. Monthly variations of entropy production in Lake Mendota in the U.S. (near Madison, Wisconsin) and those in Lake Biwa were also investigated (Aoki 1989a, 1990a). The comparative study of oligotrophic Lake Biwa (the northern basin) and eutrophic Lake Mendota leads to a hypothetical principle of entropy production for the development of living systems (Aoki 1989a, 1995, 1998), which is described in detail in Sections 8.4 and 8.5.

8.3.5   THE EARTH

In considering a large and complex system, the *systems-thinking* philosophy has been adopted as useful for the investigation of such a system. Systems-thinking, essentially the same as a holological approach by Hutchinson (Section 8.3.4), treats a large and complex system as being composed of a small number of subsystems or compartments and each subsystem is dealt with *as a whole*; that is to say, details of structure and process within subsystems are not scrutinized and each subsystem is regarded as a black box. The main concern in this approach is the pattern of networks of flows into and out of each subsystem.

The earth, a very large and complex system, is considered here according to systems-thinking, as composed of the earth's surface and the earth's atmosphere. Figure 8.8 shows the pattern of energy flow on the earth in units of [kJ cm^{-2} year^{-1}] (Aoki 1988b). Figure 8.8: (1) Energy flows associated with incident solar radiation. The incident energy flow from outer space is 1067 units, of which 320 units are absorbed by the earth's surface as direct-beam solar radiation, 224 units absorbed by the earth's surface as diffuse sky radiation, and 203 units absorbed by the atmosphere. Parts of the remaining 277 units are reflected back by the atmosphere and 43 units by the earth's surface. (2) Energy flows associated with terrestrial infrared radiation. The earth's surface emits 1045 units, of which 64 units escape to outer space through "the atmospheric window" and the remaining 981 units are absorbed by the atmosphere. The atmosphere emits 683 units to outer space and 821 units to the earth's surface. (3) Net radiation of the earth's surface and of the atmosphere: net sums of the values in (1) and (2). (4) Energy flows due to heat conduction and due to evaporation of water. Heat conduction transports 75 units of sensible heat, and evaporation transports 245 units of latent heat from the earth's surface to the atmosphere.

Based on the energetic data shown above, detailed patterns for entropy flow and entropy production in the earth's surface and in the earth's atmosphere are calculated and shown in Figure 8.9 (Aoki 1988b); units are [J cm^{-2} year^{-1} K^{-1}]. The entropy production in the earth's surface is 2064 units and that in the atmosphere is 2236 units. The net entropy flows from the earth's surface to the atmosphere transported by infrared radiation, heat conduction, and evaporation of water are in the ratios 1.0:0.4:1.4. There is an extensive circulation of entropy between the earth's surface and the atmosphere. The ratio of the entropy of incoming radiation to the earth to that of outgoing radiation from the earth is 1:18; the earth amplifies incoming radiation entropy by 18 times.

8.3.6   REMARKS REGARDING THE CALCULATIONS

In the above examples, the net entropy flows into the plant leaves, the animals, the humans, the lakes, the atmosphere, and the earth's surface are shown to be all negative. That is, they absorb "negative entropy" from the surroundings in the sense of Schrödinger (1944). This is the physical basis for organized structure and function of organisms, ecosystems, and the earth to be maintained, according to Schrödinger (1944).

The entropy productions in the above examples turn out to be all positive. This indicates that the Second Law of Thermodynamics certainly holds in the above cases, as shown in Equation (2). This is against the arguments made earlier that the Second Law may not be applied to living systems. If the Second Law of Thermodynamics does not hold for living systems, they can be perpetual-motion machines of the second kind, because the Second Law is the principle of the impossibility of constructing those machines (Thomon's Principle or Ostwald's Principle). As far

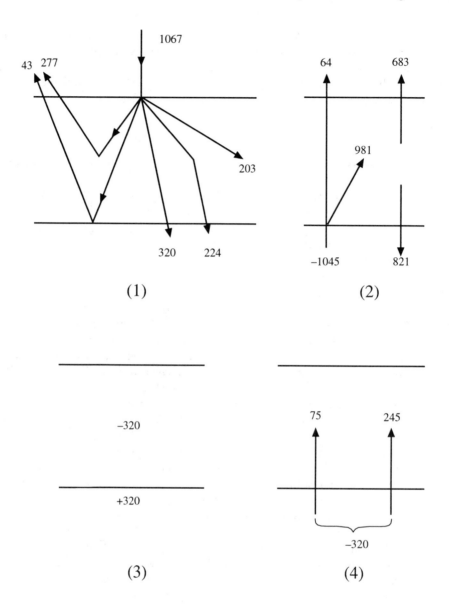

FIGURE 8.8 Energy flow in the earth (two-compartment model of the earth). Units are [kJ cm^{-2} year^{-1}].

as the above cases show, it is highly probable that the Second Law holds for living systems, and hence they cannot be any perpetual-motion machines of the second kind.

The above entropic study of plant leaf, lizard, and humans shows that the effect of mass exchange is small compared with the total. The reason is as follows: the effect of mass exchange in organized systems should, in general, be small enough in order to maintain structural and functional integrity of systems; otherwise (i.e., if mass exchange is very extensive) it would be difficult to keep sustainable organized configuration of matter, and the integrity and the organization of systems would be destroyed. Hence, the effect of mass exchange will be small in living systems which are well-organized and have orderly biological functions.

As shown in the above examples, entropy production can be numerically estimated for a large variety of natural complex systems. The next important problem to be examined is how entropy production changes with time for living systems in general (the answer for human life span is

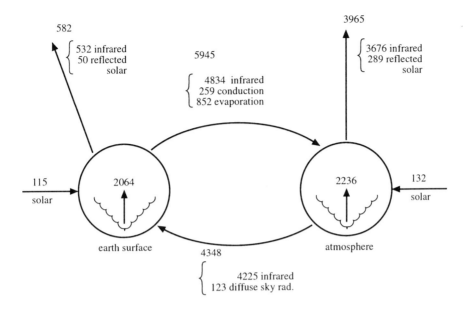

FIGURE 8.9 Entropy flow and entropy production in the earth's surface and the atmosphere. Units are [J cm^{-2} year^{-1} K^{-1}].

already given Section 8.3.3). This subject will be clarified in the comparative study of lake ecosystems in the following sections.

8.4 COMPARATIVE STUDY OF ENTROPY PRODUCTION IN LAKE ECOSYSTEMS

Along the line of research simply described in Section 8.3.4, entropy production in lake ecosystems is examined in detail in the present section (Aoki 1998). The outline of the methods of calculation is provided in Appendix A2. For the physical basis of the methods, the reader is referred to Aoki (1982) and literature cited therein.

8.4.1 LAKE BIWA AND LAKE MENDOTA

Consider two lakes as examples: Lake Biwa and Lake Mendota. Lake Biwa is located at 34°58′ –35°31′N, 135°52′–136°17′E (near Kyoto, Japan) and consists of a northern basin (the main part) and a southern basin (the smaller part). The former is oligotrophic and the latter is nearly eutrophic (in 1970s; data of that period are used in the present study). Only the northern basin is considered here. Lake Biwa is the most studied lake in Japan, and its energy budgets (i.e., energy flow and heat storage) were studied in detail by Ito and Okamoto (1974) and Kotada (1977).

From the data given by Kotoda (1977), monthly entropy productions per surface area in Lake Biwa are obtained using the methods shown in Appendix A2 similar to a previous paper (Aoki 1990a). The results, entropy production per m^2 of lake surface plotted against absorbed solar radiation energy per m^2 of lake surface, are shown in Figure 8.10.

Lake Mendota, another example, is located at 43°04′N, 89°24′W (near Madison, Wisconsin, U.S.A.), and is a eutrophic lake. It is the most studied lake in the U.S.A., and its energy budgets were investigated by Dutton and Bryson (1962) and Stewart (1973). From their data, monthly entropy productions per surface area in Lake Mendota are obtained by similar methods employed in the previous paper (Aoki 1989a). The results are shown in Figure 8.11.

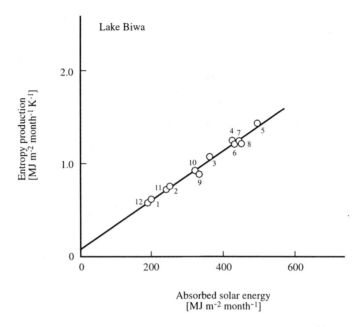

FIGURE 8.10 Monthly entropy production in the northern basin of Lake Biwa per m² of the lake surface plotted against monthly solar radiation energy absorbed by 1 m² of the lake surface. The numbers near the circles are the months.

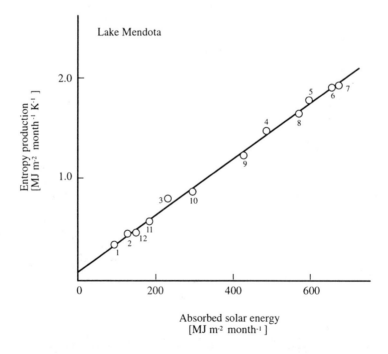

FIGURE 8.11 Monthly entropy production in Lake Mendota per m² of the lake surface plotted against monthly solar radiation energy absorbed by 1 m² of the lake surface. The numbers near the circles are the months.

Figures 8.10 and 8.11 show that the entropy production in month j [denoted as $(\Delta_i S_j)$] is a linear function of the absorbed solar radiation energy in month j (denoted as E_j):,

$$(\Delta_i S)_j = a + bW_j \tag{4}$$

The second term on the right-hand side of Equation (4) is the entropy production dependent on solar radiation energy. This term is mainly due to absorption of solar radiation by water, dissolved organic matter, and suspended solid in the lake water followed by subsequent conversion of solar radiation energy mostly to heat energy; this process causes entropy production. (Contributions to this term from photosynthesis and light respiration of phytoplankton will be very small compared with the above mentioned one.) The first term on the right-hand side of Equation (4) is the entropy production independent of solar radiation energy. This term is assumed, for the moment, to be mainly due to liberation of heat energy by respiration of organisms in the lake; this process leads to entropy production.

The entropy production per year is given by

$$\sum_{j=1}^{12} (\Delta_i S)_j = 12a + b \sum_{j=1}^{12} E_j \tag{5}$$

and entropy production per year per one unit of absorbed solar radiation energy is

$$\frac{\sum_{j=1}^{12} (\Delta_i S)_j}{\sum_{j=1}^{12} E_j} = \frac{12a}{\sum_{j=1}^{12} E_j} + b. \tag{6}$$

Values of each term in Equations (5) and (6) are easily obtained from the data in Figures 8.10 and 8.11.

Consider a vertical water column from the lake surface to the bottom, the cross section of which is 1 m². This water column consists of a light zone (euphotic zone) and a dark zone (aphotic zone).

Now, it is possible to make a comparative study between the two lakes. Table 8.1 shows, for both lakes, total and solar energy-dependent entropy productions (per year per MJ of absorbed solar radiation energy per m³ of the lake water), and entropy productions independent of solar radiation energy (per year per m³ of the lake water); they are obtained from the data in Figures 8.10 and 8.11.

As already stated, the entropy production dependent on solar radiation in the light zone (euphotic zone) is the measure of activity due to absorption of solar radiation energy by water, dissolved organic matter, and suspended solid followed by subsequent conversion of solar energy to heat energy. Hence, the values of entropy production dependent on solar radiation in the light zone in Table 8.1 will be related to the amount of dissolved organic matter and suspended solid per m³ of lake water in the light zone. In fact, the average amount of suspended solid (SS) in the light zone in Lake Biwa is 1.3 [g m^{-3}] and that in Lake Mendota is 1.9 [g m^{-3}] (National Institute for Research Advancement 1984), and the ratio of the amount of SS in Lake Mendota to that in Lake Biwa is 1:5. Also, the average amount of dissolved organic carbon (DOC) in the light zone in Lake Biwa is 1.6 [gCm^{-3}] (Mitamura and Saijo 1981) and that in Lake Mendota is 3.3 [gCm^{-3}] (Brock 1985); the ratio of DOC in Lake Mendota to that in Lake Biwa is 2:1. These ratios are consistent with the ratio of entropy production dependent on solar radiation in Lake Mendota to that in Lake Biwa shown in Table 8.1. Thus, the larger the amounts of SS and DOC, the more the entropy production is dependent on solar radiation. The entropy production dependent on solar radiation gives a kind

TABLE 8.1

Comparison of Entropy Productions in Lake Biwa and in Lake Mendota

Lakes	Total (in Whole Water Column)	Solar Energy Dependent (in Light Zone)	Solar Energy Independent (in Whole Water Column)
Lake Biwa	0.07	0.13	19
Lake Mendota	0.24	0.31	69
Lake Mendota/Lake Biwa	3:7	2:3	3:6

Total and solar energy dependent entropy productions (per year per MJ of absorbed solar radiation energy per m^3 of the lake water) are shown, respectively, in the first and in the second column, and entropy productions independent of solar radiation energy (per year m^3 of the lake water) are in the third column. Units are [kJ K^{-1} m^{-3} $year^{-1}$]. Ratios of the values for the two lakes are shown in the last row.

of *physical* measure for the amount of dissolved organic matter and suspended solid in the lake water by means of reactions to incident solar radiation.

The entropy production independent of solar radiation energy shown on the third column of Table 8.1 will be the measure of activity of respiration of organisms and distributed over the whole water column from the surface to the bottom. Hence, it will correlate with the amount of organisms in the whole water column. In fact, the average amount of total plankton plus zoobenthos in the whole water column in Lake Biwa is 0.16 [gCm^{-3}] (Sakamoto 1975) and that in Lake Mendota is 0.62 [gCm^{-3}] (Brock 1985); the ratio of the amount of plankton plus zoobenthos in Lake Mendota to that in Lake Biwa is 3:9. This ratio is consistent with the result on entropy production independent of solar radiation shown in Table 8.1. The larger the amount of organisms, the more the entropy production is independent of solar radiation. The entropy production independent of solar radiation energy will represent a *physical* measure of the degree of respiration of organisms in lake water.

Thus, entropy productions due to light-absorption (physical process) and respiration (biological process) are separately estimated and are compared with each other for the oligotrophic and eutrophic lakes.

The values in the first column of Table 8.1 are total entropy productions per year per MJ of absorbed solar radiation energy per m^3 of the average whole water column.

8.4.2 ENTROPY PRODUCTION IN ECOLOGICAL SYSTEMS

As shown in Table 8.1, the entropy productions in eutrophic Lake Mendota are larger than those in oligotrophic Lake Biwa in any of the categories considered (i.e., due to light absorption, respiration, and total). Therefore, it may be possible to propose that each entropy production of the above three categories in a eutrophic lake will generally be larger than those in an oligotrophic lake. Eutrophication in a lake is a directional process: the process tends to proceed with time from oligotrophy to eutrophy in most of present lake ecosystems surrounded by the environment full of organic matter (anthropogenic restoration is not considered here). Hence, the entropy production of the three categories in a lake will increase with time accompanying the process of eutrophication; this may be called the entropy law for eutrophication, and has already been proposed by Aoki (1989a, 1990a).

Will this tendency, however, be applied to the overall span of ecological process? Hyper-eutrophic lakes represent the ultimate senescent stage of the eutrophication process; they become very vulnerable to external perturbations and the entire algal biomass often catastrophically collapses and dies, resulting in the elimination of entire populations (phytoplankton, fish, and zooplankton) (Barica 1993). Also, the productivity of organic matter in a senescent lake is lower than that in a eutrophic lake as already shown in the classical work by Lindeman (1942). These facts suggest that entropy production will decrease with time in the senescent stage of ecological process (i.e., in the course of hypereutrophication), because entropy production is a measure of the degree of various activities in natural systems including productivity in ecological systems as stated in Section 8.2. This trend of decreasing entropy production appears to coincide with the minimum entropy production principle by Prigogine [Equation (3)]. However, it should be emphasized that they are quite different in nature: although Prigogine's principle holds near to only thermodynamic equilibrium, the trend proposed here (the decrease of entropy production with time) is applicable even for processes far from equilibrium, based on the ecosystem observations.

Thus, the entropy production in ecological systems will have the tendency of early increase and later decrease. This is in accord with the "multi(three)stage" hypothesis proposed by Aoki (1989a) for the life span of the ecological process in general: entropy production increases with time in an early stage (growing stage) of the ecological process (the entropy law for eutrophication stated above holds in this stage), is kept almost stationary or shows oscillation in an intermediate stage, and decreases in a later stage (senescent stage) of the ecological process. Thus, the ecological process will consist of multiphases in entropy production and not be unidirectional.

It is noteworthy that the multiphase tendency of entropy production proposed above is parallel to the behavior of entropy production in the human life span showing an early rise and a later fall (Section 8.3.3). Thus, in this respect, there will be a parallelism in entropy production between organisms and ecosystems.

8.5 ENTROPY PRINCIPLE IN LIVING SYSTEMS

The multi(two or three)phase tendency in entropy production in the life span of humans (Section 8.3.3) and of ecological systems (Section 8.4.2) leads to a hypothetical *entropy principle* in living systems from organisms to ecosystems. That is, entropy production in a living system, which is open and far from equilibrium, consists of two or more phases: an early increasing stage, a later decreasing stage, and an intermediate stage, as shown schematically in Figure 8.12. In Figure 8.12,

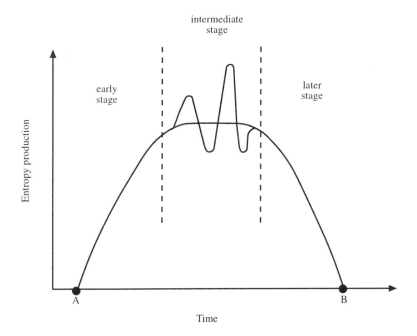

FIGURE 8.12 A hypothecial entropy principle for living systems (schematic profile). Scales are arbitrary. This tendency is also applied to exergy destruction. Explanations are given in the text.

point A is the beginning of life in living systems (e.g., fertilization in organisms, or formation of the lake in lake ecosystems) and point B is the end (death) of their lives. The early increasing stage is the phase of development and growth of the systems, the later decreasing stage is the phase of senescence and decay of the systems, and the intermediate stage is the transitional one (it may be stationary or oscillating). This entropy principle will be universal for the development of living systems from organisms to ecosystems, which have two opposing phases: birth–growth and senescence–death (Aoki 1989a, 1995, 1998). The validity of this principle can be examined by comparative study of various organisms and ecosystems as developed in the previous section.

The three-stage character of the entropy principle shown in Figure 8.12 is in accord with the classical hypothesis by Lindeman (1942) on productivity of organic matter in lake ecosystems. Lindeman presented a similar hypothetical figure on the time-course of productivity, based on the comparison of senescent Cedar Bog Lake and eutrophic Lake Mendota. Because entropy production is a measure of the degree of physical, chemical, and biological activities in natural systems including productivity in ecosystems as stated in Section 8.2, the entropy principle shown above can be regarded as a kind of generalization and thermodynamical abstraction of Lindeman's classical hypothesis on productivity of organic matter in lake ecosystems.

8.6 EXERGY PRINCIPLE IN LIVING SYSTEMS

Thus far, we have considered entropy production, which is a measure of the degree of activity and irreversibility of natural process as already stated (Section 8.2). Another thermodynamic quantity that is closely related to irreversibility is exergy or equivalently availability. Exergy is a quantity that takes into account both quantity and quality of energy, and represents the maximum available work that can be extracted from a system and its environment when the system passes from a given state to a state in equilibrium with the environment (e.g., Moran 1982). It is also a measure of the departure of the state of a system from its reference environment (e.g., Moran 1982). It should be noted, however, that exergy of living systems cannot be quantified (Aoki 1993; Maansson and McGlade 1993).

Contrary to entropy that is always produced by irreversibility of process, exergy is always destroyed by irreversibility (Moran 1982). The Gouy–Stodola theorem (Bejan 1982) shows that the amount of entropy produced by irreversibility (entropy production $\Delta_i S$) is proportional to the amount of exergy destroyed by irreversibility (exergy destruction $|\Delta_i A|$):

$$|\Delta_i A| \ = \ T^* \Delta_i S, \tag{7}$$

where T^* is the temperature of the environment of a system. Thus, the magnitude of exergy destruction $|\Delta_i A|$ is parallel to that of entropy production. By use of the Gouy–Stodola theorem, exergy destruction (not exergy itself) can be quantified if the value of entropy production is known, and vice versa. The entropy law for eutrophication stated in Section 8.4.2 can now be read as: exergy destruction increases with time in the course of a eutrophication process. This may be called the exergy law for eutrophication.

Likewise, the behavior of entropy production in the life span of living systems shown in Figure 8.12 (the entropy principle) can be applied to exergy destruction: exergy destruction consists of an early increasing stage, a later decreasing stage, and an intermediate stage. This is the exergy principle in the life span of living systems. This has already been proposed for humans (Aoki 1991) and is extended to general living systems including ecological systems (Aoki 1996, 1998).

APPENDIX

SAMPLE METHODS FOR CALCULATION OF ENTROPY FLOW AND ENTROPY PRODUCTION

A1   WHITE-TAILED DEER (AOKI 1987c)

Consider a white-tailed deer of 50 [kg] weight on a maintenance diet in a standing posture during a winter night at an air temperature $T_a = -20$ [°C]. The effective temperature of the sky (T_{sky}) and that of the ground (T_{grd}) are obtained from T_a by use of the empirical formulae. The downward and upward entropy flux become, respectively,

$$S_\downarrow = \frac{4}{3} \sigma(T_{sky})^3 = 1.075 \ [\text{Jm}^{-2}\text{s}^{-1}\text{K}^{-1}] \tag{A1.1}$$

$$S_\uparrow \frac{4}{3} \sigma(T_{grd})^3 = 1.218 \ [\text{Jm}^{-2}s^{-1}\text{K}^{-1}]. \tag{A1.2}$$

where σ is the Stefan–Boltzmann constant [as to radiation entropy, the reader is referred to Planck (1959)]. The radiation profile of the deer is assumed to be 0.85, the infrared absorptivity of the deer is assumed to be 1, and the total surface area of the deer of 50 [kg] is $S_t = 1.703$ [m²]. The infrared radiation entropy absorbed by the deer becomes

$$S_{rad}(\text{in}) = 0.85 \times S_t \times \frac{1}{2}(S_\downarrow + S_\uparrow)$$

$$= 1.660 \ [\text{Js}^{-1}\text{K}^{-1}]. \tag{A1.3}$$

From the surface temperature of the deer (T_s), the infrared entropy emitted by the deer is estimated as

$$S_{rad}(\text{out}) = 0.85 \times S_t \times \frac{4}{3}\sigma(T_s)^3$$

$$= 1.822 \ [\text{Js}^{-1}\text{K}^{-1}]. \tag{A1.4}$$

The convective entropy loss is obtained from the observed value of convective heat loss (E_{cnv}),

$$S_{cnv} = E_{cnv}/T_s = 0.209 \ [\text{Js}^{-1}\text{K}^{-1}]. \tag{A1.5}$$

Similarly, the evaporative entropy loss is given by

$$S_{evp} = E_{evp}/T_b = 0.066 \ [\text{J s}^{-1}\text{K}^{-1}], \tag{A1.6}$$

from the observed values of the evaporative heat loss (E_{evp}) and the body temperature of the deer (T_b). The conductive entropy loss to ingested food is

$$S_{cnd} = E_{cnd}/T_b = 0.022 \ [\text{Js}^{-1}\text{K}^{-1}], \tag{A1.7}$$

where E_{cnd} is the observed conductive heat loss to ingested food.

Net entropy flow into the deer from its surroundings is

$$\Delta_e S = S_{rad}(in) - [S_{rad}(out) + S_{cnv} + S_{evp} + S_{cnd}]$$
$$= -0.46[Js^{-1} K^{-1}]. \qquad (A1.8)$$

If we assume that the deer is in a steady state in entropy, as should be in the long term, entropy production ($\Delta_i S$) should occur within the deer and result in no change of entropy content of the deer,

$$\Delta_i S = -\Delta_e S = +0.46 \; [Js^{-1}K^{-1}]. \qquad (A1.9)$$

Thus, the values of entropy flow and entropy production in the deer have been quantitatively estimated. For more detailed arguments, the reader is referred to the original paper (Aoki 1987c).

A2 LAKE ECOSYSTEM (AOKI 1987d, 1989a, 1998)

Energy flows between a lake and its surroundings are due to direct solar radiation (E_{dr}), scattered (diffuse) solar radiation (E_{sc}), reflected solar radiation (E_{rf}), downward infrared radiation ($E_{i\downarrow}$), upward infrared radiation ($E_{i\uparrow}$), evaporation of water (E_{evp}), and heat conduction–convection (E_{con}). The energy balance equation is written as the change in energy content of a lake (ΔQ) equal to the energy inflow minus energy outflow,

$$\Delta Q = (E_{dr} + E_{sc} - E_{rf}) + (E_{i\downarrow} - E_{i\uparrow}) + (-E_{evp} \pm E_{con}). \qquad (A2.1)$$

Each term of Equation (A2.1) has been measured for various lakes.

Associated with these energy flows, there are corresponding entropy flows, which are expressed in the following.

A2.1 Direct Solar Radiation

The daily solar energy incident on a unit area of a horizontal surface just outside the earth's atmosphere is given by (e.g., Gates 1980; Sellers 1965)

$$Q_e = \frac{86400}{\pi} e_1 \left(\frac{\overline{d}}{d}\right)^2 (H - \tan H) \sin\phi \sin\delta \; [J \, m^{-2} day^{-1}] \qquad (A2.2)$$

where e^1 is the solar constant (e.g., Liou 1980),

$$e_1 = 1353 \; [Jm^{-2}s^{-1}], \qquad (A2.3)$$

$\overline{d}$ and d are, respectively, the mean and instantaneous distances between the earth and the sun; H is the half-day length expressed in radian; ϕ is the latitude of the observation point; and δ is the solar declination.

Likewise, the daily solar entropy incident on a unit area of a horizontal surface just outside the earth's atmosphere is given by

$$Q_s = \frac{86400}{\pi} s_1 \left(\frac{\overline{d}}{d}\right)^2 (H - \tan H) \sin\phi \sin\delta \; [J \, m^{-2} day^{-1}K^{-1}], \qquad (A2.4)$$

where s_1 is "the solar constant of second kind" (Aoki 1983), which represents the solar entropy flux incident on a plane perpendicular to incident solar radiation at the top of the atmosphere. The value for s_1 is (Aoki 1983).

$$s_1 = 0.3132[\text{Jm}^{-2}\text{s}^{-1}\text{K}^{-1}]. \tag{A2.5}$$

From Equations (A2.2) and (A2.4),

$$Q_s = Q_e \times \frac{s_1}{e_1}. \tag{A2.6}$$

The energy of direct solar radiation incident on a horizontal plane of the earth's surface is written as

$$E_{\text{dr}} = \rho Q_e, \tag{A2.7}$$

where ρ is the transmissivity of solar radiation in the atmosphere. Likewise, the entropy of direct solar radiation incident on a horizontal plane of the earth's surface is

$$S_{\text{dr}} = \rho Q_s. \tag{A2.8}$$

From Equations (A2.6), (A2.7), and (A2.8), we obtain

$$S_{\text{dr}} = E_{\text{dr}} \times \frac{s_1}{e_1}. \tag{A2.9}$$

Thus, it is possible to calculate S_{dr} from the observed value for E_{dr} by use of the known values of e_1 and s_1. Equation (A2.9) can be applied for any time interval, i.e., not only for the daily but also the monthly or the annual direct solar entropy incident on the earth's surface.

A2.2   Diffuse Solar Radiation

Suppose that a beam of diffuse solar radiation is incident on an element of area $d\sigma$ on a horizontal plane of the earth's surface through a solid angle $d\Omega = \sin\theta\, d\theta\, d\phi$ in a direction forming an angle θ with the normal to the area $d\sigma$ (ϕ is an azimuthal angle of the incident radiation beam). Let the specific intensity (Planck 1959) of the diffuse solar energy radiation be denoted as K_1. The radiation energy incident on $d\sigma$ in time dt through $d\Omega$ is, by definition, $K_1 dt d\sigma \cos\theta\, d\Omega$. Now, we assume, for simplicity, that diffuse solar radiation comes from all directions in the sky hemisphere with equal intensity and hence K_1 is independent of (θ, ϕ). Integrating $K_1\, dt\, d\sigma \cos\theta\, d\Omega$ by $d\Omega$, we obtain the total radiation energy incident on $d\sigma$ in time dt,

$$K_1 dt d\sigma \int_0^{2\pi} d\phi \int_0^{\pi/2} \cos\theta \sin\theta d\theta = \pi K_1 dt d\sigma. \tag{A2.10}$$

Thus, the energy of diffuse (scattered) solar radiation incident per unit area of the earth's surface is expressed as

$$E_{\text{sc}} = \pi K_1. \tag{A2.11}$$

By use of Equation (A2.11), we can calculate K_1 from the observed value for E_{sc}. Even if K_1 is dependent on (θ, ϕ) as will really be the case, Equation (A2.11) will remain valid by interpreting K_1 as representing a kind of average value.

Next, let the specific intensity of solar energy radiation in extraterrestrial space be K_0. Let us assume that extraterrestrial solar radiation is blackbody radiation of the temperature $T_0 = 5760$ [K] (Aoki 1983). Then, by use of the Planck's formula (Planck 1959), K_0 is given by

$$K_0 = \int_0^\infty \frac{2h}{c^2} \frac{1}{e^{h\nu/kT_0} - 1} \nu^3 d\nu = \frac{1}{\pi} \sigma T_0^4 = 0.63 \times 10^6 \text{ [GJ m}^{-2} \text{ year}^{-1}], \quad (A2.12)$$

where σ is the Stefan–Boltzmann constant.

Since extraterrestrial solar radiation with K_0 is scattered by particles in the atmosphere and then becomes diffuse solar radiation with K_1, we may write K_1 as proportional to K_0, as in Aoki (1982),

$$K_1 = \varepsilon K_0 = \varepsilon \frac{1}{\pi} \sigma T_0^4. \quad (A2.13)$$

Thus, in this model, we can regard diffuse solar radiation as graybody radiation emitted by the body of the temperature T_0 and of the emissivity ε (Aoki 1982). Therefore, the specific intensity of diffuse solar entropy radiation is obtained as (Aoki 1982)

$$L_1 = \frac{1}{\pi} \frac{4}{3} \varepsilon \sigma T_0^3 X(\varepsilon), \quad (A2.14)$$

where $\varepsilon = (K_1/K_0)$ and

$$X(\varepsilon) = \frac{45}{4\pi^4} \frac{1}{\varepsilon} \int_0^\infty y^2 [(x + 1) \ln (x + 1) - x \ln] dy, \quad (A2.15)$$

$$x = \frac{\varepsilon}{e^y - 1} \quad (A2.16)$$

From Equations (A2.11), (A2.13), and (A2.14), we find that

$$L_1 = \frac{4}{3} \frac{K_1}{T_0} X(\varepsilon) = \frac{1}{\pi} \frac{4}{3} \frac{E_{sc}}{T_0} X(\varepsilon). \quad (A2.17)$$

From a similar discussion as in Equation (A2.11), the entropy of diffuse solar radiation incident per unit time on a unit area of the earth's surface is given by

$$S_{sc} = \pi L_1. \quad (A2.18)$$

From Equation (A2.17) and (A2.18), we obtain

$$S_{sc} = \frac{4}{3} \frac{E_{sc}}{T_0} X(\varepsilon). \quad (A2.19)$$

Since $\varepsilon = (K_1/K_0)$ is known from T_0 [Equation (A2.12)] and E_{sc} [Equation (A2.11)], we can thus find the value S_{sc} from the observed value for E_{sc} and from T_0.

A2.3 Diffusely Reflected Solar Radiation

Suppose that solar radiation is reflected by an element of area $d\sigma$ on a horizontal plane of the earth's surface and is emitted into a solid angle $d\Omega = \sin\theta\, d\theta d\phi$ in a direction forming an angle θ with the normal to the area $d\sigma$ (ϕ is an azimuthal angle of reflected radiation). Let the specific intensity (Planck 1959) of the reflected solar energy radiation be denoted as K_1. The radiation energy reflected by $d\sigma$ in time dt into $d\Omega$ is $K_1\, dt\, d\sigma \cos\theta\, d\Omega$, by definition. We assume that reflection is diffuse and that K_1 is independent of (θ,ϕ).

Integrating $K_1\, dt\, d\sigma \cos\theta\, d\Omega$ by $d\Omega$, we obtain the total radiation energy reflected diffusely by $d\sigma$ in time dt:

$$K_1 dt d\sigma \int_0^{2\pi} d\phi \int_0^{\pi/2} \cos\theta \sin\theta = \pi K_1 dt d\sigma. \tag{A2.20}$$

Thus, the energy of solar radiation reflected diffusely per unit time by a unit area of the earth's surface is given by

$$E_{rf} = \pi K_1. \tag{A2.21}$$

When K_1 is dependent on (θ,ϕ), the specific intensity K_1 in Equation (A2.21) should be interpreted as representing a kind of average value.

In a similar manner as in Section A2.2, let us express K_1 as proportional to K_0 (Aoki 1982):

$$K_1 = \varepsilon K_0 = \varepsilon \frac{1}{\pi}\sigma T_0^4, \tag{A2.22}$$

where K_0 is the specific intensity of extraterrestrial solar radiation and $T_0 = 5760$ [K] is the temperature of the sun. Thus, we can regard diffusely reflected solar radiation as graybody radiation emitted by the body of the temperature T_0 and of the emissivity ε (Aoki 1982). Therefore, the specific intensity of diffusely reflected entropy radiation is obtained as (Aoki 1982)

$$L_1 = \frac{1}{\pi}\frac{4}{3}\varepsilon\sigma T_0^3 X(\varepsilon), \tag{A2.23}$$

where $\varepsilon = (K_1/K_0)$ $X(\varepsilon)$ is given by Equations (A2.15) and (A2.16). From Equations (A2.21) to (A2.33), we have

$$L_1 = \frac{4}{3}\frac{K_1}{T_0}X(\varepsilon) = \frac{1}{\pi}\frac{4}{3}\frac{E_{rf}}{T_0}X(\varepsilon). \tag{A2.24}$$

Similar to Equation (A2.21), the entropy of solar radiation reflected diffusely per unit time by a unit area of the earth's surface is given by

$$S_{rf} = \pi L_1. \tag{A2.25}$$

From Equations (A2.24) and (A2.25), we obtain

$$S_{rf} = \frac{4}{3} \frac{E_{rf}}{T_0} X(\varepsilon).$$ (A2.26)

Since $\varepsilon = (K_1/K_0)$ is known from T_0 [Equation (A2.12)] and E_{rf} [Equation (A2.21)], we can thus find the value S_{rf} from the observed value for E_{rf} and from T_0.

A2.4 Infrared Radiation Emitted by Lake Surface

The lake surface can be regarded as an emitter of graybody radiation of a temperature T_w and of an emissivity ε_w (e.g., Sellers 1965). The entropy flux of graybody radiation emitted by a graybody of the temperature T_w and of the emissivity ε_w is written as [see Equation (A2.14) or (A2.23), and the entropy flux $= \int_0^{2\pi} d\phi \int_0^{\pi/2} L_1 \cos\theta d\theta = \pi L_1$]

$$S_{i\uparrow} = \frac{4}{3} \varepsilon_w \sigma T_w^3 X(\varepsilon_w).$$ (A2.27)

The infrared emissivity of lake surface is taken as $\varepsilon_w = 0.94$ according to Sellers (1965). Thus, the entropy flux of infrared radiation emitted by the lake surface can be calculated by use of the observed value of temperature of the lake surface T_w.

A2.5 Infrared Radiation Incident on Lake Surface

Let us assume that the atmosphere can be regarded as a graybody of an effective temperature T_a and of an emissivity ε_a, as in Aoki (1988b). Then the energy flux of infrared radiation emitted by the atmosphere is $E_{i\downarrow} = \varepsilon_a \sigma T_a^4$ (e.g., Aoki 1988b). Since the atmosphere absorbs 94% of infrared radiation emitted by the earth's surface (e.g., Battan 1979) and the absorptivity equals the emissivity (Kirchhoff's law), we can put the infrared emissivity $\varepsilon_a = 0.94$. The effective temperature T_a can be obtained from ε_a and the observed infrared radiation energy $E_{i\downarrow}$,

$$T_a = [E_{j\downarrow} / \varepsilon_a \sigma]^{1/4}.$$ (A2.28)

The entropy flux of infrared radiation emitted by the atmosphere and incident on the lake surface is given by, like Equation (A2.27),

$$S_{j\downarrow} = \frac{4}{3} \varepsilon_a \sigma a T_a^3 X(\varepsilon_a) = \frac{4}{3} \frac{E_{j\downarrow}}{T_a} X(\varepsilon_a),$$ (A2.29)

the value of which is obtained from the known values $E_{i\downarrow}$, T_a and ε_a.

A2.6 Evaporation of Water and Heat Conduction–Convection

The entropy flux from the lake surface to the atmosphere due to evaporation of water is

$$S_{evp} = E_{evp} / T_w,$$ (A2.30)

where E_{evp} is the latent heat flux from the lake surface and T_w is the temperature of the lake surface.

Likewise, the entropy flux between the lake surface and the atmosphere due to heat conduction–convection is given by

$$S_{con} = E_{con} / T_w, \tag{A2.31}$$

where E_{con} is the heat flux due to conduction–convection.

Thus, the values of S_{evp} and S_{con} are obtained from observed data E_{evp}, E_{con}, and T_w.

A2.7 Entropy Production

The net entropy inflow into the lake is obtained as

$$\Delta_e S = (S_{dr} + S_{sc} - S_{rf}) + (S_{i\downarrow} - S_{i\uparrow}) + (- S_{evp} \pm S_{con}). \tag{A2.32}$$

The change in entropy content of the lake is (Aoki 1989a, 1990a)

$$\Delta S = (\text{the change in heat storage in the lake}) / T_m, \tag{A2.33}$$

where T_m is the mean temperature of lake water. Thus the value of ΔS is estimated from observed data of T_m and the change in heat storage in the lake water.

The entropy balance equation is expressed as

$$\Delta S = \Delta_e S + \Delta_{iS}, \tag{A2.34}$$

where $\Delta_i S$ is the entropy production within the lake. The entropy production is the production rate of entropy by irreversibility of processes, and is nonnegative according to the Second Law of Thermodynamics [Equation (2)]. Because almost all processes occurring in the natural world are irreversible, the entropy production is a measure of the magnitude of activity of physical, chemical, and biological processes in the natural world, as stated in Section 8.2.

From Equation (A2.34), the entropy production can be obtained if values of the entropy flow ($\Delta_e S$) and the change in entropy content (ΔS) are estimated.

The methods of calculation in the Appendix are based on simplified models in order to give clear and pertinent pictures on the subject in a simple way. More complicated and detailed treatments may be possible, but are left for future study. This is according to well-established methodologies in physics.

REFERENCES

Aoki, I., Radiation entropies in diffuse reflection and scattering and application to solar radiation. *J. Phys. Soc. Jpn.,* 51:4003–4010, 1982.

Aoki, I., Entropy productions on the earth and other planets of the solar system. *J. Phys. Soc. Jpn.* 52: 1075–1078, 1983.

Aoki, I., Entropy budgets of deciduous plant leaves and a theorem of oscillating entropy production. *Bull. Math. Biol.,* 49:449–460, 1987a.

Aoki, I., Entropy budgets of soybean and bur oak leaves at night. *Physiol Planta.* 70:293–295, 1987b.

Aoki, I., Entropy balance of white-tailed of white-tailed deer during a winter night. *Bull. Math. Biol.,* 49:321–327, 1987c.

Aoki, I., Entropy balance in Lake Biwa. *Ecol. Modell.,* 37:235–248, 1987d.

Aoki, I., Eco-physiology of a lizard (*Sceloporus occidentalis*) from an entropy viewpoint. *Physiol. Ecol. Jpn.,* 25:27–38, 1988a.

Aoki, I., Entropy flows and entropy productions in the earth's surface and in the earth's atmosphere. *J. Phys. Soc. Jpn.,* 57:3262–3269, 1988b.

Aoki, I., Holological study of lakes from an entropy viewpoint-Lake Mendota. *Ecol. Modell.*, 45:81–93, 1989a.

Aoki, I., Entropy flow budget of conifer branches. *Bot. Mag., Tokyo*, 102:133–141, 1989b.

Aoki, I. Entropy flow and entropy production in the human body in basal conditions. *J. Theor. Biol.*, 141:11–21, 1989c.

Aoki, I., Monthly variations of entropy production in Lake Biwa. *Ecol. Modell.*, 51:227–232, 1990a.

Aoki, I., Effects of exercise and chills on entropy production in human body. *J. Theor. Biol.*, 145:421–428, 1990b.

Aoki, I., Entropy principle for human development, growth and aging. *J. Theor. Biol.*, 150:215–223, 1991.

Aoki, I., Entropy physiology of swine—a macroscopic viewpoint. *J. Theor. Biol.*, 157:363–371, 1992.

Aoki, I., Inclusive Kullback index—a macroscopic measure in ecological systems. *Ecol. Modell.*, 66:289–299, 1993.

Aoki, I., Entropy production in human life span: a thermodynamical measure for aging. *Age,* 17:29–31, 1994.

Aoki, I., Entropy production in living systems: from organisms to ecosystems. *Thermochim. Acta,* 250:359–370, 1995.

Aoki, I., Entropy production and exergy destruction in lakes. Presentation at Ecological Summit 96, Copenhagen (Abstract Book 49.1), 1996.

Aoki, I., Entropy and exergy in the development of living systems: a case study of lake-ecosystems. *J. Phys. Soc. Jpn.,* 67:2132–2139, 1998.

Barica, J., Ecosystem stability and sustainability: a lesson from algae. *Verh. Int. Verein. Limnol.,* 25:307–311, 1993.

Battan, L. J., *Fundamentals of Meteorology.* Prentice-Hall, Englewood Cliffs, NJ, 1979.

Bejan, A., *Entropy Generation through Heat and Fluid Flow.* Wiley, New York; (1996) *Entropy Generation Minimization.* CRC Press, Boca Raton, FL, 1982.

Birge, E. A., The heat budgets of American and European lakes. *Trans. Wis. Acad. Sci. Arts Lett.* 18:166–213, 1915.

Brock, T. D., *A Eutrophic Lake; Lake Mendota, Wisconsin.* Springer, New York, 1985.

Dutton, J. A. and R. A. Bryson, Heat flux in Lake Mendota. *Limnol. Oceanogr.,* 7:80–97, 1962.

Gates, D. M., *Biophysical Ecology.* Springer, New York, 1980.

Hutchinson, G. E., *A Treatise on Limnology.* Vol. I. Wiley, New York, 1957.

Hutchinson, G. E., The lacustrine microcosm reconsidered. *Am. Sci.,* 52:334–341, 1964.

Ito, K. and Okamoto, I., Time variation of water temperature in Lake Biwa-ko (VIII). *Jpn. J. Limnol.,* 35:127–135 (in Japanese with English summary), 1974.

Kotoda, K., A method for estimating evaporation based on climatological data. *Bull. Environ. Res. Cent. Univ. Tsukuba,* 1:53–66 (in Japanese), 1977.

Landsberg, P. T., The fourth law of thermodynamics. *Nature,* 238:229–231, 1972.

Lindeman, R. L., The trophic-dynamic aspect of ecology. *Ecology,* 23:399–418, 1942.

Liou, K-N., *An Introduction to Atmospheric Radiation.* Academic Press, New York, 1980.

Maansson, B. AA. and J. M. McGlade, Ecology, thermodynamics and H. T. Odum's conjectures. *Oecologia* 93:582–596, 1993.

Mitamura, O. and Y. Saijo, Studies on the seasonal changes of dissolved organic carbon, nitrogen, phosphorus and urea concentrations in Lake Biwa. *Arch. Hydrobiol.,* 91:1–14, 1981.

Moran, M. J., *Availability Analysis: A Guide to Efficient Energy Use.* Prentice-Hall, Englewood Cliffs, NJ, 1982.

National Institute for Research Advancement, *Data Book of World Lakes.* National Institute for Research Advancement, Tokyo, 1984.

Nicolis, G. and Prigogine, I., *Self-Organization in Nonequilibrium Systems.* Wiley, New York, 1977.

Planck, M., *The Theory of Heat Radiation.* Dover, New York, 1988; American Institute of Physics, New York, 1959.

Sakamoto, M., Trophic relation and metabolism in ecosystem. In: Mori, S. and Yamamoto, G. (eds.) *Productivity of Communities in Japanese Inland Waters (JIBP Synthesis,* Vol. 10):405–410. University of Tokyo Press, Tokyo, 1975.

Schrödinger, E., *What is Life?* Cambridge University Press, London, 1944.

Sellers, W. D., *Physical Climatology.* The University of Chicago Press, Chicago, 1965.

Stewart, K. M., Detailed time variations in mean temperature and heat content of some Madison lakes. *Limnol. Oceanogr.,* 18:218–226, 1973.

CHAPTER 9

In the introduction to this chapter, exergy is used as the core concept which explains evolution. Species evolution has caused increasingly better use of resources (exergy and material). Species as well as ecosystems as a whole therefore tend to progress toward more complex dissipative structure producing more complex behavior. Interacting species in a common ecosystem evolve toward specialisation, speciation, synergy, complexification, diversity, and more efficient use of exergy and material resources. This leads the author to present his orientation theory: self-organising systems develop a set of emergent objectives (basic orientors) that are identical across all systems in normal environments. They are defined to have the following six properties: normal state, sparse resources, variety, fluctuation, change, and the influence from other systems. The orientors correspond to six properties of the system: existence, effectiveness, freedom of action, security, adaptability, and coexistence. The six properties of the system have a one-to-one relationship with the properties of the environment.

The relationship between the orientors and exergy is explained in the chapter as follows: better fitness for more participants in a system requires more dissipative structure that again requires more exergy throughput as well as exergy accumulation. Since the exergy flow of ecosystems is limited, increasingly better utilisation is to be expected. Ecosystems as a whole therefore move in the direction of using all available exergy gradients (in accordance with the Ecological Law of Thermodynamics, presented in Chapter 13).

Methods to quantify the orientors are also presented. As quantified, environmental indicators are used in the following nine concepts: resource availability, environmental variety, environmental reliability, environmental stability, existence, effectiveness, freedom of action, security, and adaptability. These indicators have been used in simulations. Without presenting the details of a simulation with a minimal ecosystem, the results are summarised in general conclusions that are in agreement with everyday observations and general systems knowledge: it is beneficial to be a generalist and cautious and to train and learn.

9 Exergy and the Emergence of Multidimensional System Orientation

Hartmut Bossel

CONTENTS

9.1 Introduction: Evolution and the "Intelligent" Use of Exergy 194
9.2 Emergent Goal Functions: Basic Orientors as System Responses
to Basic Environmental Properties 196
    9.2.1 Properties of the Environment 197
    9.2.2 Basic Orientors 197
    9.2.3 Properties of Orientors 198
    9.2.4 Orientor Satisfaction and Exergy 198
    9.2.5 Orientors as Implicit Goal Functions 198
9.3 Quantification of Environmental Indicators and System
Orientors for a Minimal Ecosystem 199
    9.3.1 Environmental Indicators 199
        9.3.1.1 Resource Availability 199
        9.3.1.2 Environmental Variety..200
        9.3.1.3 Environmental Reliability 200
        9.3.1.4 Environmental Stability 200
        9.3.1.5 Measures of Animat Orientor Satisfaction 201
        9.3.1.6 Existence 201
        9.3.1.7 Effectiveness...202
        9.3.1.8 Freedom of Action 202
        9.3.1.9 Security..203
        9.3.1.10 Adaptability..203
9.4 Simulations with a Minimal Ecosystem: Emergence of Intelligence
and Value Orientation 204
    9.4.1 Sample Results..204
    9.4.2 Emergence of Basic Value Orientations 206
    9.4.3 Emergence of Individual Differences in Value Orientation 206
9.5 Conclusions 207
References 208

9.1 INTRODUCTION: EVOLUTION AND THE "INTELLIGENT" USE OF EXERGY

The global ecosystem is made up of an ensemble of interacting local and regional ecosystems, each composed of biotic and abiotic subsystems. The evolution of these systems is constrained by physical and system laws and by the basic properties of their environment, including the constraints of exergy, material, and information flows. Sustainability (persistence) of a system in its environment therefore requires respecting these constraints. The conclusion is general: the very fact of its persistence demonstrates that a system has successfully adapted to its operating conditions. Evolution has forced it to respect physical and system laws and the basic properties of its environment.

The adaptation of a system to its environment is reflected in its structure, including its nonmaterial, cognitive structure. This system structure determines its behavior, and hence the adaptive response to its particular environment. System structures of material systems are dissipative: they require exergy and material flows for their construction, maintenance, renewal, and reproduction.

The dissipative structures of the global ecosystem are constructed and maintained by a finite rate of exergy input (solar energy) and a finite stock of materials. The global ecosystem is therefore forced to recycle all of its essential material resources. The development of local ecosystems is constrained by the local rate of exergy flux (solar radiation input) and by the local rate of material recycling (weathering rate, absorption rate, decomposition rate, etc.) that it produces.

Evolution favors those species or (biotic) subsystems of the ecosystem that have "learned" to use available resources more efficiently and effectively than their competitors. This learning is embedded in their genetic code, and is manifested in the dissipative structures they construct. Both will increase in complexity as a species evolves, increasing the specific exergy content of both genetic material and dissipative structure (Jørgensen et al., 1993). At the ecosystem level, species evolution will cause increasingly better use of (exergy and material) resources (Ecological Law of Thermodynamics (ELT), Jørgensen and Nielsen 1994). Species as well as ecosystems as a whole therefore tend to progress toward more complex dissipative structure producing more complex behavior.

Interacting species in a common ecosystem coevolve in the direction of increasing fitness of each individual species. Evolution of ecosystems therefore proceeds in the direction (arrow of time) of specialization, speciation, synergy, complexification, diverisity, and more efficient use of exergy and material resources. This development becomes manifest in the corresponding emergent properties (Müller 1996): exergy degradation, recycling, minimization of output, efficiency of internal flows, homeostasis and adaptation, diversity, heterogeneousness, hierarchy and selectivity, organization, minimization of maintenance and costs, and storage of available resources. These properties are not limited to ecosystems; they are a general feature of "living systems" (Miller 1978), including human organizations.

In particular, ecosystems will therefore build up in the course of their development as much dissipative structure as can be supported by the available exergy gradient (ELT). Available opportunities will eventually be found out by the processes of evolution, and will then be utilized. This fact can be used to predict the direction in which ecosystems will develop under given conditions (Jørgensen 1994).

The ability to respond successfully to environmental challenges can be interpreted as "intelligent behavior," defined by van Heerden (1968) as: "Intelligent behavior is: to be repeatedly successful in satisfying one's ... needs in diverse, observably different, situations on the basis of past experience."

Ashby (1962) more specifically considered the development of intelligence in response to the challenge of keeping the system state within certain limits in a particular environment: "... there is no difficulty, in principle, in developing synthetic organisms as complex, and as intelligent as we please. But we must notice two fundamental qualifications: first, their intelligence will be an adaptation to, and a specialization toward, their particular environment, with no implication of validity for any other environment such as ours and secondly, their intelligence will be directed toward keeping their own essential variables within limits. They will be fundamentally selfish."

Ashby expressly links this environmentally determined intelligence to the normative (and fundamentally selfish) task of "keeping their own essential variables within limits." In our terms, this means learning a behavioral (value) orientation that is not merely focused on the acquisition of exergy itself, but on securing this acquisition even under adverse circumstances.

The same point has been made in more detail by the orientation theory. Orientation theory (Bossel 1977, 1992, 1994, 1998) is based on the concept that, while adapting to their specific environment, self-organizing systems develop a set of emergent objectives (basic orientors) that are identical across all systems in "normal" environments. Corresponding to the six basic properties of normal environments: normal state, sparse resources, variety, fluctuation, change, and other systems, systems learn to simultaneously pay attention to corresponding orientors: existence, effectiveness, freedom of action, security, adaptability, and coexistence. These orientors emerge as reflections of the environment in the function of the system. Systems that do not pay balanced attention to these orientors exhibit pathological behavior and eventually fail for lack of fitness. The range of application of these concepts covers self-organization as well as the design of systems. It ranges from biological systems to social systems, psychology, and the design of technical or political systems. In addition to policy analysis applications (e.g., Hornung 1988), the concepts have previously been used to simulate "intelligent" behavior of a policy maker in Forrester's world model (Bossel and Strobel 1978).

It is possible to study the evolution of such "normative intelligence" by simulation of the evolution of the cognitive structure of artificial animals (animats) in an artificial environment. Wilson (1985) developed his "animat" specifically to study the development of intelligence at an elementary level. Using algorithms proposed by Holland (1975, 1976, 1980, 1984, 1992), his animat was able to learn a set of behavioral rules that enabled it to avoid obstacles and find "food" in a complex two-dimensional environment. Wilson's objective was to contribute to the understanding of natural and artificial intelligence. In Wilson's study, the "needs" mentioned by van Heerden were essentially limited to somatic needs (collection of "food," avoidance of "injury" in collisions).

We have used Wilson's animat (with some modifications) to study the emergence of orientor attention in an evolutionary system (Krebs and Bossel 1997). In our study, we introduced more specific definitions of "needs" (i.e., basic value orientation) to analyze results. However, there was no explicit optimization of needs satisfaction, i.e., no explicit objective function. As in Wilson's study, the animat was trained mainly by exergy payoff (i.e., food). Behavioral rules that directly or indirectly led to success eventually became more dominant in the knowledge base.

The results were then analyzed to determine the emergence of multidimensional objectives in the course of the adaptation process. Wilson's study demonstrated that the animat would learn to attend to its survival needs with "respectable," but not "perfect," performance. This indicates that optimization for efficiency is not the only objective emerging in the animat's learning phase, but that considerable emphasis (and exergy expenditure) is given to other behavioral dimensions required for survival. Results of these simulations will be analyzed in an orientation theory framework. In order to do this, the properties of the animat's environment must be expressed by corresponding indicators. To express the current state of orientor satisfaction, quantifiable expressions for the basic orientors must be defined.

In this contribution, the role of exergy in shaping system development will first be discussed, stressing the multidimensionality of the associated behavioral problem for the system. This requires multidimensional system orientation and response. The concepts of orientation theory and their connection to the system's exergy balance are then outlined (Section 9.2). To demonstrate the applicability of the concepts, quantifiable indicator and orientor sets are then defined for an evolutionary system in a simple environment (Section 9.3). These are used to evaluate results of computer simulations and to assess the relative performance of individuals with different normative orientations, i.e., basic orientor emphasis (Section 9.4). Conclusions are then drawn from the analysis and the experimental results, in particular concerning the relevance and significance for exergy analysis and ecosystem research (Section 9.5).

9.2   EMERGENT GOAL FUNCTIONS: BASIC ORIENTORS AS SYSTEM RESPONSE TO BASIC ENVIRONMENTAL PROPERTIES

Basic concepts can be introduced by visualizing a simple animal with limited vision in a simple environment. The animal requires exergy for self-organization, motion, harvesting food, and maintenance. The environment provides food in certain locations, usually associated with obstacles that must be avoided since they have an exergy cost. In Section 9.4, results of corresponding computer experiments will be discussed.

In a stable environment where sufficient (regenerating) food is distributed in a completely regular pattern, evolutionary adaptation would eventually lead to optimization of an animal's movements in a regular grazing pattern, with a single objective: optimum exergy uptake, and use. The regular grazing pattern reflects the complete certainty of the next step, which the animal learns by accumulating and internalizing experience in a "cognitive structure" aiding its limited vision.

In more complex and diverse environments, the animal, because of its limited vision, may not know for several steps which situation it will encounter next. It will therefore have to develop decision rules that have greater generality and are applicable to (and will be reinforced by) different motion sequences with different outcomes. In addition to the requirement of harvesting and using exergy resources effectively and efficiently, another objective is now implicitly added: to secure food under the constraint of incomplete information, i.e., a security objective. Note that this is an emergent property that is not explicit in the reward system (which still rewards only food uptake). Failure to heed this implicit security objective will reduce food uptake and may endanger survival. On the other hand, the pressure to "play it safe" will occasionally mean giving up relatively certain rewards. In other words, efficiency is traded for more security, and both are now prominent normative orientations (goals, values, interests) incorporated in the cognitive structure.

Orientation theory (Bossel 1977, 1994, 1998) deals in a more general way with the emergence of behavioral objectives (orientors) in self-organizing systems in general environments. The proposition is that if a system is to survive in a "normal" environment characterized by a given normal environmental state, sparse resources, variety, unreliability, change, and other systems, it must be able to physically exist in (be compatible with) this environment, effectively harvest necessary resources, freely respond to environmental variety, protect itself from unpredictable threats, adapt to changes in the environment, and interact productively with other systems. These essential orientations emerge in the course of the system's evolution in its environment. The terms *existence, effectiveness, freedom of action, security, adaptability,* and *coexistence* are used to characterize the six basic orientors of system development.

The basic orientor proposition has three important implications:

1. If a system evolves in a normal environment, then that environment forces it to implicitly or explicitly ensure minimum and balanced satisfaction of each of the basic orientors.
2. If a system has successfully evolved in a normal environment, its behavior will exhibit balanced satisfaction of each of the basic orientors.
3. If a system is to be designed for a normal environment, proper and balanced attention must be paid to the satisfaction of each of the basic orientors.

The third implication has particular relevance for the creation of programs, institutions, and organizations in the sociopolitical sphere, among other things. Note that for a specific system in a specific environment, each orientor will have a specific meaning; i.e., "security" of a nation is a multifaceted objective set with very different content from the "security" of an individual. However, the systems-theoretical background for satisfaction of the "security" orientor is the same in both cases.

9.2.1 PROPERTIES OF THE ENVIRONMENT

In addition to the physical constraints of exergy and material flows, ecosystem and species development is determined by the general properties of the environment (Bossel, 1977, 1994, pp. 233 ff.).

1. *Normal environmental state.* The actual environmental state can vary around this state in a certain range.
2. *Scarce resources.* Resources (exergy, matter, information) required for a system's survival are not immediately available when and where needed.
3. *Variety.* Many qualitatively different processes and patterns occur in the environment constantly or intermittently.
4. *Reliability.* The normal environmental state fluctuates in random ways, and the fluctuations may occasionally take it far from the normal state.
5. *Change.* In the course of time, the normal environmental state may gradually or abruptly change to a permanently different normal environmental state.
6. *Other systems.* The behavior of other systems changes the environment of a given system.

9.2.2 BASIC ORIENTORS

If evolution enforces fitness of (surviving) species, then these systems must reflect the properties of their environment in their structure. More generally, the basic properties of the environment require corresponding basic system features. Since the basic environmental properties are independent of each other, a similar set of independent system features must exist, and it must find expression in the concrete features of the system structure.

The independence of the six general properties of the environment requires separate "attention" to each of them by the system. The set of criteria demanding this attention is termed *basic orientors*, since they provide (consciously or unconsciously) behavioral orientation to the system. A viable system must ensure minimum and balanced attention to each of the basic orientors (Bossel 1977, 1994, pp. 233 ff.). There is a one-to-one relationship between the properties of the environment and the basic orientors of systems.

1. *Existence.* Attention to existential conditions is necessary to insure the basic compatibility and immediate survival of the system in the normal environmental state.
2. *Effectiveness.* In its efforts to secure scarce resources (exergy, matter, information) from, and to exert influence on its environment, the system should on balance be effective.
3. *Freedom of action.* The system must have the ability to cope in various ways with the challenges posed by environmental variety.
4. *Security.* The system must have the ability to protect itself from the detrimental effects of variable, fluctuating, unpredictable, and unreliable environmental conditions.
5. *Adaptability.* The system should be able to change its parameters and/or structure in order to generate more appropriate responses to challenges posed by changing environmental conditions.
6. *Coexistence.* The system must modify its behavior to account for behavior and interests (orientors) of other systems.

Obviously, the system equipped to secure better overall orientor satisfaction will have better fitness, and will therefore have a better chance for long-term survival and sustainability. In persistent systems or species, these orientors will be found as emergent goal functions (or systems interests).

9.2.3 Properties of Orientors

While developed here with an eye on evolutionary processes, the orientor concept applies equally well to other systems (human organizations, technical systems, personal development, etc.). The same set of environmental properties applies, and the same set of orientors is required in response.

Each of the orientors stands for a unique requirement. Attention (conscious or unconscious) must therefore be paid to all of them, and the compensation of deficits of one orientor by over-fulfillment of other orientors is not possible. Fitness forces a multicriteria response, and comprehensive (conscious or unconscious) assessments of system behavior and development must also be multicriteria assessments.

In the assessment and orientation of system behavior, we deal with a two-phase assessment process where each phase is different from the other: Phase 1: First, a certain minimum satisfaction must be guaranteed separately for each of the basic orientors. A deficit in even one of the orientors threatens long-term survival. The system will have to focus its attention on this deficit. Phase 2: Only if the required minimum satisfaction of *all* basic orientors is guaranteed is it permissible to try to raise system satisfaction by further improving satisfaction of *individual* orientors.

Characteristic differences in the behavior of otherwise very similar systems (humans, political, or cultural groups) can often be explained by differences in the relative importance attached to different orientors (i.e., emphasis on "freedom," or "security," or "effectiveness," or "adaptability") in Phase 2 (i.e., when minimum requirements for all basic orientors have been satisfied in Phase 1).

9.2.4 Orientor Satisfaction and Exergy

Better orientor satisfaction (better fitness) for more participants in a system requires more dissipative structure, which requires more exergy throughput as well as exergy accumulation. Since the exergy flow of ecosystems is limited (capture of solar radiation by photoproduction), increasingly better utilization is to be expected. This saturates at maximum exergy flow utilization for the ecosystem as a whole. Ecosystems as a whole therefore move in the direction of using all available exergy gradients (see ELT). For organisms in the ecosystem, this implies development tendencies toward specialization (using previously unused gradients), more complex structure (greater use efficiency), larger individuals (less maintenance exergy required per biomass unit), mutualism, etc. For species development this translates into a principle of maximum exergy use efficiency. On the basis of these principles, prediction of development trends in ecosystems is possible (Jørgensen 1994).

9.2.5 Orientors as Implicit Goal Functions

The selection for better fitness in evolutionary processes favors systems (organisms) with better coping ability. Aspects of the behavioral spectrum of a system that improve coping ability (basic orientors) can be understood as implicit goal functions: existence, security, effectiveness, freedom, adaptability, coexistence. (These orientors imply other goals such as efficient resource use, maximum exploitation of energy gradients, etc.)

The existence of these implicit goal functions does not imply teleologic or teleonomic development toward a given goal (where the final product is specified, May, 1974). These goal functions do not determine the exact future states of the system at all; they only pose constraints on choices (or evolutionary selection). The process and its rules are known; the product is unknown. The spectrum of (qualitatively different) possible future development paths and sustainable states remains enormous. The shape of the future, and of the systems that shape it, cannot be predicted this way. All one can say with certainty, however, is that (1) all possible futures must be continuous developments from the past, and (2) paths with better orientor satisfaction are more likely to succeed in the long run (if options to change paths have not been foreclosed).

9.3 QUANTIFICATION OF ENVIRONMENTAL INDICATORS AND SYSTEM ORIENTORS FOR A MINIMAL ECOSYSTEM

Orientation theory is not just a conceptual framework for understanding system evolution and behavior under the exergy availability constraint. It also allows quantitative and comparative analysis of system performance under different environmental conditions.

Orientation theory concepts will now be applied to analyze behavior and performance of an artificial animal (animat) in different environments (Krebs and Bossel 1997). The animat experiment contains all components necessary for a study in the basic orientor framework. It is assumed that the system is physically able to exist in this environment. The orientor "existence" is therefore satisfied and will not be considered further. Also, we deal with a single animat only and do not consider interactions with other individuals, i.e., the "coexistence" orientor plays no role. This means that we only have to deal with the four orientors "effectiveness," "freedom of action," "security," and "adaptability," corresponding to the four environmental properties "resource availability," "variety," "reliability," and "environmental change." For these properties as well as for the orientors, quantiative indicators have to be defined that reflect the particular problem setting.

In the framework of systems theory, the animat is a self-organizing system depending on an exergy supply (food) from its environment. The exergy balance is therefore of particular significance, and it will be essentially captured in the effectiveness orientor. A positive exergy balance is a necessary condition for survival. Attention to the other orientors is mandatory to ensure a positive exergy balance even under adverse environmental conditions.

For the analysis, environmental indicators and measures of orientor satisfaction will be defined for a simple artificial animal (modification of Wilson's 1985 animat). These measures have been used to assess behavior and performance in computer simulation experiments; corresponding results are discussed in Section 9.4. The same approach can be applied to much more complex environments and systems (e.g., Bossel and Strobel 1978, Hornung 1988, Bossel 1994, 1998).

Wilson's animat can move in one-step increments in eight directions of a two-dimensional environment containing food F and trees T. F-encounters are rewarded by food (i.e., exergy required for survival), T-encounters mean collision and exergy loss. The animat can only "see" one step ahead. This limited image of its environment is reflected in its "sense vector." Initially, it moves at random, but its cognitive structure gradually develops on the basis of past experience (Holland's 1975, 1992 genetic algorithm is used for this). For a given sense vector, it learns (or develops) a certain behavioral rule (i.e., movement in a particular direction). Exergy availability in the environment, and the exergy balance of the (simulated) organism turn out to be the crucial elements defining the multicriterial orientor space.

9.3.1 ENVIRONMENTAL INDICATORS

9.3.1.1 Resource Availability

For its exergy supply the animat depends on the consumption of food F in its environment. The F objects are arbitrarily distributed in the environment. Resource availability depends (1) on the total amount available, and (2) on the distribution pattern. The total distance traveled to harvest all resources (with correspondent energy consumption for motion) may be long if they are uniformly distributed, or quite short, if food is concentrated in a small area of the environment. The indicator e for resource availability is therefore defined by relating the total amount of resources m_{total} to the total harvest distance $d_{harvest}$.

$$e = m_{total}/d_{harvest}$$

"Total harvest distance" in a given animat environment is here computed simply by starting at one F, moving from there to the nearest neighboring F until all Fs have been visited once, and summing all the steps taken. The animat will have to respond to scarce resources by developing effective search grazing strategies (i.e., minimizing harvest distance; effectiveness orientor).

9.3.1.2   Environmental Variety

The variety of the environment is defined by the number of distinct patterns it presents to the perception of the animat, i.e., the number of potential sense vectors. (Note that variety thus defined is a property of the environment relative to the system.) Variety is calculated by summing up the number of distinct sense vectors n_{sense} that may be "seen" by the animat from any position in the environment. The indicator f for environmental variety is defined by relating this to the number of occupied positions in the environment $n_{position}$:

$$f = n_{sense} / n_{position}$$

Given the same number of nonempty spaces (occupied positions), an environment with a greater number of different patterns (distinct sense vectors) will have greater environmental variety f. To cope with environmental variety, the animat must evolve suitable condition–action pairs for all distinct sense vectors, and must learn to generalize appropriately. Greater environmental variety generally requires more behavioral rules, or more general rules, for dealing with a greater number of different sense vectors (i.e., freedom of action orientor, cf. Ashby's 1958 Law of Requisite Variety).

9.3.1.3   Environmental Reliability

Certain environmental factors may be subject to random fluctuations and therefore be unpredictable. Random changes make the environment more or less reliable for the system, and it will have to develop appropriate responses to it. A reliability indicator has to be defined that describes this particular property of the environment.

In order to introduce the aspect of reliability into the animat model, a third kind of object is added. With a given probability $p_{uncertain}$, this object looks occasionally like food (F object), but it is actually a tree (T object) with an associated collision penalty. If $p_{uncertain}$ equals 0, all objects are what they appear to be. If $p_{uncertain}$ is greater than zero, the animat cannot completely trust its senses; the environment is more of less unreliable. The reliability indicators s is therefore defined by

$$s = 1 - p_{uncertain}$$

The animat will have to respond to an unreliable environment by developing behavioral rules representing more cautious strategies (i.e., security orientor).

9.3.1.4   Environmental Stability

Most environments change more or less gradually, if only as a result of coevolution with the systems inhabiting them. Change can affect existential conditions, resource availability, environmental variety, environmental reliability, environmental stability, and the number and types of systems with which the organism has to interact. The different rates of change of the different environmental properties have different meanings for the system and must be accounted for separately. Therefore change cannot usually be expressed in one simple indicator.

In the case of the animat, environmental change means change in the set of potential sense vectors and their reliability. Some expression of a rate of change r_{envir} of the environment, or of change per unit time, must be used to define a corresponding indicator. Conversely, a time constant

T_{envir} characteristic of the "residence time" of the average sense vector in the environment could be introduced, where

$$r_{\text{envir}} = 1/T_{\text{envir}}$$

This change rate is meaningful only if defined by reference to a time period T_{system} characteristic of the system (e.g., the average time between food discoveries). An indicator a of environmental stability could then be defined (for $T_{\text{system}} < T_{\text{envir}}$) by

$$a = 1 - r_{\text{envir}} T_{\text{system}} = 1 - (T_{\text{system}}/T_{\text{envir}}) = 1 - (r_{\text{envir}}/r_{\text{system}})$$

If the characteristic time of the environment T_{envir} is much greater than that of the system T_{system}, the environmental stability (with respect to the system) is high. (Equivalently, if the rate of change of the environment r_{envir} is much smaller than the rate of adaptation of the system r_{system}, the environmental stability is high). The animat will have to respond to a changing environment by an appropriate adaptation rate (i.e., adaptability orientor).

9.3.1.5 Measures of Animat Orientor Satisfaction

The animat is trained in a constant environment defined by given values of the environmental indicators e, f, and s, and $a = 1$. From a randomly chosen initial position, it moves in single steps according to its behavioral rules until food is discovered. The process is then repeated from a different, randomly selected location. At each step, learning takes place in the bucket-brigade process (Holland 1984, 1992) via competition, taxation, and reward distribution. Genetic rule generation (Holland 1975, 1992) introduces new concepts by crossover and mutation at longer (adjustable) intervals, and rule creation is implemented in the rare cases where no applicable behavioral rule is available. After initial growth and rapid changes of the knowledge base of the animat, i.e., of the set of behavioral rules, this set becomes more or less stationary: an optimal solution has been reached.

Behavioral rule evolution is determined by several random factors, and this is reflected in the final behavioral rule sets for different simulation runs: although the overall performance of animat individuals may be similar, their behavioral rule sets will be very different. A systematic comparison of the behavioral rule sets seems impossible, and it is even more difficult to analyze which behavior may be caused by what (chain of) behavioral rules. We therefore compare system performance in its different aspects by using exergy-based measures of orientor satisfaction. These have to be defined using relevant parameters of animat performance. The orientor measures are defined such that values <1 indicate a threat to survivial, while values ≥ 1 show adequate orientor satisfaction.

9.3.1.6 Existence

The existence orientor is concerned with the fundamental compatibility of system and environment, i.e., the physical ability of a system to survive in the normal state of a given environment (e.g., it is satisfied for a herring in seawater at a temperature of 5°C with sufficient oxygen and plankton for food, but not at 100°C).

Satisfaction of the existence orientor can be assessed by a simple checklist of the necessary conditions c_n ($n = 1 \dots N$) for survival. The existence orientor satisfaction ξ is true ($\xi = 1$) if all necessary conditions are true ($c_n = 1$), and false ($\xi = 0$) if one or more of the conditions are not satisfied ($c_n = 0$),

$$\xi = c_1 \bullet c_2 \bullet \dots \bullet c_n \bullet \dots \bullet c_N$$

The animat is assumed to have been designed for its particular environment. Furthermore, the food supply in this environment is sufficient for survival. In this case therefore the existence satisfaction measure $\xi = 1$.

9.3.1.7 Effectiveness

The survival of any (open) system in its environment depends on its ability to secure the necessary resources, in particular, exergy. It is clear that this requires an average rate of exergy uptake that must be equal or greater than the total average rate of exergy loss from the system (entropy generation). This relationship is improved if the rate of exergy loss is decreased by more efficient exergy use, but it should be noted that survival only requires that the necessary resources be effectively secured, and not efficently used. Effectiveness therefore should not be confused with efficiency.

An effectiveness measure for the animat is defined by considering its exergy balance. It gains exergy at a rate e_{gain} by encounters with F (providing food), and it loses exergy at a rate e_{loss} in collisions with T (trees), in moving around, and for the self-organization processes of rule generation in its cognitive unit. The effectiveness measure ε is therefore defined by relating net gain rate e_{net} to exergy availability rate e_{avail},

$$\varepsilon = 1 + (e_{gain} - e_{loss}) / e_{avail}$$

$$\varepsilon = 1 + e_{net}/e_{avail}$$

The exergy availability rate is given by resource availability e and speed of motion: $e_{avail} = e \bullet v_{motion}$. Obviously, if the measure of effectiveness ε drops below 1, the net exergy balance of the organism is negative, and survival is at stake. The better the ratio of net exergy gain vs. exergy availability, the better is the satisfaction of the effectiveness orientor. Effectiveness in harvesting exergy is therefore crucial for the survival of the organism, and the remaining orientors will be defined in terms of it.

9.3.1.8 Freedom of Action

The system's ability to cope with environmental variety is reflected in the freedom of action available to it, i.e., the behavioral variety embodied in its cognitive structure. A crude measure of freedom of action is the number of behavioral rules: in a more demanding environment, more rules are generated. However, the number of rules is not a reliable indicator of behavioral variety. A few general rules, or rule chains, may provide much more freedom of action than a large number of specialized rules. In view of these difficulties in analyzing the rule set, an experimental approach was used to measure freedom of action directly and reliably. A similar experimental approach was used to measure the security dimension.

To determine measures for the freedom of action and security orientors, the animat's cognitive unit is "frozen" when it has reached a more or less steady state after extensive trainging in the given environment. All processes for generating and reinforcing behavioral rules are deactivated. There is no learning; the rule set remains fixed.

The experimental procedure for determining the animat's freedom of action consists of increasing the environmental variety and observing the animat's resulting performance. Starting with the initial (training) environmental variety f_{normal}, environmental variety is increased by adding randomly generated new patterns of T and F objects to the normal environment. Because the animat is now unable to learn an effective response to new environmental challenges, the effectivenss orientor ε eventually drops below 1 (the exergy balance becomes negative) at a critical level of environmental variety $f_{critical} \geq f_{normal}$. The freedom of action measure φ is therefore defined by the ratio of these two values,

$$\varphi = f_{critical}/f_{normal}$$

If the animat just survives in its normal environment and cannot cope with additional variety, then $f_{critical} = f_{normal}$ and $\varphi = 1$. If its freedom of action φ exceeds that required in the normal environment, it can survive increasing variety up to a point where environmental variety reaches the critical value $f_{envir} = f_{critical}$.

9.3.1.9 Security

The system's ability to cope with environmental uncertainty is reflected in the satisfaction of the security orientor. Again, this property cannot be deduced directly from the set of behavioral rules; it has to be determined by experiment.

The procedure again employs a frozen cognitive unit as described for the freedom of action orientor. In this set of experiments, environmental reliability is gradually reduced from an initial value s_{normal} by introducing random uncertainty into the recognition of T objects, where some of them now look like Fs, but still cause T collisions (see the section on environmental reliability). When environmental reliability is reduced to a critical value $s_{critical}$, the animat can no longer cope with this environment, and its effectiveness drops below 1 ($\varepsilon < 1$, negative exergy balance). The security measure σ is therefore defined by the ratio of s_{normal} vs. $s_{critical}$.

$$\sigma = s_{normal}/s_{critical}$$

If the security measure σ exceeds 1 in the normal environment, then the animat can survive a less reliable environment up to the point where environmental reliability has dropped to the critical value $s_{envir} = s_{critical}$.

9.3.1.10 Adaptability

The ability to adapt the cognitive system to a changing environment is expressed by the adaptability measure. By definition, this must capture the performance of the learning process. This means that — in contrast to the measurement of the freedom of action and security orientors — the learning process must be part of the experiment.

The learning phase of the animat (in a constant environment) typically takes some 10^4 to 10^5 food discovery cycles. Realistic environmental changes would occur over similar or even much shorter time intervals. If gradual changes were introduced into the simulations, processes of regular learning would be empirically inseparable from processes of adaptation (to change).

Adaptability is a function (1) of the cognitive abilities of the animat, and (2) of the maginitude and complexity of environmental change challenging the system per unit time. A direct measure of adaptability is the time it takes a system to adapt its behavior to a given new environment to the point where its survival is again assured ($\varepsilon \geq 1$, positive exergy balance). We can formulate a nondimensional measure of adaptability α by relating the time required for adaptation of the system T_{system} to the time constant of environmental change T_{envir}, or equivalently, the rate of system change r_{system} to the rate of change of the environment r_{envir},

$$\alpha = T_{envir}/T_{system} = r_{system}/r_{envir}$$

If $\alpha \geq 1$, i.e., if either adaptation time T_{system} is sufficiently small or environmental change sufficiently slow (T_{envir} large), the adaptability orientor is satisfied. However, if either the environment changes too quickly or the time required for adaptation (to this particular change) is too long (or both), the system can no longer adapt completely, and survival is threatened.

9.4 SIMULATIONS WITH A MINIMAL ECOSYSTEM: EMERGENCE OF INTELLIGENCE AND VALUE ORIENTATION

The ability to develop a cognitive system reflecting its environment makes Wilson's (1985) animat a suitable vehicle for investigating goal function emergence and value orientation. Holland's (1975, 1992) algorithms (rule competition, bucket-brigade, genetic algorithm, rule creation) are very effective processes that seem to capture the essentials of "real" processes found in the evolution of organisms and ecosystems. Note that the bucket-brigade very effectively builds up a cognitive model that enables anticipatory behavior (Rosen 1979): since rewards flow back to earlier rules leading to later pay-off, and activation of the initial rules in a pay-off chain means that the system suspects possible pay-off and anticipates the near future, i.e., it has an internal model of the results of its actions under the given circumstances.

The results of our experiments are more fully reported in Krebs and Bossel (1997). The following presents an example and summarizes major conclusions.

9.4.1 Sample Results

The animats were trained for "survival" in an environmental space of 60 columns by 30 rows (1800 possible positions). The properties of the training environments are summarized in Table 9.1. Food and trees were distributed in different distinct patterns of Fs and Ts.

Resource availability in both environments was similar, but their environmental variety differed by a factor of more than two. Under one set of training conditions, considerable unreliability (25%) was introduced in E3. Since the animat's training depends on a number of random factors, each animat develops a different cognitive system (classifiers and decision rules), even though final performance may be similar. In order to show general trends despite these individual differences, mean values were generated over populations of 50 individuals. The coginitive systems tended to stabilize after about 10^5 cycles with adequate survival performance. (General restuls are summarized below.)

One remarkable result from these experiments is that individuals achieve comparable performance in a given training environment with very different cognitive systems and, in particular, with different orientor emphasis. While this does not provide any particular advantage in the training environment, it provides distinct fitness advantages if the animat is moved to a different environment.

Figure 9.1 shows the orientor strengths for three particular individuals: a generalist (A) stressing freedom of action, a specialist (B) focusing on effectiveness, and a cautious type (C) emphasizing security.

TABLE 9.1
Properties of Training Environments

	Simple Environment E3	Complex Environment E4
Occupied positions	78	83
Food	24	21
Harvesting steps (1 cycle)	134	132
Distinct patterns	8	22
Sense vectors	32	73
Resource availability e	0.179	0.159
Environmental variety f	0.41	0.88
Environmental reliability s	1 (0.75)	1
Environmental stability a	1	1

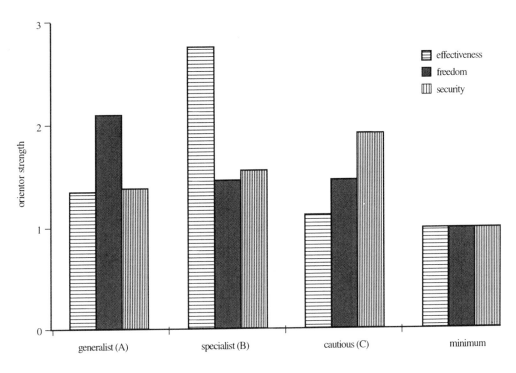

FIGURE 9.1 Orientor emphasis in three different animat individuals.

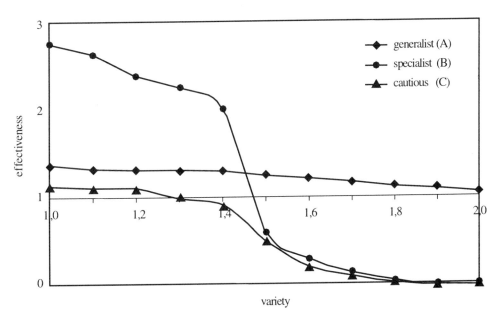

FIGURE 9.2 Performance of three different animat individuals for increasing environmental variety.

Figure 9.2 shows the performance of these types as the variety of the (training) environment is increased (by adding new patterns of food F and trees T). Here the generalist has a distinct advantage: its effectiveness remains above the critical level (=1) even for a large increase in environmental variety. The specialist can survive a medium amount of additional variety, while the cautious type can only cope with a small increase.

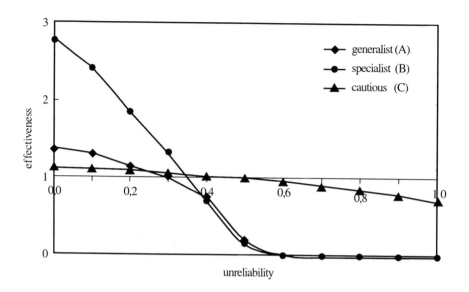

FIGURE 9.3 Performance of three different animat individuals for increasing environmental unreliability.

Figure 9.3 shows the performance of the three types for increasing unreliability of the environment (achieved by converting more and more Fs into Ts, while still appearing as Fs). In this case, the cautious type has a distinct advantage, while the generalist fares worst.

9.4.2 EMERGENCE OF BASIC VALUE ORIENTATIONS

The animat experiments confirm the basic proposition of orientation theory: In a self-organizing system, balanced multidimensional value orientation emerges as a result of fitness selection. The emergent basic orientors (basic value dimensions, system interests) are a reflection of the properties of the system's environment. Moreover, in "normal" environments, the same value dimensions emerge in all self-organizing systems, irrespective of their specific composition or their specific environment. The experiments with an individual animat assumed implicit satisfaction of the existence orientor, and therefore dealt with the four orientors: effectiveness, freedom of action, security, and adaptability.

In the animat experiments (and similarly, in real life), implicit and (more or less) balanced multidimensional attention to the basic orientors emerges from the simple one-dimensional mechanism of rewarding success in the given environment. Thus, in the course of its evolutionary development in interaction with its environment, the system evolves a complex multidimensional behavioral objective function from the very unspecific requirement of fitness. Conversely, this also means that balanced attention to the emergent basic orientors is necessary for system viability and survival — they would not have emerged unless important for the viability of the system.

9.4.3 EMERGENCE OF INDIVIDUAL DIFFERENCES IN VALUE ORIENTATION

"Balanced attention" still leaves room for individual differences in the relative emphasis given to the different orientors. Individuals belonging to the populations used in the animat experiments evolve significant differences in value emphasis (e.g., specialist, generalist, cautious type). These individual variations, while not significantly reducing performance in the standard training environment, provide comparative advantage and enhanced fitness when resource availability, variety, or reliability of the environment change. They also result in distinctly different behavioral styles. However, pathological behavior will follow if orientor attention becomes unbalanced (e.g., dominant emphasis on effectiveness, or freedom of action, or security, or adaptability), even if the system should still manage to secure the resources for survival.

Changing orientor attention involves trade-offs. For example, greater satisfaction of security can be bought only at the cost of reduced satisfaction of other orientors, in particular effectiveness.

Training of the animat in different environments, the performance of animat individuals in environments that differ from their training environments, and the simulation of adaptive learning in a changing environment, lead to some general conclusions that are in full agreement with everyday observations and general systems knowledge.

- Generalists have a better survival chance than others if moved to an environment of greater variety.
- Cautious types have a better survival chance than others if moved to a less reliable environment.
- Training in more unreliable and/or more diverse environments increases satisfaction of the security and/or freedom of action orientors at the cost of the effectiveness orientor.
- Training in an uncertain environment teaches caution and improves fitness in a different environment.
- Learning caution (better satisfaction of the security orientor) takes time and decreases effectiveness, but increases overall fitness.
- Investment in learning (energy cost of learning in the animat) pays off in better fitness: the learning investment is (usually) much smaller than the pay-off gain.

9.5 CONCLUSIONS

The central role of exergy for the understanding of behavior and evolution of organisms and ecosystems has been borne out by this study. Important attributes of the environment as well as the system must be formulated in terms of exergy. The implications go beyond thermodynamic considerations and reach into the development of cognitive structure and the emergence of normative objectives (goal functions, values, system interests) in evolving systems.

Applying van Heerden's (1968) definition of intelligence (Section 9.1), animat individuals develop intelligent behavior in an even wider sense than suspected by Wilson (1985), and more closely in line with the visionary statement by Ashby (1962) (Section 9.1); they also develop a complex objective function (balanced attention to basic orientors), or value orientation. Serious attention to basic values (existence, effectiveness, freedom, security, adaptability, and coexistence) is therefore an objective requirement emerging in and characterizing self-organizing systems. These basic values are *not* subjective human inventions; they are objective consequences of the process of self-organization in response to normal environmental properties.

Self-organizing systems are dissipative systems by their very nature: the processes of maintenance and organization of structure require that the system must be able to maintain a resource throughput (exergy, materials, information). The system's exergy balance therefore plays a central role, which is also reflected in the orientors: Given satisfaction of the existence orientor, satisfaction of the effectiveness orientor is essential for survival. Satisfaction of the freedom of action and security orientors are preconditions for longer-term effectiveness (and survival), while adaptability in turn ensures effectiveness, long-term freedom of action, and security. The system of basic orientors therefore exhibits a hierarchical structure of urgency, with different time constants associated with the different orientors (existence: immediate; effectiveness: short term; freedom of action and security: longer term; adaptability: long term).

Since the cognitive system with its inherent value emphasis is a function of the environment to which the system has adapted, it follows that an adaptive, self-organizing system will also undergo value change (change in relative value emphasis) if the system has to adapt to an environment with different or changing environmental properties (e.g., a less reliable environment will require more caution, i.e., value change to security emphasis).

Orientation theory in conjunction with a simulated organism (animat) has been found to be useful in analyzing and understanding processes of self-organization and system evolution, and in particular of the evolution of cognitive structure and resultant behavior. The concepts are also applicable and useful to the design of (complex) systems that have to cope with a given environment. Such systems must be designed with a cognitive unit that allows balanced attention to the full (multidimensional) set of basic orientors; one-dimensional criteria (e.g., cost-effectiveness, maximization of gross national product, etc.) will normally be insufficient and lead to pathological behavior.

The approach therefore has particular significance for path studies of potential future development of human systems (Bossel 1998). In such studies, the time development of relevant system indicators (state variables) must be mapped on the basic orientor dimensions in order to determine the potential contribution to system development, and to its potential for survival and development.

Quantification of a system's orientor satisfactions and comparison of the orientor stars (polar diagrams of orientor satisfaction) for competing systems provides a more comprehensive picture and fuller understanding of system performance than an aggregate measure of fitness (such as net energy accumulation over a given time period). In particular, although the behavioral rule sets developing in adaptive artificial intelligence systems (such as Wilson's animat) cannot be evaluated and compared directly with respect to their performance, the orientor mappings offer a means of objective performance evaluation and comparison.

Numerical assessment and comparison of system performance requires quantification of (1) measures of environmental properties, and (2) measures of the orientor dimensions. Both measures are system dependent. The same environment means different things to different organisms, depending on their sensors and their system structure: a meadow has a different composition of environmental properties for a cow than for a honeybee. Similar distinctions are required with respect to orientors: the security of a nation requires a different objective function than that of an individual. However, definition and quantification of suitable criteria and measures is usually possible, even in complex societal contexts (for example, Bossel and Strobel (1978), Hornung (1988), Bossel (1994), Bossel (1998)).

REFERENCES

Ashby, W. R., Requisite variety and its implications for the control of complex systems. *Cybernetica*, 1: 83–99, 1958.

Ashby, W. R., Principles of the self-organizing system. In: H. von Foerster and G. W. Zopf (Eds.): *Principles of Self Organization*. Pergamon Press, New York, 1962.

Bossel, H., Orientors of nonroutine behavior. In: H. Bossel (Ed.): *Concepts and Tools of Computer-Assisted Policy Analysis*, Birkhaeuser Verlag, Basel, 227–265, 1977.

Bossel, H., Real-structure process description as the basis of understanding ecosystems and their development. *Ecol. Modell.*, 63: 261–276, 1992.

Bossel, H., *Modeling and Simulation*. A. K. Peters, Wellesley, MA, and Vieweg, Wiesbaden, 1994.

Bossel, H., *Earth at a Crossroads: Paths to a Sustainable Future*. Cambridge University Press, Cambridge, U.K., 1998.

Bossel, H. and Strobel, M., Experiments with an "intelligent" world model, *Futures*, June 1978, 191–212, 1978.

Holland, J. H., *Adaptation in Natural and Artificial Systems*, University of Michigan, Ann Arbor, 1975.

Holland, J. H., Adaptation. In: R. F. Rosen (Ed.): *Progress in Biology IV*. Academic Press, New York, 1976.

Holland, J. H., Adaptive algorithms for discovering and using general patterns in growing knowledge-bases. Int. *J. Policy Analysis Information Syst.*, 4, 217–240, 1980.

Holland, J. H., Genetic algorithms and adaptation. In: O. G. Selfridge, R. L. Rissland, and M. A. Arbib (Eds.): *Adaptive Control of Ill-Defined Systems*. Plenum, New York, 1984.

Holland, J. H., *Adaptation in Natural and Artificial Systems,* MIT Press, Cambridge, MA (revised edition of 1st Ed. University of Michigan, Ann Arbor, 1975), 1992.

Hornung, B. R., Grundlagen einer problemfunktionalistischen Systemtheorie gesellschaftlicher Entwicklung-Sozialwissenschaftliche Theoriekonstruktion mit qualitativen, computergestuetzten Verfahren. Europaeische Hochschulschriften, Reihe XXII, Soziologie, Bd. 157, Peter Lang, Frankfurt, 1988.

Jørgensen, S. E., Models as instruments for combination of ecological theory and environmental practice. *Ecol. Modell.,* 75/76: 5–20, 1994.

Jørgensen, S. E., Mejer, H. F., and Nors-Nielsen, F., Emergy, environ, exergy and ecological modelling. *Ecol. Modell.,* 77: 99–109, 1995.

Jørgensen, S. E. and Nielsen, S. N., Models of the structural dynamics in lakes and reservoirs. *Ecol. Modell.,* 74: 39–46, 1994.

Krebs, F. and Bossel, H., Emergent value orientation in self-organization of an animat. *Ecol. Modell.,* 96: 143–164, 1997.

Mayr, E., Teleological and teleonomic: A new analysis. *Bost. Stud. Phil. Sci.,* 14: 91–117, 1974.

Miller, J. G., *Living Systems.* McGraw-Hill, New York, 1102, 1978.

Müller, F., Ableitung von integrativen Indikatoren zur Bewertung von Ökosystemzuständen für die Umweltökonomische Gesamtrechnung. Ökologiezentrum, Universität Kiel, 130, 1996.

Rosen, R., Anticipatory systems in retrospect and prospect. *General Systems Yearbook,* 24, 11–23. 1979.

van Heerden, P. J., *The Fountain of Empirical Knowledge.* Wistik, Wassenaar, The Netherlands, 1968.

Wilson, S. W., Knowledge growth in an artificial animal. In: J. J. Grefenstette (Ed.): *Proceedings of the First International Conference on Genetic Algorithms and Their Applications.* Lawrence Earlbaum, Pittsburgh, PA., Morgan Kaufmann, San Mateo, 16–23, 1985.

Make it visible and it will
disappear in the crowd,
make it probable and you will
be able to distinguish it

CHAPTER 10

This chapter focuses on the use of thermodynamics on systems far from thermodynamic equilibrium. It applies entropy and exergy calculations to these systems. Equation (10) for entropy is essential in these calculations. Equation (12) is used for exergy. This latter equation can be used to interpret the contributions to exergy, namely, from the total biomass and from Kullback's measure of information. This implies again that the exergy maximum principle (also named ELT in Chapter 9, the Ecological Law of Thermodynamics) is an increase of biomass and/or an increase of the information contained in a biomass unit. There are limitations to how much the biomass can increase, while the information contained in the biomass in principle can increase to an unlimited extent. The author shows that the adoption of the maximum exergy principle implies that a certain level of diversity is maintained.

The energy needed to cause environmental destruction can be described by a flow of artificial energy (fossil fuel, electricity, etc., denoted energy load) and by a second term associated with a flow of chemical elements to maintain the molar concentrations inside the system (denoted chemical load). The latter term coincides with the exergy as demonstrated in the chapter. The author concludes that entropy (or exergy destruction) can be used as a measure of environmental degradation. He also concludes that the maximum exergy principle is one of the possible corollaries of the thermodynamic extreme principles.

10 Thermodynamics and Ecology: Far from Thermodynamic Equilibrium

Yuri M. Svirezhev

CONTENTS

10.1 Introduction 213
10.2 The Physical Approach: Direct Calculation of the Entropy
      and the "Entropy Pump" Hypothesis 214
10.3 Systems Far from Thermodynamic Equilibrium..219
10.4 Exergy and Entropy: Exergy Maximum Principle..221
10.5 Exergy and Information 222
10.6 Entropy Production and Chemical Load 224
10.7 Average Rate of Entropy Production and Equilibrium in Average 225
10.8 Conclusion...226
References 227

10.1 INTRODUCTION

"Thermodynamics is full of highly scientific terms. *Entropy, thermal death of the Universe, ergodicity, statistical ensemble* — all these words sound impressive in any order. But, placed in the appropriate order, they can help us find the solution of urgent practical problems. The problem is how to find this order ..."

(from table-talks in Moscow)

"... nobody knows what entropy is in reality, that is why in the debate you will always have an advantage."

John von Neumann, 1948

Many studies are known which attempt to apply (directly or indirectly) thermodynamic concepts and methods in theoretical and mathematical ecology for the macroscopic description of biological communities and ecosystems. Such attempts can be divided into two classes.

The *first* class includes the direct transfer of such fundamental concepts as entropy, the First and Second Laws of Thermodynamics, Prigogine's theorem, etc. into ecology. The literature on this subject is enormous. The most recent publications are Weber et al. (1988), Jørgensen (1997), Schneider and Kay (1994).

The *second* class includes some attempts to use the *methods* of thermodynamics, like the Gibbs statistical method. In the 1940s a very elegant method for the construction of the formal statistical mechanics was proposed by Khinchin. This method can be applied to a wide class of dynamical systems, in particular, to Volterra's "prey–predator" system (Kerner, 1957, 1959; Alexeev, 1976). Unfortunately, none of these results can be interpreted satisfactorily from the ecological point of view (Svirezhev, 1976).

Strictly speaking, there are no principal prohibitions to applying thermodynamic concepts to such physical–chemical systems as ecological systems. The problem is the following: there is no direct homeomorphism between the models (in a broad sense) in thermodynamics and the models in ecology. For example, the model of ideal gas (the basic model of thermodynamics) cannot be applied directly to a population or, moreover, to a biological community. The macroscopic state of the ideal gas is an additive function of the microscopic states of molecules. The stable structure of a biological community is the consequence of interactions between populations rather than the function of characteristics of individual species, etc. It is appropriate to mention the well-known ecological paradox: the diversity of a community is maximal when the distribution of species is uniform, i.e., when there are no abundant and rare species, and there is no structure.

But in spite of this, I look at the problem of the application of thermodynamic ideas to ecology with optimism. I think that if we could manage to formulate the thermodynamic ecological model correctly, and if we were able to correctly formulate the concept of the thermodynamic system in relation to ecosystems, the use of these formulations in ecology would be very fruitful.

10.2   THE PHYSICAL APPROACH: DIRECT CALCULATION OF THE ENTROPY AND THE "ENTROPY PUMP" HYPOTHESIS

From the viewpoint of thermodynamics, any ecosystem is an open thermodynamic system. The climax of the ecosystem corresponds to the dynamic equilibrium (steady state), when the entropy production in a system is balanced by the entropy flow from the system to the environment. This work is being done by the "entropy pump."

Let us consider one unit (hectare, m^2, etc.) of the Earth's surface which is occupied by a natural ecosystem (i.e., meadow, steppe, forest, etc.) and it is maintained in the dynamic equilibrium (steady state). For instance, such sort of equilibrium may be an ecosystem climax. There is a natural periodicity in such a system (1 year); during this period, the internal energy of the ecosystem is increased by a value of gross primary production (which can be expressed either in calories or in joules). One part of this production is used for respiration with the further transformation into heat, while another part (the net primary production), on the one hand, turns into litter and other forms of soil organic matter, and, on the other hand, is taken by consumers and for compensation of the respiratory losses. But, since the system is at dynamic equilibrium, an appropriate part of dead organic matter in litter and soil has to be decomposed (releasing a place for a "new" dead organic matter from annual net primary production). The "old" dead organic matter has "to be burned," so that the chemical energy of it is transformed into heat. The temperatures and pressures in the ecosystem and its environment are assumed to be equal, i.e., we consider an isothermic and isobaric process.

In the theory of open systems the total variation of entropy

$$dS(t) = d_iS(t) + d_eS(t), \tag{1}$$

where $d_iS(t) = dQ(t)/T(t)$, $dQ(t)$ is the heat production caused by irreversible processes inside the system and $T(t)$ is the current temperature (in K) at a given point of the Earth's surface. The value $d_eS(t)$ corresponds to the entropy of exchange processes between the system and its environment.

In fact, total heat production = heat emission of the plant metabolism (heat emitted during the process of respiration, $R_v(t)$) + heat emission of the consumer's metabolism ($R_c(t)$) + heat emission of the decomposition of "dead" organic matter ($D(t)$). Really $R_v(t) \gg R_c(t)$ so that the total metabolism of the ecosystem is equal to the metabolism of its vegetation, in practice.

Integrating (1) in respect to a natural period (one year), we get

$$S(t + 1) - S(t) = \int_t^{t+1} \frac{R_v(\tau) + D(\tau)}{T(\tau)} d\tau - \delta_e S, \tag{2}$$

where $\delta_e S = \int_t^{t+1} \frac{d_e S}{d\tau} d\tau$. Using the mean value theorem, the integral in (2) can be rewritten in the form

$$\int_t^{t+1} \frac{R_v(\tau)}{T(\tau)} d\tau = \frac{1}{T(\theta_1')} \int_t^{t+1} R_v(\tau) d\tau = \frac{1}{T(\theta_1')} [P_0(t) - P_0^n(t)], \qquad \theta_1', \theta_2' \in ([t, t+1])$$
$$\tag{3}$$

$$\int_t^{t+1} \frac{D(\tau)}{T(\tau)} d\tau = \frac{1}{T(\theta_2')} \int_t^{t+1} D(\tau) d\tau = \frac{1}{T(\theta_2')} \tilde{D}_0(t),$$

The values $P_0(t)$ and $P_0^n(t)$ are annual gross (total) and net primary productions; the temperatures $T(\theta_1')$, $T(\theta_2')$ are some "mean" annual temperatures where they may be different. In all these values the notation "t" means a number of corresponding year.

The ecosystem state changes inside in 1-year intervals but if these changes are periodical each and all the mean characteristics of the ecosystem (i.e., by averaging inside each 1-year interval) are not changed from year to year, then we can consider this state of the ecosystem as stationary. The characteristic time will be equal to 1 year.

Therefore, since the system is in dynamic equilibrium (steady state) then the "burned" part of the dead organic matter $D_0(t) = P_0^n(t)$ and, moreover, in (1)

$$S(t + 1) - S(t) = 0.$$

Then

$$\delta_e S(t) = \frac{1}{T(\theta_1')} P_0(t) + \left[\frac{1}{T(\theta_1')} - \frac{1}{T(\theta_2')} \right] P_0^n(t). \tag{4}$$

Considering the integral

$$\int_t^{t+1} R_v(\tau) d\tau = \int_t^{t+1} r(\tau) p_0(\tau) d\tau = r(\theta_3') \int_t^{t+1} p_0(\tau) d\tau = r(t) P_0(t), \qquad \theta_3' \in [t, t+1],$$

where $r(t)$ and $p_0(t)$ are the current respiration coefficient and total primary production inside the 1-year interval, respectively, $r(q_3^t) = r(t)$ is a "mean" respiration coefficient so that $P_0^n(t) = [1 - r(t)] P_0(t)$, we also get

$$\delta_e S(t) = \left\{ \frac{1}{T(\theta_1')} \left[\frac{2 - r(t)}{1 - r(t)} \right] - \frac{1}{T(\theta_2')} \right\} P_0^n(t). \tag{4'}$$

According to our assumption, the entropy $\delta_e S(t)$ must be "sucked out" by the solar "entropy pump" (in accordance with the steady-state condition). Consequently, the power of this pump at some point of the Earth is equal to $\delta_e S(t)$.

Note that the process of the formation of a new biomass and its next degradation represents a complex chain of multiple chemical and biochemical reactions. In our approach, which is based on the Hess theorem (see, for instance, Rubin, 1967), we assume that the total entropy produced by this chain depends only on the thermodynamic characteristics of the initial and final elements, i.e., it is determined only by the chain input and output.

The entropy which is produced by the destruction of chemical structure of biomass is considerably less than the "heat" entropy. Hence, we can neglect the former (for the calculation of the balance).

We mentioned above that, in the general case, $T(\theta_1') \neq T(\theta_2')$. In order to estimate these values we must calculate the integrals in (3). For this we must know, first, how the respiration, R_v, and the decomposition, D_v, depend on the temperature, and, second, the seasonal dynamics of the temperature. In fact, these dependencies differ from each other, since the respiration is determined by metabolism of plants and the decomposition occurs by the activity of decomposers (worms, nematodes, microbes, etc.). It is known that the time-peaks for respiration and decomposition are distinguished; the first peak occurs earlier than the second one which occurs closer to autumn (Volobuev, 1953). As a consequence, $T(\theta_1^T) > T(\theta_2^T)$ It is obvious that for such calculations we need detailed information about intraseasonal dynamics of the metabolic and decomposition processes. It is absolutely unrealistic to have information at global and regional levels. However, if we assume that both the photosynthesis and the respiration, and also the decomposition of dead organic matter similarly depend on the current temperature then we get $T(\theta_1') \approx T(\theta_2') = T(t)$. Apparently the best approximation for this "mean" temperature will be the *mean active temperature*, i.e., the arithmetic mean of all temperatures above 5°C that is a good indicator of biological activity. The next approximation may be the mean temperature of a vegetation period. At last the annual mean temperature is not the best possible approximation.

If we accept this assumption, then Formulas (4), (4′) are written in the form

$$\delta_e S = \frac{P_0}{T} = \frac{1}{1-r} \frac{P_0^n}{T}. \tag{5}$$

Let us assume that the considered area is influenced by anthropogenic pressure, i.e., a flow of artificial energy (W) into the system takes place. We include in this notion (the flow of artificial energy) both the direct energy flow (fossil fuels, electricity, etc.) and the inflow of chemical elements (pollution, fertilisers, etc.). The latter is the so-called "grey energy" flow.

We suppose that the first inflow dissipated inside the system is transformed directly into heat and, moreover, it modifies the plant productivity. The second inflow, changing the chemical state of environment, also modifies the plant productivity.

Let the gross production of the ecosystem under anthropogenic pressure be P_1 (in caloric units or joules). When repeating the previous arguments we learn that the entropy production within this "disturbed" ecosystem is equal to

$$\delta_i^1 S = \frac{1}{T} [W + P_1(W)]. \tag{6}$$

We make a very important assumption and suppose that a part of the entropy released at this point by the entropy pump is still equal to $d_e S = P_0/T$. Really we assume that the power of the local entropy pump corresponds to a natural ecosystem that would be situated in that location. In other words, the climatic, hydrological, soil, and other environmental local conditions are organised in such a way that only a natural ecosystem that is specific, namely for this combination, can be at the steady state without an environmental degradation. This is the entropy pump hypothesis.

We assume that despite anthropogenic perturbation, the disturbed ecosystem is again tending to a steady state. If we accept it we must also assume that the transition from natural to anthropogenic ecosystem is performed sufficiently fast so the "adjustment" of the entropy pump could not change. Therefore, regarding the other part of the entropy,

$$\delta S \; = \; \sigma \; = \; \delta_e^1 S - P_0 / T \tag{7}$$

should be compensated by the outflow of entropy to the environment, and only the unique way exists. *This compensation can occur only at the expense of environmental degradation ($\sigma > 0$),* that resulted, for instance, from heat and chemical pollution, and mechanical impact on the system.

We assume also that the "natural" and "anthropogenic" ecosystems are connected by *the relation of succession.*

I would like to include a few words about the relation of succession. Let us assume that the anthropogenic pressure has been removed. The succession from the anthropogenic ecosystem toward a natural one has started. The next stage of this succession would be a "natural" ecosystem in our sense. Really, if the anthropogenic pressure has been weak, the "natural" ecosystem (in our sense) will be typical of a "wild" ecosystem existing in this locality.

On the other hand, if the anthropogenic ecosystem is an agroecosystem, surrounded by forest, a grass-shrubs ecosystem (not a forest) will be successionally close to its "natural" ecosystem. And finally, we can define a *successionally closed ecosystem* (i.e., an ecosystem, successionally closed to an ecosystem under anthropogenic stress) as the first stage of succession of an "anthropogenic" ecosystem where the anthropogenic stress is removed.

What is a "dynamic" sense of the "successional closeness"? Why do we need the concept?

The point is that in this approach we can compare only close steady states, their vicinities must be intersected significantly, and the time-scale of a quasi-stationary transition from a natural to anthropogenic ecosystem and vice versa must be small (in comparison to the time-scale of succession).

Let us consider the nonlocal dynamics of the system. We stop the flux of artificial energy (i.e., the anthropogenic energy and chemical fluxes) into the ecosystem. As a result, if the ecosystem is not degraded, a succession would take place at the site that tends toward the natural ecosystem type specific for the territory (grassland, steppe, etc.). This is a typical reversible situation. Under severe degradation, a succession would also take place, but toward another type of ecosystem. This is quite natural, since the environmental conditions have been strongly perturbed (for instance, as a result of soil degradation). This is an irreversible situation. So, the "successional closeness" concept means that we remain in a framework of "reversible" thermodynamics. And if there is no input of artificial energy, the steady state for a given site (locality) will be presented by the natural ecosystem, as the local characteristics of the entropy pump correspond exactly to the natural type of ecosystem.

Nevertheless, there is a small error. When we discussed a successionally closed system above, we assumed implicitly that any stage of the succession is a dynamic equilibrium. Since a succession is a transition process between two steady states, this statement is incorrect, but insofar that we can suggest that the time-scale of ecological succession is much greater than the time-scale of anthropogenic processes, we can consider a succession as a thermodynamically quasi-stationary process (simultaneously, we remain inside the model of equilibrium thermodynamics). However, if we want to construct a thermodynamic model of succession, we should drop the hypothesis on quasi-stationary transition.

Returning to (6) and (7) we get the formula for σ.

$$\sigma T \; = \; W + P_1 - P_0. \tag{8}$$

The values in (8) are not independent. For instance, P_1 depends on W. Since we are not able to estimate this correlation within a framework of theory of thermodynamics, we have to use an empirical correlation.

If we remember the sense of the σ then it is obvious that the σ value can be used as the criterion for environmental degradation or as the entropy fee that has to be paid by society (really, suffering from the degradation of environment) for modern industrial technologies.

Of course, there is the other way to balance the entropy production within the system. For instance, we can introduce an artificial energy and soil reclamation and pollution control (or, generally, ecological technologies). Using the entropy calculation we can estimate the necessary investments (in energy units).

We know that our attempt to apply thermodynamic concepts to analyse ecosystems is not original. For instance, we may refer to the brilliant work of Ulanowicz and Hannon (1987) in which the direct method for calculation of entropy fluxes inside ecosystems was developed. In our approach we use a different (indirect) method for similar calculations based on the "entropy pump" hypothesis. Its correctness depends on a plausibility of this one. Since a direct test is a very complex task we say that it is plausible if we obtain coincidences between theoretical conclusions and real observations (for instance, in agricultural systems). Nevertheless we shall try once again to argue our point of view using a classic and standard thermodynamic method of consideration. In other words we shall develop a *thermodynamic model of vegetation*.

From the point of view of our model the whole plant is presented in the form of one "leaf." The change of entropy caused by photosynthesis (and it is a serious oversimplification) is described by Formula (1) where the term d_eS describes the processes of the sunlight assimilation minus unavoidable energy losses, the term d_iS describes the production of inner entropy caused by a heat dissipation of the vegetation cover accompanying its life functioning, and the term dS corresponds to the total increase of plant biomass caused by photosynthesis.

We assume that the whole heat produced in leaves by spontaneous metabolic reactions is output into the environment at a rate equal to the rate of entropy production d_iS/dt. The main processes that output the heat are the heat radiation and the evapotranspiration. The first process gives

$$\left(\frac{d_iS}{ds}\right)_{q_W} = q_T \frac{\delta T}{T^2} ,$$

and the second one gives

$$\left(\frac{d_iS}{dt}\right)_{T.q_T} = \gamma q_W \frac{\delta p_m}{T} ,$$

so that

$$\frac{d_iS}{dt} = q_T \frac{\delta T}{T^2} + \gamma q_W \frac{\delta p_m}{T} , \qquad (*)$$

where T is the mean temperature of the "leaf–air" system, δT is the temperature difference between leaves and surrounding air, q_T is the corresponding heat flow, $\gamma = 600$ cal/g of water is the heat of vapour formation for water, q_W is the rate of evapotranspiration, and δp_m is the difference of specific humidity between leaf-stomata and air. The energy balance will be

$$P_n = (Phar - E_l) - T \frac{d_iS}{dt} , \qquad (**)$$

where P_n is the net primary production (in caloric units), *Phar* is the solar photoactive radiation, i.e., the energy of assimilated sunlight, and E_l is the energy losses accompanying the absorbtion and photosynthetic assimilation of solar energy.

The value d_iS/dt was obtained experimentally (using formula (*)) for a coniferous forest in Northern Russia (see Rubin, 1967) and it was equal to 3.3 kcal/m²*K*year. Therefore the value of so-called *dissipative function*

$$\beta = T*d_iS/dt = 3.3*285 = 940 \text{ kcal/m}^2*\text{year.}$$

The mean annual net production for trees was equal to 4320 kcal/m²*year. Then the amount of utilised solar energy, i.e., the total (gross) annual production is equal to $P = P_n + \beta = 4320 + 940 = 5260$ kcal/m²*year. The direct measurements show that *Phar* = 534,420 kcal/m²*year. Using these data we may estimate, for instance, the efficiency coefficient for total photosynthesis $\eta_{ph} = 5260/534,420 \approx 1\%$ and also the respiration coefficient $r = 940/5260 \approx 20\%$. All these values are very similar to realms. Note that we took into account the evapotranspiration in our calculation.

Since this area must be in a steady state, then the dead organic matter, which is equivalent to the created biomass, must be decomposed and the corresponding amount of heat must be emitted to the environment. And finally, the annual entropy production will be equal to

$$dS = \frac{1}{T}(P_n + Td_iS) = \frac{P}{T}$$

(compare to Formula (5)).

10.3   SYSTEMS FAR FROM THERMODYNAMIC EQUILIBRIUM

Before introducing some special concepts like the exergy, etc. we must remember that all these concepts consider the ecosystem as a system far from thermodynamic equilibrium. The "basic" variable for this theory is a rate of the entropy production, or a rate of the energy dissipation, the so-called *dissipative function* (β). Immediately a series of questions arises, dealing with the behaviour of the dissipative function β.

1. How can we calculate the dissipative function β for a system far from thermodynamic equilibrium if we do not know the appropriate kinetic equations?
2. What can we calculate in this case?
3. What kinds of "thermodynamic" statements can be formulated in this case?

We attempt to answer these questions.

Let us assume that we have the system of ordinary differential equations, which describe the dynamics (kinematics) of the considered system:

$$\frac{dC_i}{dt} = f_i(C_1,...,C_n), \qquad i = 1,...,n; \quad C_i \in (\mathbf{P}^n,) \tag{9}$$

where $\mathbf{P}^n$ is a positive orthant of state space and $\{Ci(t)\}$ is a vector of state variables (the positive orthant is a domain of state space in which all the state variables are nonnegative).

We also assume:

1. The system (9) has a single stable equilibrium point $\{C_{i*}\}$, so that
2. $C_i \Rightarrow C_{i*}$ for $t \Rightarrow \infty$ and for any $C_i, C_{i*} \in P^n$.

Actually, we do not have a good description for the system of differential equations (9).

In the best case we have a time series of observations recorded in the course of transition from an initial state $\{C_{io}\}$ toward the stable equilibrium point $\{C_{i*}\}$. It corresponds to some solution $\mathbf{C} = \mathbf{C}(\mathbf{Co}, t)$ of an unknown basic system of differential equations (9). Note that the problem of the reconstruction of (9) is a typically incorrect problem (in a mathematical sense). Nevertheless, we can calculate the *total amount of dissipated energy*,

$$L = \int_{t_0}^{\infty} \beta(t)dt,$$

for the transition $\mathbf{C}(t_0) \Rightarrow \mathbf{C}^*(\infty)$ in one special case.

Let the system dynamics be a movement in a potential field with the chemical potentials $\mu_i = \mu_{io} + RT \ln C_i$, $i = 1,...,n$, where C_i are molar concentrations of corresponding components and R is the gas constant. We assume $\mu_{10} = \mu_{20} = ... = \mu_{n0}$, i.e., the components C_i are substances of *identical* (or close) origin. Affinity for reaction (transition) $C_i \Rightarrow C_i^*$ is equal to $A_{i,i*} = \ln(C_i/C_{i*})$. The initial values are arbitrary so that we can consider any point $\mathbf{C}(t)$ (except the singular point $\mathbf{C}^*$) as an initial point.

Since the affinity for transition $C_i \Rightarrow C_i^*$ is equal to $A_{ii*} = \ln(C_i/C_i^*)$ then

$$L = \int_{t_0}^{\infty} \beta(t)dt = \int_{t_0}^{\infty} \sum_{i=1}^{n} \beta_i(t)dt = \sum_{i=1}^{n} \int_{t_0}^{\infty} \ln \frac{C_i(t)}{C_i^*} \frac{dC_i}{dt}$$

$$= \sum_{i=1}^{n} \int_{C_i(t)}^{C_i(\infty)} (\ln C_i - \ln C_i^*)dC_i = \sum_{i=1}^{n} \left\{ -C_i \ln \frac{C_i}{C_i^*} + (C_i - C_i^*) \right\}.$$

We omit the factor RT in the expression for β, which is inessential when we consider an isotherm process. And finally

$$L = \sum_{i=1}^{n} L_i = -\sum_{i=1}^{n} \left[C_i \ln \frac{C_i}{C_i^*} - (C_i - C_i^*) \right]. \tag{10}$$

You can see that $L < 0$ for any $C_i > 0$, except $C_i = 0$, when $L = 0$. It means that the *value of total change for entropy of an open system far from thermodynamic equilibrium*, when it passes from some nonsteady state to stable dynamic equilibrium, *is a negative value*. Also, this value does not depend on characteristics of this transition.

Certainly, *spontaneous* processes, when a system tends to a stable dynamic equilibrium after small fluctuations, which appear inside a system, *are accompanied by an increase of entropy*. In our case the transition $\mathbf{C} \Rightarrow \mathbf{C}^*$ is not spontaneous *but forced*, it depends on interaction between the system and its environment.

It is obvious that the decrease of entropy (in similar transition processes) is a result of free energy consumption (by the system) from the environment. It is, in turn, a result of exchange processes, giving a negative contribution into the entropy production.

"Revenons à nos moutons," and let us remember

$$\beta = \frac{dS}{dt} = \frac{d_eS}{dt} + \frac{d_iS}{dt},$$

where $\beta_e = d_eS/dt$ is a result of exchange between the system and its environment, and $\beta_i = d_iS/dt$ is a result of internal spontaneous irreversible processes inside the system.

For "quasi-equilibrium" systems (the domain of linear thermodynamics, the "Prigogine World"), β (and $|\beta_e|$) is minimal; therefore, in similar open systems, when all the transition processes have been finished and some steady, "quasi-equilibrium" state is established, the *"total energy store" must be minimal.*

In our case, for open systems far from thermodynamic equilibrium, the "total energy store" increases at the expense of exchange between the system and its environment in the course of forced transition to the stable dynamic equilibrium (steady state), so that $L < 0$.

10.4 EXERGY AND ENTROPY: EXERGY MAXIMUM PRINCIPLE

Let us suppose that the right-hand sides of Equations (9) depend on some parameters $\alpha_1,...,\alpha_m$, so that

$$\frac{dC_i}{dt} = f_i(C_i,...,C_n;\alpha_1,...,\alpha_m), \qquad i = 1,...,n. \tag{11}$$

The vector of parameters α describes the state of the environment. It is obvious that the steady state $\mathbf{C}^*$ depends on α. We consider the following *Gedankenexperiment*:

1. Let the current state of environment be described by the vector α^1, then $\mathbf{C}^* = \mathbf{C}^*(\alpha^1)$.
2. We change the environment from state α^1 to state α^2 very quickly in comparison with the actual time of the system.
3. We spend the energy (work) E^{12} to realise this change.
4. After this change the state $\mathbf{C}^*(\alpha^1)$ ceases to be a stationary state, and the system starts to evolve toward a new stationary (steady) state $\mathbf{C}^*(\alpha^2)$.

If we calculate the dissipative energy for this transition, we get

$$L^{12} = -\sum_{i=1}^{n}\left\{C_i^*(\alpha^1)\ln\frac{C_i^*(\alpha^1)}{C_i^*(\alpha^2)} - C_i^*(\alpha^1) - C_i^*(\alpha^2)\right\}. \tag{12}$$

Since we cannot cancel the action of the Second Law, $E^{12} \geq -L^{12}$ and min $E^{12} = -L^{12}$. We shall consider an extreme case and assume that $E^{12} = -L^{12}$.

If we assume that the vector $\bullet^1$ corresponds to the current state of the environment (the biosphere), the vector α^2 corresponds to some *prebiological* environment, and $\mathbf{C}^*(\alpha^2) = \mathbf{C}^0$ are equal to concentrations of biogenic elements in some prebiological structures, we see immediately that E^{12} is S. E. Jørgensen's *exergy* (Jørgensen, 1992). Therefore,

$$\text{Exergy } (Ex) = \sum_{i=1}^{n}\left\{C_i\ln\frac{C_i}{C_i^s} - (C_i - C_i^0)\right\}. \tag{13}$$

(Note that we omitted the factor RT.)

Let us remember that the exergy is equal to the work necessary for such a transformation of the system environment, so that in this new environment the system evolves to a prebiological state. The latter can be considered as a thermodynamic equilibrium, i.e., it corresponds to the death of the system. In other words, the exergy is a necessary energy in order to *kill* the system.

Note that the work cannot be done directly on the system; it must be done on its environment; i.e., we cannot destroy the system directly. In order to do this, we must change the environment in a hostile way (for the system). For this, a mechanical work may be used, and the system will be mechanically destroyed. This may be a result of change of the chemical status of environment (for instance, by pollution). Therefore, Jørgensen's *exergy maximum principle* postulates that this work must be *maximal*.

Let us consider the impact of a poisonous substance from this point of view. Let the mortal concentration of poison be equal to c^m. In order to "poison" the living system we must increase the poison concentration from the basic concentration in a "normal" environment, c^*, to the mortal concentration, c^m. For this we must perform a work against the chemical potential $RT \ln(c^m/c^*)$. It is equal to (for one volume unit)

$$W_{ch} = RT[C^m \ln(c^m/c^*) - (c^m - c^*)].$$ (14)

If the exergy of volume unit of the living system is equal to Ex_v then, in accordance to our postulate, $Ex_v = W_{ch}$. Since the basic concentration for normal condition, c^*, is usually small, then the term $\ln(c^m/c^*)$ would be sufficiently large, and, as a result, the exergy, Ex_v, will also be sufficiently large.

Finally I would like to call your attention to the following. There is a principal difference between the types of *Gedankenexperiment* in classic thermodynamics and here. If, in classic thermodynamics, we perform the work on the system in order to change the state of the system, then, in nonequilibrium thermodynamics, to obtain the same result, we must perform the work on the system environment.

10.5 EXERGY AND INFORMATION

Introducing the new variables

$$p_i = C_i \Big/ \sum_{i=1}^{n} C_i, \qquad \sum_{i=1}^{n} C_i = A,$$

where A is the total amount of matter in the system, we can rewrite formula (13) in the form

$$Ex = A \sum_{i=1}^{n} p_i \ln \frac{p_i}{p_i^0} + \left\{ A \ln \frac{A}{A_0} - (A - A_0) \right\}.$$ (15)

The vector $\mathbf{p} = \{p_1,...,p_n\}$ describes the *structure* of the system, i.e., p_i are *intensive* variables. The value A is an *extensive* variable.

The value

$$K = \sum_{i=1}^{n} p_i \ln \frac{p_i}{p_i^0}$$

is the so-called *Kullback measure*, which is popular in information theory; $K \geq 0$, ($K = 0$ *iff* $\mathbf{p} = \mathbf{p_0}$).

What is the exact meaning of the Kullback measure? (See, for instance, Kullback, 1959.) Suppose that the initial distribution $\mathbf{p}^0$ is known. Then we have some additional information, and, in consequence, the distribution is changed from $\mathbf{p}^0$ to $\mathbf{p}$. So, $K(\mathbf{p}, \mathbf{p}^0)$ is the *measure of this additional information*. Note that K is a specific measure (per unit of matter). Then the product

$A*K$ can be considered as a measure of the *total amount of information* for the whole system, which has been accumulated in the process of transition from some *reference state* corresponding to a thermodynamic equilibrium, i.e., some *prevital* state, to the current state of living matter.

We can present the expression for the exergy in the form,

$$Ex = Ex_{inf} + Ex_{mat}, \quad \text{where}$$

$$Ex_{inf} = A * K(\mathbf{p}, \mathbf{p}^0) \geq 0,$$ (16)

$$Ex_{mat} = A * \ln \frac{A}{A_0} - (A - A_0) \geq 0,$$

i.e., as the sum of two terms: the first is a result of structural changes inside the system, and the second is caused by a change of total mass of the system.

If we accept Jørgensen's *exergy maximum principle* then we must postulate that the exergy must increase during the system evolution, i.e., $dEx/dt \geq 0$ along the system trajectory. From (16) we have

$$dEx / dt = dEx_{inf} / dt + dEx_{mat} / dt = A \frac{dk}{dt} + K \frac{dA}{dt} + \ln \frac{A}{A_0} \frac{dA}{dt}$$

$$= A \left\{ \frac{dK}{dt} + \left(K + \ln \frac{A}{A_0} \right) \frac{1}{A} \frac{dA}{dt} \right\} \geq 0.$$

Denoting $\ln A/A_0 = \xi$ and taking into account $A > 0$, we get the evolutionary criterion in the form

$$\frac{dK}{dt} + (K + \xi) \frac{d\xi}{dt} \geq 0.$$ (17)

If the positiveness of $d\xi/dt$ means an increase of the total biomass in the course of evolution then the positiveness of dK/dt can be interpreted as an increase of the information contained in a biomass unit. It is obvious that if the volume (mass) and its information contents increase, then the exergy also increase. If the total biomass is not changed ($d\xi/dt = 0$) then the system can evolve only if the information contents of the biomass are increasing. On the other hand, the information content can be decreasing ($dK/dt < 0$), but, if the total biomass is growing sufficiently fast ($d\xi/dt \gg 1$), then the exergy is growing and the system is evolving. The evolution is carried out if the biomass is decreasing, but the information contents of the biomass is growing (sufficiently fast). And last, there is one paradoxical situation when the exergy is increasing while the total biomass and its information contents are decreasing. If A, A_0 then $\xi < 0$, and $\xi(d\xi/dt) > 0$. From (17) we have

$$\xi \frac{d\xi}{dt} \geq \left| \frac{dk}{dt} \right| + K \left| \frac{d\xi}{dt} \right|.$$ (18)

this inequality can be realised if $|\xi| \gg K$ and $|dK/dt| \ll 1$, i.e., the information contents is sufficiently low ($K \ll 1$) and the process of its further decrease is very slow. Since the condition $A \ll A_0$ must be fulfilled in order to $|\xi| \gg 1$, then we can say that the system "paid" its own evolution by its own biomass.

In the vicinity of thermodynamic equilibrium, at the initial stage of evolution $K \cong \xi \sim 0$. Then from (17) we get $dK/dt > 0$, i.e., at the initial stage of evolution the system must increase the information contents of its own biomass in order to evolve.

It is interesting that the exergy maximum principle possesses certain selective properties. To clarify this we consider the system with constant total biomass (ξ = const, $d\xi/dt$ = 0). Then the state with exergy maximum corresponds to the maximum of K. We see that

$$\max_{\mathbf{p}} K = \max_i [\ln(1/p_i^0)]$$

This means that if the system is increasing its own exergy (without change of its biomass) in the course of its evolution, then the system will eliminate all the components except one, namely, the element with minimal initial concentration. Usually in a prebiological state (Jørgensen's reference state) this is some living element or substance. In other words, the system which is increasing its own exergy selects among its components those presented in a minimal quantity at the beginning of evolution.

But among these elements there are those that are necessary for life maintaining and the system must retain them. How to solve this contradiction? We can do this to introduce some constraints. We are required to maintain a certain level of system diversity. Formally it means (in this partial case) that

$$\max \sum_{i=1}^n p_i \ln(p_i/p_i^0) \text{ under the constraint } H = -\sum_{i=1}^n p_i \ln p_i = \text{const}$$

(and naturally $\sum_{i=1}^n = 1$). The exergy must be increasing but not arbitrary, maintaining some certain level of diversity. There is a deep analogy with Fisher's fundamental theorem of natural selection (see, for instance, Svirezhev and Passekov, 1990).

10.6   ENTROPY PRODUCTION AND CHEMICAL LOAD

Let us remember Formula (8) describing the "degradative" part of the entropy in a disturbed ecosystem, and present the value W as a sum of two terms: $W = W_f + W_{ch}$. If the first term (W_f) can be associated to a direct flow of artificial energy such as electricity, fossil fuels, etc. (energy load then the second term (W_{ch}) is associated to an inflow of chemical elements that maintain molar concentrations C_i ($i = 1,...,n$) inside the system (chemical load). We assume (as above) that the inflow W_f is dissipated inside the system and transformed into heat. Then we can rewrite (8) in the form

$$\sigma T = W_f + W_{ch} + P_1 - P_0. \tag{19}$$

To calculate W_{ch} let the system, described only by the chemical concentration $\mathbf{C}(t) = \{C_i(t), i = 1,...,n\}$ as state variables, move from an initial point $\mathbf{C}(t_0) = \mathbf{C}^0$ toward a stable equilibrium $\mathbf{C}$ which is maintained by the pressure of chemical load. We assume that the basic concentrations $\mathbf{C}^0$ correspond to a natural (wild) ecosystem. We assume also that if the chemical inflow would be stopped then the system would evolve to a natural state. In this case the maintenane of concentration $\mathbf{C}$ inside the system means that we are performing a work against the chemical potential. As a result the entropy is produced inside the system with the rate

$$\frac{d_i S}{dt} = R \sum_{i=1}^n A_{i,i_0} q_i = R \sum_{i=1}^n \ln\left(\frac{C_i}{C_i^0}\right) q_i. \tag{20}$$

Here A_{i,i_o} are the affinities ($i = 1,...,n$), and q_i are the flows of chemical elements into the system. Therefore

$$W_{ch} = RT \sum_{i=1}^{n} \ln \left(\frac{C_i}{C_i^0} \right) q_i. \tag{21}$$

10.7 AVERAGE RATE OF ENTROPY PRODUCTION AND EQUILIBRIUM IN AVERAGE

Let an anthropogenic pressure, described as the inflow $W(t) = W_f + W_{ch}$, start to act on the natural ecosystem at the initial moment t_0. If only the entropy pump "sucks out" the entropy from the system, then the rate of entropy production will be

$$\frac{dS}{dt} = \frac{1}{T} \{ W_f(t) + r(t)P_i(t) + D(t) \} + R \sum_{i=1}^{n} \ln \left(\frac{C_i}{C_i^0} \right) \frac{dC_i}{dt} - \frac{P_0}{T}, \tag{22}$$

where P_1 and P_0 are the gross primary production of anthropogenic and natural ecosystems, respectively, D is the rate of decomposition for dead organics, and r is the respiration coefficient. After integrating, we get

$$S(t) = S_0 + \frac{1}{T} \left\{ \int_{t_0}^{t} W_f dt + \int_{t_0}^{t} rP_1 dt + \int_{t_0}^{t} D dt + RT \sum_{i=1}^{n} \left[C_i \ln \left(\frac{C_i}{C_i^0} \right) - (C_i - C_i^0) \right] - P_0(t - t_0) \right\}. \tag{23}$$

We assume that the direct energy inflow (W_f) dissipates inside the system and the chemical inflow realises the transition $C^0 \Rightarrow C(t)$ working against the chemical potentials $\mu_i \sim \ln C_i^0$. We assume also that the temperature T and the productivity P_0 are constant in the course of the transition.

We consider a quasi-stationary process in the sense that $(1 - r(t))P_1(t) \approx D(t)$, i.e., there is a dynamic equilibrium between the formation of new biomass and the decomposition of dead organics. Let the period of an anthropogenic stress be equal to $t - t_0 = \tau$. Then the average rate of the entropy production

$$\left\langle \frac{dS}{dt} \right\rangle = \frac{S(t) - S_0}{\tau} = \frac{1}{T} \{ \langle W_f \rangle + \langle P_1 \rangle + \langle W_{ch} \rangle - P_0 \}, \tag{24}$$

where

$$\langle W_f \rangle = \frac{1}{\tau} \int_0^\tau W_f(\tau + t_0) d\tau, \quad \langle P_1 \rangle = \frac{1}{\tau} \int_0^\tau P_1(\tau + t_0) d\tau,$$

$$\langle W_{ch} \rangle = \frac{RT}{\tau} \sum_{i=1}^{n} \left[C_i \ln \left(\frac{C_i}{C_i^0} \right) - (C_i - C_i^0) \right].$$

These values are also some average characteristics. Let us consider the following thermodynamic models.

Model 1. We consider an average equilibrium assuming that the equilibrium condition $dS/dt = 0$ holds in some average sense, i.e., $\langle dS/dT \rangle = 0$. We assume also that the unique way to provide the last equality is to change the productivity so that

$$\langle P_1 \rangle = P_0 - \langle W_f \rangle - \langle W_{ch} \rangle. \tag{25}$$

Thus we have the thermodynamic model for reduction of the ecosystem productivity under anthropogenic pressure.

Model 2. Sometimes we have an experimental (observed) dependence with the productivity P_1 on W_f and $\mathbf{C}$. Then calculating the average $\langle P_1 \rangle$ we can check the equality (25). If $\langle P_1 \rangle < \langle P_0 \rangle - \langle W_f \rangle - \langle W_{ch} \rangle$ then we can assume that there is, in addition to the reduction of productivity, another mechanism which sucks the entropy excess. Such a mechanism would be a degradation of the environment. The entropy measure of the degradation is equal to

$$\langle \sigma \rangle T = \langle W_f \rangle + \langle W_{ch} \rangle + \langle P_1 \rangle - P_0. \tag{26}$$

(See also Section 10.2.)

Model 3. Let the dynamic of the anthropogenic system (with respect to the chemical load) be "impulse"; i.e., at the end of each year the system is spontaneously returned to the initial (natural) state so that at the beginning of each next year, the system is starting from the state. This situation is typical for agroecosystems that are not far from natural ones. This spontaneous transition is accompanied by the entropy production described by (25) for t = 1 year. Since the averaging interval is equal to 1 year, we can consider the average values as the annual ones. If we assume also that from year to year the system is at the equilibrium in average then we can introduce the following entropy measure for an environmental degradation (analogous to (26))

$$\sigma T = W_f + W_{ch} + P_1 - P_0, \tag{27}$$

where

$$W_{ch} = RT \sum_{i=1}^{n} \left[C_i \ln \left(\frac{C_i}{C_1^0} \right) - (C_i - C_i^0) \right].$$

It is interesting that the value W_{ch} formally coincides with the exergy of S. E. Jørgensen (1992).

10.8   CONCLUSION

The chapter demonstrates how to apply the concepts and methods of classical (and nonclassical) thermodynamics to ecological problems. Ecosystems are far from thermodynamic equilibrium. When we try to calculate the entropy by a direct method, we have difficulty solving the problem. It becomes unrealistic. I suggest the hypotheses of the entropy pump and calculation of the entropy production for ecosystems under anthropogenic stress by some circuitous way. Calculated circuitously the entropy can be used as a measure of environmental degradation under anthropogenic impact (for instance, intensive agriculture).

Recently S. E. Jørgensen suggested the exergy concept and based the optimal parametrisation for ecosystem models on this method. The nonequilibrium thermodynamics point of view proved that the exergy concept is one of possible corollaries of the thermodynamic extreme principles.

REFERENCES

Alexeev, V. V., Biophysics of living communities. *Uspekhi Fisicheskikh Nauk,* 120(4): 647–676, 1976.

Jørgensen, S. E., *Integration of Ecosystem Theories: A Pattern.* Second Edition, Kluwer Academic Publishers, Dordrecht, 383, 1997.

Kerner, E. H., A statistical mechanics of interacting biological species. *Bull. Math. Biophys.,* 19: 121–146, 1957.

Kerner, E. H., Further considerations on the statistical mechanics of biological associations. *Bull. Math. Biophys.,* 21: 217–255, 1959.

Khinchin, A. Ja., *Mathematical Foundations of Statistical Mechanics.* Gostekhizdat, Moscow, 167, 1943.

Kullback, S., *Information Theory and Statistics.* Wiley, New York, 395, 1959.

Rubin, A. B. *Thermodynamics of Biological Processes.* Moscow State University Press, Moscow, 240, 1967.

Schneider, E. D. and Kay, J. J., Complexity and thermodynamics. Toward a new ecology. *Futures,* 26(6): 626–647, 1994.

Svirezhev, Yu. M., Vito Volterra and the modern mathematical ecology. In: Volterra, V., *Mathematical Theory of Struggle for Existence.* Nauka, Moscow (the postscript to the Russian translation of this book), 1976.

Svirezhev, Yu. M., Entropy as a measure of environmental degradation. *Proc. Int. Conf. on Contaminated Soils,* add. volume: 26–27, Karlsruhe, Germany, 1990.

Svirezhev, Yu. M. and Passekov, V. P., *Fundamentals of Mathematical Evolutionary Genetics.* Kluwer Academic Publishers, Dordrecht, 395, 1990.

Ulanowicz, R. E. and B. M. Hannon, Life and the production of entropy. *Proc. R. Soc. London B,* 232: 181–192, 1987.

Volobuev, V. P., *Soils and Climate.* Izdatel'stvo AzAN, Baku, 252, 1953.

Weber, B. H., Depew, D. J., and Smith, J. D., (eds.), *Entropy, Information and Evolution: New Perspectives on Physical and Biological Evolution.* MIT Press, Cambridge, MA, 173, 1988.

CHAPTER 11

The core of this chapter is formulated as the Laws of Ergodynamics. The first law describes the situation where the considered system settles down in a steady state of ergodynamic equilibrium: maximum energy input and least specific entropy production. Translated to exergy: capture as much exergy as possible from the exergy input (from solar radiation) and utilise this exergy input as efficiently as possible, relative to the size of the structure to be maintained, as the input is limited. It agrees with the conclusions in Chapters 9, 10, 12, 13, and 14, although the author of this chapter has used other "tools" to reach the same conclusions.

The second ergodynamic statement is concerned with the situation where the system moves further away from thermodynamic equilibrium. The more the system moves away from thermodynamic equilibrium, the more energy flux is needed to maintain the system. At the same time there is an increase in internal energy of the system and an increase in order. This is in parallel with the development of ecosystems discussed in Chapters 9, 10, 13, and 14. The system will increase the exergy captured from the energy through-flow (<solar radiation), which implies that the system must increase its structure, costing more for maintenance. When the system has reached the level where it captures as much energy as possible, it moves toward least specific entropy production to better utilise (gradually) the energy input.

In the first part of the chapter, observations leading to six principles are presented. The six principles lead to the above presented Laws of Ergodynamics. The observations are based on a wide range of ecosystems and biological components that can be seen from the following examples: fish population in arctic lakes, islands in the Indian Ocean, whales from the Baffin Bay, European virgin forests, dominant species, and interspecific interactions.

The second part of the chapter is devoted to tests of the hypotheses, the two Laws of Ergodynamics. The author is able to support the hypotheses by looking into

1. Succession,
2. The production/biomass ratio or rather the reciprocal of the ratio, the turnover time, which is increasing with the development of ecosystems,
3. r- and K-selection,
4. Stability and diversity — the lack of complete correlation between these two concepts can be explained from the ergodynamic laws,
5. Global energy flow, which is completely in accordance with the principles presented in this chapter,
6. Darwinian evolution and selection, which is a parallel to the support of the maximum exergy principle in Chapters 9, 10, and 14.

The author offers a coherent treatment of the use of thermodynamics on ecosystems based upon a wide range of ecological observations. His results are consistent with many of the other chapters in this volume, but they are self-containing and are based on the utilisation of other "tools" combined with observations.

This chapter is a condensed version of a book with the same title to be published shortly by the author of this chapter.

11 Imperfect Symmetry: Action Principles in Ecology and Evolution

Lionel Johnson

CONTENTS

11.1 Premises 232
    11.1.1 Introduction 232
    11.1.2 The Stepping Stones of Rosetta 232
    11.1.3 The Probative System 234
    11.1.4 Are Arctic Systems Unique? 240
        11.1.4.1 Giant Land Tortoises of Aldabra Atoll 240
        11.1.4.2 Bowhead Whales of Baffin Bay–Davis Strait 244
        11.1.4.3 The Orange Roughy from the Deep Waters off New Zealand 244
        11.1.4.4 European Virgin Forests 245
11.2 Strong Inference 246
    11.2.1 The Autonomous Lake: Closed and Open Systems 246
    11.2.2 Categories of Ecosystems 247
    11.2.3 Characteristics of the Dominant Species 247
    11.2.4 Interspecific Interactions 248
11.3 The Action Hypothesis 249
    11.3.1 The Homeorhetic System 249
    11.3.2 Onsager's Reciprocal Relations 250
    11.3.3 Action Principles 251
    11.3.4 Biological Systems 251
    11.3.5 Origins 253
    11.3.6 The Move Away from Thermodynamic Equilibrium 253
    11.3.7 Broken Symmetry 254
11.4 Testing the Hypothesis 255
    11.4.1 Succession 255
    11.4.2 The Production/Biomass Ratio 259
    11.4.3 *r*- and *K*-Selection 260
    11.4.4 Action, the Binding of Energy and Time, is the Inherent
        Stabilizing Property of Living Things 262
    11.4.5 Stability 262
    11.4.6 Diversity 263
    11.4.7 Stability and Diversity 268
11.5 Global Energy Flow 269
11.6 Evolution and Natural Selection 270
    11.6.1 Evolution 270

1-56670-272-0/01/$0.00+$.50
© 2001 by CRC Press LLC

11.6.2 Natural Selection..275
11.7 Conclusion..276
11.8 Darwin's Legacy 277
Acknowledgments...278
References 278

11.1 PREMISES

11.1.1 INTRODUCTION

> The job of science is not to collect a body of general notions. It is to produce theories.
>
> Rigler 1975a

Whether or not physical principles govern the fundamental behaviour of organisms is a question of long standing, for biological systems have proved remarkably opaque to attempts to bring them within the fold of thermodynamics. Nevertheless, all organisms and ecosystems represent energy transfer through materials; therefore the laws and principles of thermodynamics should be applicable. This being so the principles should be discoverable; if not, then there are unknowable aspects that preclude further scientific consideration. If we adopt the former stance, then solving the cryptogram presented by the emergent patterns of the natural world is extremely important. Without understanding the fundamental mechanisms, we will experience great difficulty in our attempts to ameliorate the present sad state of the world ecosystem.

11.1.2 THE STEPPING STONES OF ROSETTA

> There is a general consent that primeval nature, as in the uninhabited forest or the untilled plain, presents
> a settled harmony of interaction among organic groups which is in strong contrast with the many serious
> maladjustments of plants and animals found in countries occupied by man.   Forbes 1880

The observations on which the thesis to be presented is founded resulted from the investigation of lakes in the Canadian Arctic (Figure 11.1). I believe they provided the connecting link that allowed the biological cryptogram to be deciphered, as the Rosetta stone allowed translation of Egyptian hieroglyphs. These lakes, essentially pristine in nature, range in size from several hectares to some of world rank. They were shown to contain fish populations which, in the great majority of cases, exhibited a structural pattern of great interest and significance.

Canadian Arctic lakes are relatively young in geological terms, having been formed during the retreat of the most recent Pleistocene ice sheet between 10,000 and 6000 BP; since that time they have been sequestered from the changing outside world by their geographic inaccessibility and hostile climate. Given their recent formation and the harsh environmental conditions of short open water period, low temperature, and low primary productivity, it is no surprise to find that their ecosystems are relatively simple. These lakes are among the few ecosystems that, up to the present time, have remained largely free from human interference other than possible worldwide effects of global warming and atmospheric transfer of pollutants.

The initial stimulus to developing a thermodynamic approach to ecosystem structure was occasioned by the results of surveys of the Barren Grounds large lakes carried out in the late 1950s and early 1960s by the Arctic Unit of the Fisheries Research Board. It was shown that the previously unexploited fish populations of these lakes were composed of old individuals exhibiting bell-shaped length- and age-frequency distributions. This indicates a high degree of stability in face of the environmental conditions experienced (1975, 1976).

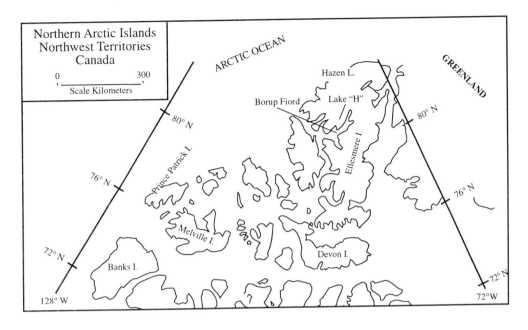

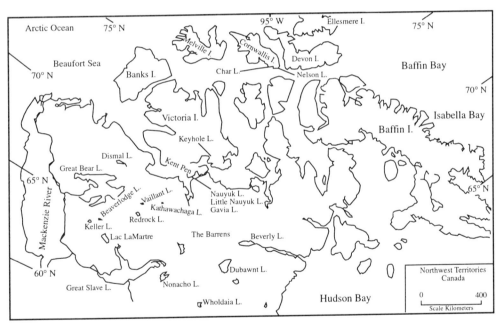

FIGURE 11.1 Map of the Canadian Northwest Territories, showing the lakes referred to in the text.

These results were in direct conflict with two tenets of the then prevailing wisdom: (1) that age and length distributions followed a negative logarithmic distribution curve (Beverton and Holt 1957; Ricker 1975), whereas the actual curves were similar to those that might be expected of a stable mammal population where size is largely genetically controlled (Figure 11.2) and (2) the high degree of constancy or stability was contrary to the theoretical postulate that simple systems tend to fluctuate, whereas diverse systems are more stable (MacArthur 1955; Elton 1966). Although now greatly revised and modified, it was, at the time, considered an important contribution to theory

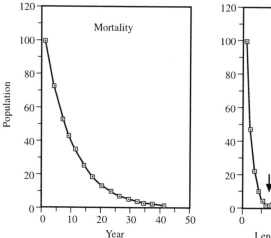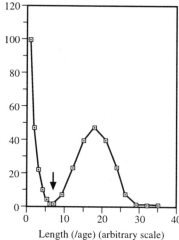

FIGURE 11.2 Diagram of the age- or length-frequency distribution, as frequently depicted in fisheries science and as found in undisturbed lakes in the Canadian Northwest Territories.

(Kolata 1974). The stability factor in Arctic lakes was particularly significant given that the lakes in question are surrounded by tundra, and tundra was then recognized as supporting the ecosystem at the simple/fluctuating pole of the diversity-stability spectrum. The major qualitative differences between the lake and tundra appeared to be (1) the "open" nature of the tundra which, during the short summer, is the feeding and breeding ground of vast numbers of migratory birds and mammals, compared with the high degree of autonomy of the lake, and (2) the buffering of climatic variability by the physical properties of water. Such beginnings demanded further investigation.

11.1.3 The Probative System

Platt (1964) concludes that the approach to analysis of complex systems most likely to be rewarding is to study the simplest system you think has the properties, in which you are interested. From this study draw strong inferences that can be tested in more complex systems. Perhaps the simplest natural autonomous ecosystems in existence are the permanently frozen lakes of Antarctica (Goldman 1970) although here the biota consists of algae and bacteria only. More interesting, more abundant, and more accessible are the lakes of the Canadian Arctic which support a more diverse biota ordered in a simple food web with a single dominant species, which is also a terminal predator.

During winter, which lasts for 9 to 10 months, many Arctic lakes are completely cut off from the external world with all tributary streams frozen solid and the lake itself encapsulated by a layer of ice up to 2 m thick. Many small lakes exist in small sparsely vegetated watersheds with most of the spring snow melt seeping through the active surface layer; hence the lakes receive only a very small input of nutrients and dissolved organic carbon, and this is balanced by almost the same amount being lost through the outlet (Rigler 1975b).

In many lakes Arctic charr (*Salvelinus alpinus*) is the dominant species and at the highest latitudes the only fish species present (Johnson 1980). In contrast to their low primary productivity, fish populations are generally of large size and numerically abundant. Predators, almost exclusively fishing birds, are relatively insignificant in their effect on the adult population, being present only during the short open water period. In larger, deeper, and more southerly lakes, two, noncompeting species, lake whitefish (*Coregonus clupeaformis*) and lake trout (*S. namaycush*), occur as codominants. In these large, deep lakes the fish populations are essentially free from predation by birds because of the large size of individual fish and their access to a refuge in deep water.

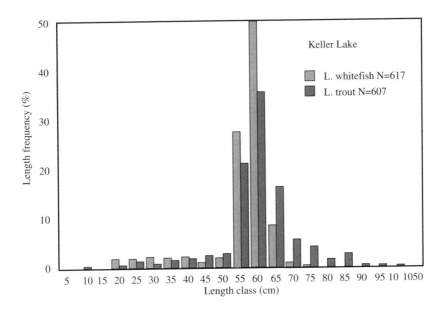

FIGURE 11.3 Length-frequency of lake trout (*Salvelinus namaycush*) and lake whitefish (*Coregonus clupeaformis*) from Keller Lake in the N.W.T., Canada.

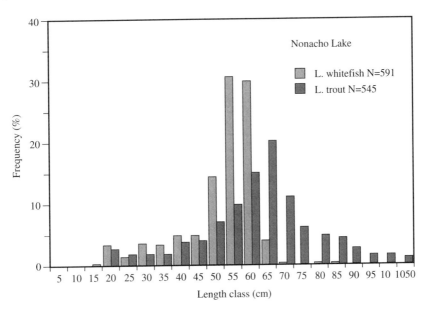

FIGURE 11.4 Length-frequency of lake trout (*Salvelinus namaycush*) and lake whitefish (*Coregonus clupeaformis*) from Nonancho Lake in the N.W.T., Canada.

The fish populations in these Arctic lakes reach a stationary state evident in the bell-shaped, or unimodal configuration of the length–frequency distribution of the sample (Figures 11.3 and 11.4). In addition some populations show a second mode of relatively larger or smaller fish than those of the main mode (Figure 11.5). This modal configuration is indefinitely retained in the absence of significant environmental change (Figure 11.6). These populations are made up of old fish; in Arctic charr the average age may be 12 to 14 years with some individuals reaching 25 years; in

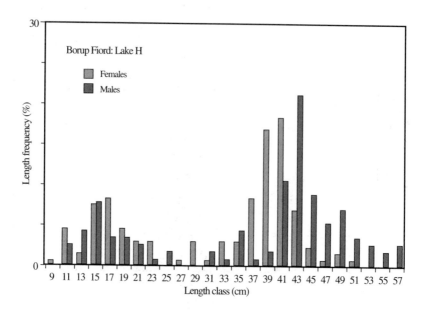

FIGURE 11.5 Length-frequency distribution of Artic charr (*Salvelinus alpinus*) from "H," Borup Fiord, Ellesmere Island, N.W.T.

the case of lake trout in the big, deep lakes, some individuals probably reach 60 to 70 years of age, possibly more. Significantly, there is only a very loose correlation between length and age, with the widest spread of age in the modal length class (Figures 11.7 and 11.8). This steady state reached in Arctic lakes was considered to be homologous with the "climax" state of vegetation theorists.

The unimodal configuration generally observed in the size distribution of fish samples is frequently interpreted as an effect of gear selectivity (Ricker 1975) or predation and, in some cases, cannibalism. Predation by birds or cannibalism effectively removes juvenile "surplus" to the requirements for replacement of the adults lost to natural mortality, effectively strengthening the modal configuration, although not its primary cause (J. Hammar, personal communication). In the present investigations the effect of gillnet selectivity was obviated by the use of gillnets with a wide range of mesh sizes and, in certain cases, by intensive fishing (Johnson 1983).

In 1962, a long-term study of Keyhole Lake (Lat. 69°22′N; Long. 106°14′W), a small lake (47 ha) on Victoria Island, was initiated (Hunter 1970). Between 1962 and 1967 the previously untouched Arctic charr population was reduced to a small remnant of its initial abundance by intensive fishing. By direct observation the initial population was thus shown to have the same well-defined unimodal configuration found in less intensively sampled lakes. In 1975, after an interval of 8 years without further disturbance, a second sampling showed that the original configuration had been regained (Figure 11.9) (Johnson 1983). A return to the original configuration necessitates the existence of constant forces fashioning the population structure.

The density of fish in Arctic lakes is generally high and comparable with that of many lakes in more southerly regions despite very low primary productivity. In a High Arctic lake, Char Lake (Lat. 74°N) at Resolute Bay, Cornwallis Island, Rigler estimated the standing stock of Arctic charr to be 10 kg.ha^{-1} with an annual primary productivity of 22.0 g.C.m^{-2}.y^{-1} (Rigler 1975b; McCallum and Regier 1984;), and in Keyhole Lake, a standing stock of 44 kg.ha^{-1} was determined by direct observation (Hunter 1970).

Further investigations in Keyhole Lake (1985–1988) (Vanriel and Johnson 1995) led to the conclusion that the feeding relationships are relatively simple, with the various trophic levels forming a hierarchical structure in which the energy distribution is pyramidal (Figure 11.10), comparable with the pyramid of numbers developed by Elton (1927) and the thermodynamic notions of Semper (1881). These "trophic levels" represent energy accumulations and do not necessarily indicate the direct

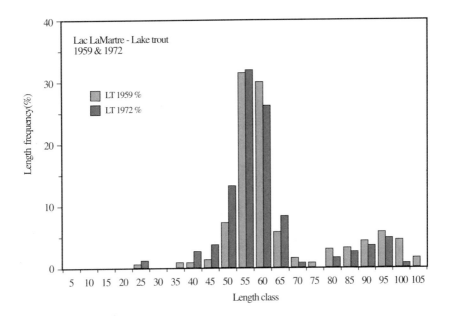

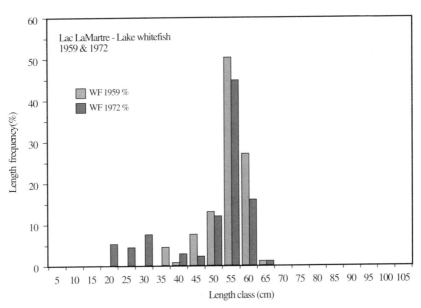

FIGURE 11.6 Length-frequency distribution of lake trout (*Salvelinus namaycush*) and lake whitefish (*Coregonus Clupeaformis*) from Lac LaMatre in 1959 and again in 1972, showing constancy of the distribution. Note the second mode of large lake trout.

transference of energy from one level to the one above; manifestly, algae do not feed on bacteria. At each ascending hierarchical level the energy density was shown to become more and more concentrated, both in time and space, through an increase in mean age, abundance, size, and uniformity, as well as an increase in energy density in the specific sense of joules per gram (Jg^{-1}).

An interesting but anomalous situation was encountered in one small lake, Gavia Lake (7 ha) on the Kent Peninsula of the northern mainland, where the population was dominated by one large Arctic charr, with three to four specimens of medium size, the rest being a rather disorganized mass of small

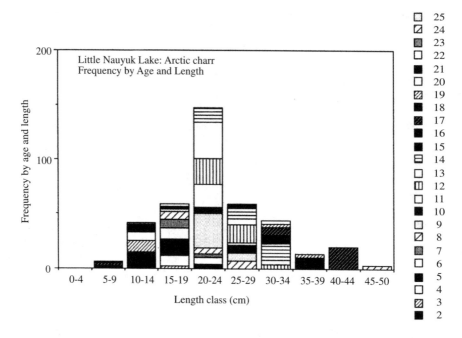

FIGURE 11.7 Little Nauyuk Lake, Artic charr (*Salvelinus alpinus*) length- and age-frequency distribution.

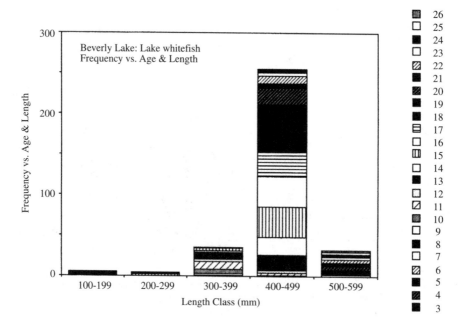

FIGURE 11.8 Beverly Lake, Lake whitefish (*Coregonus clupeaformis*).

individuals, 10 to 15 cm in length (Johnson 1994c). With the removal of the large fish, the remaining population rapidly assumed the more usual modal configuration, retaining this characteristic structure over the subsequent years of the investigation (Figure 11.11). The initial length–frequency distribution was attributed to stress, most likely the result of late winter conditions when the lake is reduced to approximately half its volume owing to ice formation. This is corroborated by a similar configuration

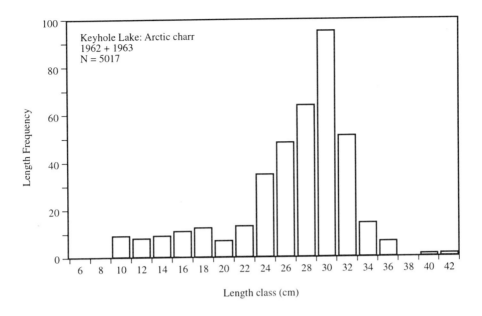

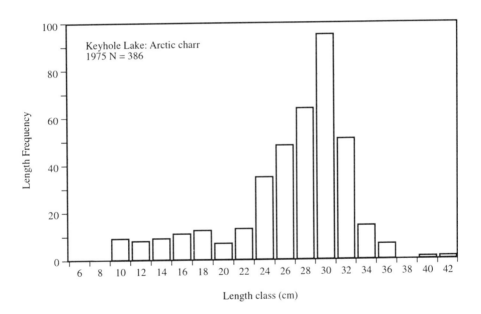

FIGURE 11.9   Keyhole Lake. Arctic charr (*Salvelinus alpinus*), length-frequency distribution in 1962 + 1963 and 1975, showing the recovery of the original population structure after reduction of the initial population to very low abundance.

observed in plants where increasing density of sowing created increasing inequality among individuals (Harper 1967) (Figure 11.12). With the removal of the large individuals the population assumed a bell-shaped length–frequency configuration that was retained until the experiment was terminated.

The evidence from Gavia Lake indicates the fine balance existing in the forces determining population structure. In unusually harsh circumstances a population evidently has the capacity to develop an internal hierarchy that is eliminated when conditions are normalized.

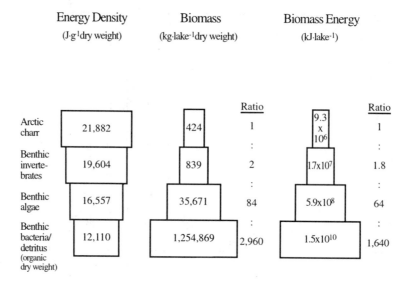

FIGURE 11.10 Keyhole Lake: energy density, biomass, and biomass energy represent accumulated energy reserves at various levels. Direct energy transfer from one level to another is not necessarily implied, i.e., algae are not consumed by many organisms, but pass through a detrital, bacterial, or fungal cycle first. (From Vanriel, P. and L. Johnson, *Ecology*, 76.6, 1741–1757, 1995. With permission.)

The Gavia Lake Arctic charr population provides an important reminder that apparently anomalous occurrences are to be expected in any attempt at generalizing biological processes.

11.1.4 Are Arctic Systems Unique?

If general inferences and conclusions are to be made concerning biological systems, it is essential to show that observed phenomena are worldwide and not specific to one region or location. Thus the question arises: are Arctic lakes unique or can similar configurations be identified in other ecosystems in various parts of the world. Unfortunately, many prime examples now exist only in anecdotal form: the unimaginable herds of bison of the North American Prairies; the immense flocks of the now extinct passenger pigeons; the large populations of the great whales in the early 19th century before whaling reduced their numbers to insignificance; the vast herds of ungulates and their associated predators on the African veldt; the now extinct moas of New Zealand; and the numberless accounts of rivers that, "in the early days," could be crossed on the backs of the ascending fish.

Nevertheless, in the appropriate circumstances of isolation, well-documented cases can be found in which the dominant species assumes a configuration in all respects comparable with the undisturbed fish population. These findings have been extensively discussed in previous publications (Johnson 1972; 1975; 1976; 1981; 1983; 1985; 1988; 1990; 1992a; 1994a; 1994b). Only four examples will be given here, one from terrestrial animals, the giant tortoises of Aldabra (*Geochelone gigantaea*); one from marine mammals, the Baffin Bay–Davis Strait bowhead whale population (*Balaena mysticetus*); one from the virgin forests of Europe; and one from the deep-sea fishery for the orange roughy (*Hoplostethus atlanticus*).

11.1.4.1 Giant Land Tortoises of Aldabra Atoll

Aldabra Atoll lies in the Indian Ocean about 500 km northwest of Madagascar and about 700 km from the African mainland. Formerly a dependency of the island of Mauritius, it is now administered by the Government of the Seychelles.

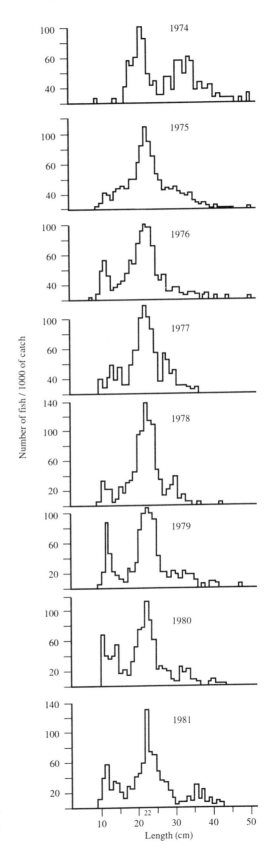

FIGURE 11.11 Gavia Lake Arctic charr (*Salvelinus alpinus*). Length-frequency distribution showing the assumption of a model distribution following removal of the few large fish in 1974. The bimodal configuration in 1974 is believed to be due to gillnet mesh selectivity. This was corrected in subsequent years.

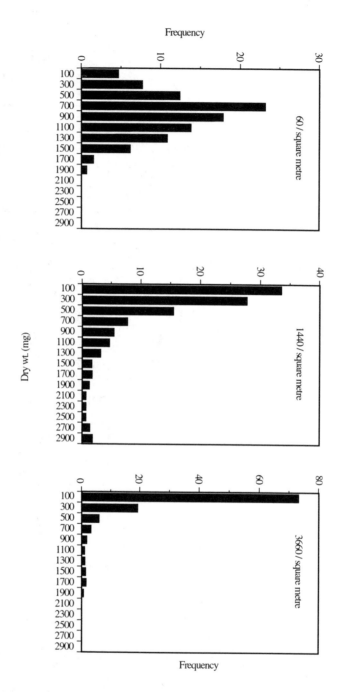

FIGURE 11.12   Showing the influence of stress on population configuration. Three different levels of planting density of *Linum usitatissimum*. As planting density increases the modal configuration is shifted to a negative logarithmic curve. (From Harper, J. L., *J. Ecol.*, 55, 247–270, 1967. With permission.)

"Aldabra is one of the least disturbed of all low-latitude islands and for historical and environmental reasons possesses an exceptionally rich and interesting fauna and flora." Stoddart and Wright 1967.

In a tropical setting and in isolation, surrounded by water rather than land, Aldabra is composed of numerous islets that are mirror images of an Arctic lake.

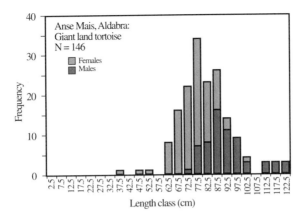

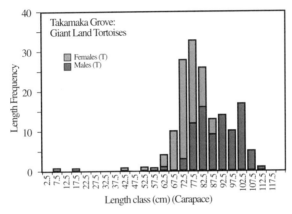

FIGURE 11.13   Aldabra Atoll, Giant land tortoise (*Geochelone gigantaea*) showing length-frequency of populations from Takamaka Grove and Ansc Mais. Aldabra Atoll. (From Gaymer, R., *J. Zool. Lond.*, 154, 341–363, 1968. With permission.)

The flora and fauna were intensively studied during the 1960s as part of an environmental impact assessment in preparation for the construction of a staging base for the British Forces (Bourn and Coe 1978; Gaymer 1968; Grubb 1971; Swingland 1977; Swingland and Coe 1978; Swingland and Lessels 1979).

Initially there were no mammals on the atoll apart from three species of bat, but cats, dogs, and rats have been subsequently introduced. The dominant terrestrial animal is the giant land tortoise that in the previous century was heavily exploited for meat and "tortoise shell." Due to political pressure applied by leading British scientists, including Charles Darwin, Richard Owen, and Joseph Hooker, the trade was curtailed and eventually stopped, allowing the tortoises to regain their former abundances (Günther, 1877, quoted by Stoddart and Wright 1967). The various islets each contain distinct tortoise populations.

All the studies on the tortoise indicate a unimodal size structure of the population (Gaymer 1968) (Figure 11.13). Adult male tortoises reach a maximum length of 100 cm as measured along the carapace; females are somewhat smaller with maximum length of about 80 cm. In the population of Middle Island the majority of tortoises have lengths between 60 to 80 cm at an age of 10 to 25 years of age reaching asymptotic size at an age between 20 and 30 years (Grubb 1971). In captivity individuals have reached ages of over 100 years. Recognizing the apparent scarcity of smaller tortoises reflected in the unimodal length–frequency pattern, Grubb remarks: "because of the paucity of small tortoises in these samples a special effort was made to measure animals of about 20 cm or less," but despite this additional effort small tortoises in expected abundances were not encountered in any of the populations.

11.1.4.2   Bowhead Whales of Baffin Bay–Davis Strait

The bowhead whale is one of the largest animals in existence, measuring 12 to 17 m in length and weighing up to 80 tonnes, although their primary food consists of the minute copepods largely of the *Calanus* species (Finley 1990). Ages are difficult to determine with a satisfactory degree of precision, but since 1981 in the Western Arctic, where legal whaling by the Inuit takes place, six "traditional" harpoon heads have been found in five separate specimens (Philo et al. 1993; George et al. 1998). This indicate ages of over 130 years, for it is at least this length of time since such technology was used by the Alaskans. Hunted for their oil and "whalebone" during the 19th century, it is estimated that some 28,000 individuals were killed in the Eastern Arctic (Finley and Darling 1990; Ross 1979) before low returns and regulations terminated the fishery. K. J. Finley personal communication) believes that bowheads are at least 20 years old before reaching sexual maturity, making them the existing animal with the longest known generation time. Based on the above, the reproductive lifespan probably exceeds 60 years. For these reasons bowhead whales along with other large whales are considered the "quintessential K-strategists" (Nerini et al. 1984). Given their immense bulk, the specific metabolic rate is likely to be low.

Although now protected from hunting for nearly 100 years, there appears to have been little increase in the population. A tiny remnant of the original population has been studied intensively by Finley (Finley 1990; Finley and Darling 1990) at Isabella Bay, Baffin Island. In this population juveniles are extremely rare, although Finley suggests that there might possibly be segregation of size groups, although no "juvenile" pods have been observed. Finley provides a length–frequency distribution to the Isabella population as observed on 28–29 September 1986 (Figure 11.14).

11.1.4.3   The Orange Roughy from the Deep Waters off New Zealand

The orange roughy (*Hoplostethus atlanticus*) is a fish that exists worldwide, occupying waters at a depth between between 700 to 900 m. Gauldie et al. (1989) believe that their habitat is bounded by the species' physiological response to differences in oceanographic conditions, particularly oxygen levels. Many of the stocks are essentially pristine, having been exploited only in recent times with

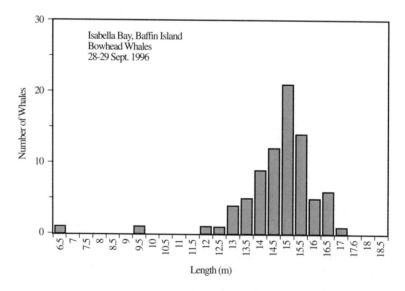

FIGURE 11.14 Baffin Bay–Davis Strait bowhead whale (*Balaena mysitcetus*) population. Length-frequency distrubutaion as measured by K. J. Finley on 28–29 Sept. 1986 at Isabella Bay, Baffin Island, N.W.T., one of the most important gathering and feeding grounds for bowhead whales. (From Finley, K. J., *Arctic*, 43.2, 137–152, 1990. With permission.)

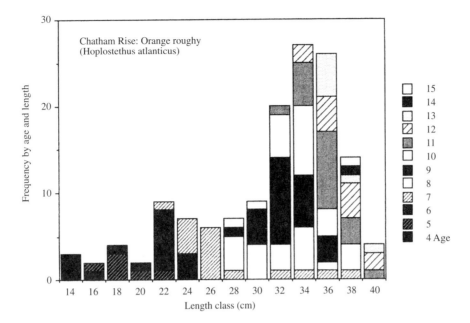

FIGURE 11.15 Chatnam Rise, New Zealand. Orange roughy (*Hoplostethus atlantiicus*). Length- and age-frequency distribution. (From Gauldie, R.W., I.F. West, and N.M. Davies, *J. Appl. Ichthyol.*, 5, 127–140, 1989. With permission.)

the development of the necessary technology. Gauldie et al. showed that there is an almost exact correspondence between the structure of these populations, characteristically "K-selected," and those of Arctic charr and other northern species. Samples of fish showed a bimodal structure with the larger individuals exhibiting great uniformity. The smaller mode was relatively poorly represented numerically and frequently absent. The size and age frequencies remained more or less stationary in the face of exploitation and declining biomass. Gauldie et al. provide numerous examples of orange roughy populations from the deep waters off New Zealand, all of which follow a similar pattern with a mode at 34 cm. One example, the orange roughy population from Chatham Rise, is provided (Figure 11.15).

These features are virtually identical with those observed during the initial stages of the fishery for three species of *Lates* and one of *Luciolates* in Lake Tanganyika (Coulter 1970).

11.1.4.4 European Virgin Forests

It is evident that forest ecosystems are a different order from the lake or terrestrial systems so far described, for the species of greatest age and biomass, generally regarded as the dominant species, is also the main primary producer. Until the arrival of man and the use of tools, vast areas of the temperate and tropical regions were covered with forest. The growth characteristics of trees and their largely inedible nature enabled them to assume dominant status and, with it, the capacity to impose "top-down" constraints on the growth of other vegetation. At the same time, the proporties of the trees and the constraints they impose determine the characteristics of the dependent animals. Food chains or food webs form starting with the leaves and flowers of trees, the subordinate vegetation or decomposing detrital material. The great variety of habitat provided by the physical structure of the forest allows for a great diversity of species and highly complex food webs.

Many of the temperate forests of the world have been modified by man so that it is now difficult to find undisturbed regions where natural conditions prevail. However, E. W. Jones in his detailed study of the virgin forest in Europe found examples of uniformly sized trees with a wide range of ages. Jones comments:

A far commoner form [than the inversed-*J*] of uneven-aged forest has a regular continuous canopy and has very much the appearance of an even-aged high forest. It is perhaps too much to say that all ages are present, but although old stems predominate, the dominant stand includes a wide range of ages. In an example of Bosnian beech forest given by Cermak the age range was 200 years, although 70% of the stems were between the limits of 170 to 200 years: the youngest stems were 90 to 100 years old. Regeneration in such forests often appears to be absent, and there is a striking deficiency of smaller stems, giving a diameter distribution which may approach that of an even-aged stand. This, however, does not mean that the forest is not reproducing itself and maintaining its present structure: the condition doubtless arises through the relatively small fraction of the life spent in growing from ground-level into the canopy when a gap is formed. Where the average length of life is 300 years only two or three gaps per hectare, each containing one or two young stems, would be sufficient to perpetuate the forest.

From Jones, E. W., *New Phytol.*, 44, 130–148, 1945. With permission.

The redwood forests of California and Oregon occupy a position comparable to the bowhead whales of Baffin Bay, in that they exhibit immense size and great age in ecosystems of relatively low diversity. They are among the most productive of temperate zone climate forests (Whittaker 1960; 1965).

11.2   STRONG INFERENCE

11.2.1   THE AUTONOMOUS LAKE: CLOSED AND OPEN SYSTEMS

A lake is an old and relatively primitive system, isolated from its surroundings. Within it matter circulates, and controls operate to produce an equilibrium comparable with that in a similar area of land. In this microcosm, nothing can be fully understood until its relationship to the whole can be seen.

Forbes 1887

The Second Law, one of the fundamentals of thermodynamics, states that energy within an *isolated* system assumes state of thermodynamic equilibrium: a state of maximum entropy in which no internal work can be done. An isolated system is one in which there is no exchange of energy, entropy, or matter with the external world. This is an idealized state that cannot exist in actuality for no materials are totally impervious to heat transfer. Manifestly, biological systems cannot exist in an isolated system for their very existence is dependent on a continuing input cycle of free energy or energy capable of doing work.

Thus I postulate that, in biology, the equivalent construct to the isolated system is a *closed ecosystem* with a regular cycle of energy input. The accepted terminology, *closed system*, is perhaps unfortunate because it is frequently confused with *isolated system*. A closed system exchanges energy and entropy, but not materials, with the external world. Thus free energy, or energy of low entropy capable of doing work, such as light, can cross the system boundaries, be converted to energy of high entropy, and then be discharged to the outside world. A closed system, in effect, functions as a sink for free energy, dissipating only heat to the external universe. Equally in biology, this is an idealized situation; thus biological systems are frequently described as *open systems* in which matter as well as energy and entropy are exchanged. However, as Dr. H. Hamilton (personal communication) points out, there is really no such thing as an open system, for such a condition as the term implies is actually a subsystem of a larger *closed* system.

At the global level, the biosphere is therefore perceived as a closed system exchanging only energy and entropy with the external universe. However, this concept has been challenged by Westbroek (1991) on the grounds that earth movement, erosion, and volcanic activity are continually changing the availability of nutrients and other factors necessary for life processes, thus materially altering the internal environment and the capacity of a system to assimilate energy. This is undoubtedly true, but this view must be modified by the fact that significant alterations in the Earth's environment, for the most part, take place relatively slowly, and catastrophic events, such as the impact of a large meteor, are usually widely spaced in time. In these circumstances the relatively rapid rate at which the biota is able to adjust and accommodate to new conditions allows a relatively stable state to be reached over long time periods. Hence, there are long periods of stasis over evolutionary time as recognized in the theory of punctuated equilibrium (Eldredge and Gould 1972; Gould and Eldredge 1977). As Slobodkin (1964) states,

> I accept as an axiom (as did Darwin and Wallace) that, baring drastic physical change or the introduction of new species, the number and kinds of animals found in any region of the world remains essentially constant from year to year.

This I interpret as meaning that over long time periods there are no discernible trends in species or population abundances. Notwithstanding, when viewed over the whole panorama of evolutionary time, the biota appears to be in a continuous state of change with the evolution of new species and the elimination of others. Evidently there are two different factors to be considered, one leading to short-term stasis, the other inducing change.

Within the biosphere all ecosystems function as subsystems of the whole although some subsystems much more closely approach the closed condition than others. The postulate is that many Arctic lakes, for long periods of time, approach the closed condition within the closed biosphere and to this extent they form microcosms of the whole.

11.2.2 CATEGORIES OF ECOSYSTEMS

It is evident that there are two basic types of ecosystems, each with many variants. One is plant dominated exemplified in the vast forests of recent times occupying relatively moist regions of the Earth. The other is the animal-dominated ecosystem as exemplified in a lake ecosystem or grasslands formed in generally less favorable climatic or edaphic regions unsuited to forest cover. The lake type has short, linear food chains which, as diversity increases, support food webs of increasing complexity. In such ecosystems the energy reserves are largely contained in the detritus (Vanriel and Johnson 1995). These reserves are important in reducing fluctuations in energy input. In the same category are grassland, where food chains are again relatively short and the main organic reserves are in the soil. As in all biological systems, the situation in grasslands in not quite as simple as it first appears. Much of the animal biomass is tied up in the primary grazing animal, but there are terminal predators such as wolves or lions that do not seem to control, to a significant extent, the size of this population, but function largely as housekeepers, maintaining the herds in good condition.

11.2.3 CHARACTERISTICS OF THE DOMINANT SPECIES

The dominant organisms, terminal in the food web in one case, but the main primary producer in the other, exhibit similar characteristics of maximum continuous biomass and greatest age. The major characteristics are large size, great age, and uniformity in size, and a low level of juveniles or potential recruits to the "establishment." The dominant species may be defined as the species that most closely approaches its maximum potential value for energy acquisition and conservation; it is the species limited only by energy availability (including in plants access to nutrients and moisture) and space; thus it is able to live out its physiological life span. It is the species that determines the overall stability and continuity of the system (Frank 1968). The dominant species is a self-regulating entity, internally modulating growth rate and abundance and imposing constraints on species lower in the hierarchy. The harsher and more pulsed the energy input, the greater the tendency to dominance by a single species.

Uniformity in size within a fish population came as an initial surprise but emerged as an important factor in attaining a state of least energy flux. Uniform distribution of energy among the largest possible group of similar individuals represents a condition of maximum energy accumulation relative to energy input. This is evident from the fact that the larger the individual the lower the specific metabolic rate (Kleiber 1961); therefore, a greater biomass can be maintained for the same energy input in a population of uniformly sized individuals, than in a population with a negative logarithmic size–frequency distribution. Similarly, the greater the average age and the more uniform the individual size, the less the demand for younger and smaller (and more energetically expensive) replacement individuals, hence the greater and more constant the total biomass.

Great variation in age within the modal size group, in a class of animals (or plants) generally considered to be of indeterminate growth pattern, indicates uniform energy distribution through symmetrically interacting individuals. This necessitates modulation of individual growth rate by interactive processes. This uniform energy distribution within a closed system thus has evident similarities with the uniform distribution of energy within an isolated system as required by the Second Law. The process of energy equipartitioning among uniformly interacting atomisms or thermodynamic engines has been described by Soodak and Iberall (1978) as *homeokinesis*. Homeokinesis thus transforms a population of individuals into a coherent, loosely integrated thermodynamic unit.

In the dominant species this unit attains maximum biomass and exhibits the most efficient energy usage, that is, the greatest biomass per unit of energy assimilated, the least specific energy dissipation, or the least specific entropy production.

11.2.4 Interspecific Interactions

Unless unique properties are assigned to the dominant species, it must be inferred that all species have the same inherent tendency to accumulate energy and delay to a maximum its passage through the system. This is confirmed by the fact that many different species in different phyla can assume dominance in the appropriate circumstances. Thus each species population has a dampening function that increases as the trophic hierarchy is ascended. However, the extent to which a species approaches a state of least dissipation is dependent on interspecific interactions, particularly the extent to which it is consumed by other species, as well as the characteristics of its life history that enable it to survive at all in a given ecosystem.

Semper (1881) was the first ecologist to consider the transfer of biological material along the food chain (heterotrophy, or different feeding, as in animals as opposed to autotrophy in plants) and the natural law that energy transformations demand energy expenditure. Hence there is a decline in the energy accumulation at each level of consumption. He also suggested that the transfer rate from prey to predator was about 10%. However, this figure, although frequently accepted as a reasonable generalization, has been shown to vary greatly depending on the organisms concerned (McIntosh 1980a; Slobodkin 1972). Whatever the actual figure, considerable energy is dissipated in work processes when it is "pumped" from a lower, less energy dense trophic level to one that is higher or more energy dense (Vanriel and Johnson 1995). In addition, undigested residues are voided in the form of food readily available to insects and microorganisms; again, this necessitates further energy dissipation. As in a river, energy is dissipated in the riffles (energy transfer), but conserved in the pools, thus the greater the complexity of the food web, the greater the number of energy transformations and the greater the energy dissipated. Hence the high-standing stock, despite the low primary productivity in Arctic lakes, is attributed to short and direct food chains.

For an ecosystem to exist, the sum of all the dissipative interactions must be less than the rate of energy input during the formative stages, or equal to the input at the steady state. Hence the "goal" of an autonomous ecosystem as a whole, as well as its parts, is a state of least energy flux and maximum biomass. As Bormann et al. (1969) and Likens et al. (1970) conclude the community tends to dampen environmental fluctuation, to stabilize its microclimate and environmental chemistry.

The observations presented above lead to the conclusion that there is a pattern of energy distribution and energy flow within an ecosystem, and the simpler and more autonomous the system, the more clearly does the pattern emerge. In the first instance, there are the general characteristics of a dominant species: large size, great age, and uniformity of individuals. Second, the general pattern of energy flow from small-sized, highly fluctuating populations to larger, longer-lived populations that impose an increasing degree of continuity and stability.

From these observations the following strong inferences may be drawn.

1. A closed ecosystem with a regular energy input assumes a steady state of maximum attainable biomass, or least energy flux, variously referred to as least specific dissipation (Prigogine 1978), or least entropy production. Energy flux is defined as the energy input required to sustain a given unit of biomass energy.

2. Each population of each species in an ecosystem tends to acquire maximum free energy and time delay to the maximum passage of that energy through the population. Differences in the capacites of populations to acquire and conserve energy lead to the formation of a hierarchy. Populations, through interactive processes, function as coherent units or holons within the hierarchy.

3. An ecosystem functions as a series of coupled "shock absorbers." Each species population, in that it time-delays energy flow to the maximum of which it is capable, imposes a dampening moment on fluctuations in energy input.

4. Energy transfer between species is energetically "expensive"; hence, the greater the diversity, the greater the number of energy transfers and the greater the energy expended in maintaining the system. Thus, in systems of low diversity, high biomass accumulates despite low primary productivity and, conversely, in systems of high diversity biomass is low relative to energy input.

5. For long periods of time, the Earth functions as a closed system within which a multiplicity of subsystems exists usually referred to as "ecosystems."

6. Species populations and ecosystems are coherent dissipative units of the whole biosphere which forms a dissipative unit at the highest level.

These features of the biosphere and its subsystems can most readily be explained on the basis of Action Principles, where Action is the quantity: energy *times* time, and is expressed in joule-seconds in the S.I. system of units.

11.3 THE ACTION HYPOTHESIS

It emerges that living matter, while not eluding the 'laws of physics' as established to date, is likely to involve 'other laws of physics' hitherto unknown, which however, once they are revealed, will form just as integral a part of this science as the former.   Moore 1989

11.3.1 THE HOMEORHETIC SYSTEM

The evident pervasiveness of the characteristics described above, throughout a wide range of ecosystems, both aquatic and terrestrial, indicates the existence of fundamental physical forces regulating energy flow. Therefore, I postulate: in a closed ecosystem with a regular energy input regime, structure develops under the influence of two countervailing force fields, one tending to increase the total energy assimilated and reduce energy flux to the minimum possible, the other tending to accelerate energy flux. These two countervailing force fields operate within a third force field, that of gravity. For an ecosystem to exist over time, it is evident that the tendency to acquire and delay energy flow must exceed the opposing force until equality is attained at the steady state or climax. The steady state is characterized by the acquisition of maximum energy from the environment and its retention for the maximum period of time possible, contingent on the various physiologies of the assembled species, the heterotrophic interactions, and the nature of the physical environment.

The ecosystem may thus be regarded as *homeorhetic* system, the essential characteristic of which is the automatic equilibration of energy flows (Waddington 1968) (Figure 11.16) provided the values

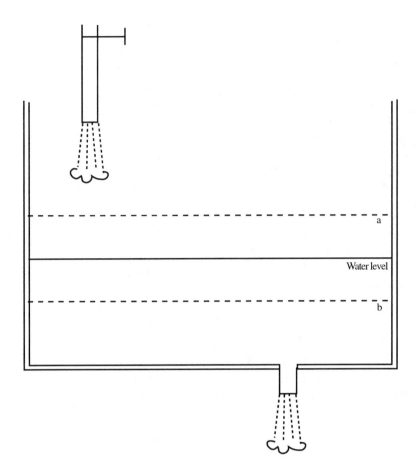

FIGURE 11.16 Diagram of a homeorhetic system. A homeorhetic system is one that automatically adjusts inflow and outflow and comes to a steady state. A faucet discharges into a tank with a drain at the bottom. If the inflow of water exceeds the capacity of the drain, the water level rises. The pressure at the drain increases as the water level rises and consequently discharge increases until a steady state is reached when inflow and outflow are equal. If the inflow is increased or the outlet reduced, the water level rises (a); conversely if the inflow is reduced or the outlet increased, the water level falls (b). The level tracks changes in inlet and/or outlet with a certain time delay. The time delay and hence the stability of the system will depend on the dimensions of the reservoir. This system can work only within a certain range of values.

remain within certain limits. This is in contradistinction to a *homeostatic* system that maintains a specific state as, for example, the carbon dioxide level of the blood.

11.3.2   Onsager's Reciprocal Relations

Onsager's reciprocal relations (1931) state that when two energy transfer processes, such as diffusion or conduction, interfere with each other, the system assumes a state of least dissipation. These relations are strictly true only in the vicinity of thermodynamic equilibrium. As a closed ecosystem reaches a steady state of least dissipation, or least specific entropy production, in accordance with Onsager's theorem, it must be assumed that there are two thermodynamic forces interfering with each other: one tending to reduce energy flux and increase the energy of the biomass, the other tending to accelerate energy flux. The problem is that an ecosystem although reaching a steady state is far from thermodynamic equilibrium and therefore Onsager's theorem would, at first sight, seem to be invalid.

11.3.3 ACTION PRINCIPLES

These features of the biosphere and its subsystems can most readily be explained on the basis of Action Principles, where Action is the quantity: energy *times* time, and is expressed in joule-seconds in the S.I. system of units.

The modern concept of *action* is founded on the work of Pierre de Fermat (1601–1665) in the 17th century. Fermat found that he could derive the laws of reflection and refraction from the premise that light passing through transparent media follows the path of least time, which is not necessarily the shortest distance between the points of observation. This became known as Fermat's Principle of Least Time. Fermat ascribed this principle to the fact that light passes through media with different refractive indices at different velocities, although it was not generally accepted at that time that light in its passage takes any time at all.

Subsequently, Maupertuis attempted to extend this principle to mechanical systems, developing a principle which defined the trajectory of a system as the path of "least action"; unfortunately he failed to define "action" adquately. After modification by Euler and Lagrange, the great Irish mathematician William Hamilton was able to generalize the concept in the Principle of Least Action. This was subsequently modified slightly by Feynman (1963) who defines *action* (*S*) as in the case of a thrown object, as the product of the difference between the potential energy (*P*) and kinetic energy at each point in the trajectory, integrated over the trajectory as a whole. *Action* thus has the dimensions of energy *times* time, and is expressed in the S.I. system of units in joule-seconds.

$$\text{Action} \;=\; S \;=\; \int_{t1}^{t2}(KE - KP)dt$$

The Principle of Least Action states that the trajectory of a system is defined by that path along which the computed *action* is least. But as pointed out by Watson (1986) this principle, strictly speaking, should be referred to as the "Principle of Stationary Time," because the action is insensitive to small deviations in the path. The Principle of Least Action is one of the foundations of modern theoretical physics (Feynman 1963; Watson 1986; Motz and Weaver 1989); the quantum of action is Planck's constant (*h*).

11.3.4 BIOLOGICAL SYSTEMS

If, in a biological system, the potential energy is considered to be equivalent to the energy in the biomass and the kinetic energy is required to maintain that biomass, the *action* can, at least in theory, be computed in an identical manner to that for physical systems. However, from field observations, *action*, the product of energy and time, proceeds in an ecosystem, not to a minimum value, but to a maximum, attained through large size, long life span, and maximum abundance.

Thus, although *action*, as defined in the physical world, is an abstract mathematical quantity, in biological systems, by contrast, it is expressed in tangible form as living substance. To maintain symmetry of terminology, biological systems may be regarded as expressing a Principle of Most Action. In the biological world an increase in action is equivalent to an increase in energy density in both time and space. This implies a general trend in living organisms toward increased mean size, increased mean life span, an increase in specific energy density ($j.g^{-1}$), and an increase in uniformity and abundance. This trend is observable in the mean of all organisms, not in all species. The trend to most action (or least specific dissipation) is a population effect and implies cohesion within the species population and, at higher levels of abstraction, within the ecosystem, and ultimately within the biosphere as a whole. As Prigogine and Stengers (1984) state,

> One of the most interesting aspects of dissipative structures is their coherence. The system behaves as a whole as if it were the site of long-range forces. In spite of the fact that interactions among molecules do not exceed a range of some 10^{-8} cm, the system is structured as though each molecule were 'informed' about the overall state of the system.

The contrary force field is a child of the Principle of Least Action. If the Principle of Least Action is universal in its application, that is, if it is a fundamental property of energy in all its forms, then it must function in the living world as well as in the nonliving. In the biological world the effect of the Principle of Least Action is comparable with that of Fermat's Principle of Least Time: energy tends to pass through the system as rapidly as the constraints allow.

Least and most action may be regarded as opposing attractors. The end point of "most action" is a state of "crystallization," a permanent state with no further change. The least action attractor is spontaneous combustion and dissolution in a shower of photons. As both states are lethal to organisms and the force fields are equal at the steady state, the system settles down in the vicinity of the midpoint between the attractors. This is also the midpoint with respect to energy flux, which is also in the point of maximum power output.

The living world thus appears to exist at the intersection of two thermodynamic force fields which interfere with each other. These force fields interact within the aegis of a third force field, that of gravity. In accordance with Onsager's relations, the system goes to a state of most action (least entropy production). Unfortunately, the application of Action Principles to biological systems, as with Onsager's relations, is considered invalid by physicists. This is because the Principle of Least Action is applicable only to energy-conserving systems: that is, systems where the total energy remains constant. Manifestly, this is not the case in biology, because systems can exist only on a continuous input of energy, feeding, as Schrodinger (1944) aptly says, negative entropy. Most action in organisms manifests the properties of negative entropy, or *ectropy* (Scheer 1970) as they are antisymmetrical to those of entropy.

The above are important objections, yet as will be seen in subsequent sections, the application of Action Principles provides such an interesting and far-reaching explanation for major biological processes, from ecosystem structure to evolution and natural selection, that I believe it is the physical laws that require extension.

I postulate that biological systems are contingent on work done. If free-energy (energy with the capacity to do work) is available to a work process, where a work process is defined as any process that raises the potential energy of a system (Scheer 1970), the work done, in accordance with the principle of least work, will be minimum relative to the energy input. In a biological system dependent on a continuous energy input this implies that the work per unit time, or rate of energy dissipation, the power of the system, tends toward a minimum value. Countervailing this trend is the trend toward accelerated dissipation (operation of the Principle of Least Time) or increased power.

Out of these biological considerations it seems possible to formulate the Laws of Ergodynamics (the dynamics of work: erg = unit of energy).

When two energy transport processes interact in the vicinity of thermodynamic equilibrium and in the presence of appropriate substrates, work may be done. A work process is defined as any process that increases the potential energy of the system. If the energy input cycle to the system is maintained at a constant level, the system comes to a steady state in which the work done is a minimum. One of the forces tends to reduce energy flux through the system (designated the Principle of Most Action), the other tends to accelerate energy flux (the Principle of Least Action). The system settles down in a steady state of ergodynamic equilibrium, which is equilibrium maintained by work done. Ergodynamic equilibrium is characterized by a state of least energy flux but maximum total energy flow (maximum energy input, least work, least specific entropy production, or least specific dissipation). The second Law of Ergodynamics states: a system at ergodynamic equilibrium, in the appropriate conditions and in the presence of appropriate substrates, may move away from thermodynamic equilibrium, provided the system remains in the vicinity of ergodynamic equilibrium. The further the system moves away from thermodynamic equilibrium, the greater the energy flux (power output) necessary to maintain the new ergodynamic equilibrium; at the same time there is an increase in internal energy of the system and an increase in order.

These laws do not conflict with established principles of thermodynamics; they are merely extensions of those laws, applicable only in specific dynamic conditions. Nevertheless, the outcome of their application, at least in the short term, is diametrically opposite to the outcome of the classical Laws of Thermodynamics: energy tends to be condensed and conserved. At the same time it is dissipated as rapidly as the constraints allow. Work is done by imposing a time delay on the passage of energy from one state to another, and the result of work is an increase in potential energy in some part of the system. The amount of work done is always the least for the conditions and constraints prevailing, and the slower it is done the more efficiently it is performed. To exist over time, the Principle of Most Action must dominate system behavior in the short term, but like its sibling, the Second Law (entropy always wins), the Principle of Least Action (energy flux tends to increase) always wins in the long run. The biological process may therefore be envisaged as a "struggle" between the principle of least work and the principle of least action. The only way that least action can win is through an increase in work done (an increase in power).

This interpretation of biological events in physical terms, I believe, meets field and laboratory observations. It is now up to physicists to establish the quantum mechanical base for these observationally derived laws. Feynman (1963), it may be noted, was also anxious to achieve this fundamental level of understanding for the equivalent observation that "if currents are made to go through a piece of material obeying Ohm's law, [a practical application of Onsager's relations] the currents distribute themselves inside the piece so that the rate at which heat is generated is as little as possible."

11.3.5 ORIGINS

To gain insight into the nature of the processes operating it is necessary to consider the circumstances envisaged as existing at the time of origin of organisms. This is contingent on the assumption that physical principles have remained constant over the intervening period.

Although organisms may have existed before the advent of photosynthesis, the predominance of and dependence upon photosynthetic organisms at the present time establishes photosynthesis as the primary engine driving the biological system. Thus the initiation of photosynthesis is a good place to begin.

Initially, a receptive atom was raised to a higher energy level by the addition of the energy in a photon to an electron in the electron shell thereby moving it to a higher orbit; thus the atom attained a higher energy state. In normal circumstances, after a short time delay of 10^{-7} to 10^{-8} sec, the electron returns to its ground state with the emission of a photon (Szent-Gyorgi 1960; 1961). However, while the molecule is in a higher energy state, it is more reactive and therefore may form compounds (incipient crystals) with other suitably reactive molecules. This is a work process in that the potential energy of the compound is increased (Scheer 1970) and the time delay before the molecule returns to the ground state is extended. Thus, close to thermodynamic equilibrium, there are two interacting energy transport processes, one tending to accelerate return to the ground state and the other delaying the return by diversion through a work process and the slow dissipation of the energy in the original photon as heat.

If the energy input is continuous the process reaches an end point and no further work is done, but if intermittent, as produced in an electric motor by a commutator, a continuous power output is obtained. The alternation of day and night in temperate and tropical regions presumably provided this commutator effect. As Anderson (1972) states "... most methods of extracting energy from the environment in order to set up a quasi-stable process involve time periodic machines such as oscillators or generators, and the processes of life work the same way."

11.3.6 THE MOVE AWAY FROM THERMODYNAMIC EQUILIBRIUM

In the initial stages this reaction occurred in the vicinity of thermodynamic equilibrium within a closed system, giving rise to the formation of protoorganisms. These protoorganisms were charged entities, transient and relatively unstable. In that biological systems are unstable and liable to change, changes occurred at the molecular level which allowed the residence time to increase thereby

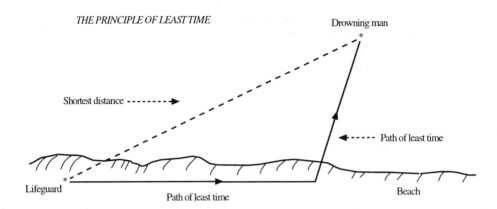

FIGURE 11.17 Diagram illustrating the Principle of Least Time. A lifeguard sees a swimmer in difficulty half a kilometer down the beach and 50 m off shore; to reach the person in distress, the lifeguard runs down the beach, being able to make faster time running than swimming, then entering the water at a point that minimizes the elapsed time to reach the swimmer.

ensuring an increase in the efficiency of the work done; conversely, changes also occurred tending to increase the rate of energy dissipation.

With lengthening life span, and provided the energy input maintained a regular cycle, the protoorganisms interacted reaching a steady state where conservation and dissipation are evenly balanced and a homeorhetic system is formed. In the initial case of a single species of protoorganisms, energy was uniformly distributed among the interacting individuals through homeokinesis.

The inherent malleability of protoorganisms and the short-term ascendancy of the Principle of Most Action initiated increased energy acquisition, longer and longer time delays, and greater stability in face of the environmental fluctuations experienced. Such changes were interspersed with changes causing more rapid energy flow, so that the system moved in an irregular stepwise manner in two directions, one toward greater total energy flow, less energy flux, and greater biomass, and the other toward increased energy flux. Each opportunity for change is accepted as it occurs in the malleable organism.

Organisms are thus trapped between two thermodynamic forces, one tending toward crystallinity and no further work, and the other toward instaneous dissipation in a shower of photons (combustion). Both states are lethal so ergodynamic equilibrium must be maintained if the system is to continue in existence (Figure 11.17).

11.3.7 BROKEN SYMMETRY

Similar protoorganisms, interacting symmetrically, thus formed a *tabula rasa*, or "clean slate," on which it was possible for signals to be encoded. As Polanyi (1968) states,

> In the light of the current theory of evolution, the codelike structure of DNA must be assumed to have come about by a sequence of chance variations established by natural selection. But this evolutionary aspect is irrelevant here; whatever may be the origin of the DNA configuration, it can function only as a code if its order is not due to potential energy. It must be as physically indeterminate as a sequence of words on a printed page.... It is this physical indeterminacy of the sequence that produces the improbability of the occurrence of any particular sequence, thereby enables — a meaning that has a mathematically determinate information content equal to the numerical improbability of the arrangement.

Thus, the symmetrically interacting organisms forming the *tabula rasa* could identify and encode environmental signals, where a signal is defined as any regularly occurring energy fluctuation. Breaking the indeterminacy and increasing the information content of the system demands an increase in the work done (Scheer 1970). Once the initial symmetry was broken, the whole array of environmental signals could be explored; in the process new energy fluctuations were created which in turn became signals.

11.4 TESTING THE HYPOTHESIS

Biological hypotheses are extremely difficult to test in a manner comparable with the procedure in physics and chemistry; the best one can often hope for in a specific test is a qualified maybe, and then only in certain circumstances. Thus it is necessary to rely on the concilience of inductions, the manner in which a wide range of observations and generalizations fit conformably within a mutually compatible and understandable whole. Thus a hypothesis forms a mental overlay placed over the reality to be tested. If the thesis fits, the general outlines will be congruent, but details may be obscure.

11.4.1 SUCCESSION

> Succession is one of the oldest, most basic, yet still in some ways, the most confounded of ecological concepts ... the apparent intractability and continual contradictions of the successional question after decades of study lead to the suspicion that there is more involved than straight forward, matter-of-fact scientific consideration.
>
> McIntosh 1980a

Succession describes the general and, on the whole, orderly changes in the biota at a particular location following environmental change or disturbance. Over time, and in the absence of further change, the ecosystem reaches a steady state frequently referred to as the climax (Clements 1916, 1936; Jones 1945; Whittaker 1951, 1953). Primary succession describes the directional and essentially irreversible changes initiated by naturally occurring erosion and sedimentation, as, for example, in delta formation and the gradual conversion of a lake to forest or the gradual afforestation of a moraine left by a retreating glacier. A transect across such a transitional zone shows the successive vegetational stages from lake to dry land and eventually forest cover. Secondary succession is the revegetation of bare soil resulting either from natural events such as a falling tree exposing bare soil, or the reversion to forest of land previously cleared by human agency. Succession is dealt with in some detail because of the light thrown on this rather obscure process by application of the Action Principles.

Succession, in undisturbed or relatively unchanging conditions, follows an orderly, but not rigid, sequence of events leading to the steady state or climax. To a large extent succession has been studied within the framework of plant ecology, although as Connell and Slayter (1977) clearly recognize, there are "... in addition to the competitive interactions between plants and sessile animals, interactions with herbivores, predators and pathogens [that] are of critical importance to the course of succession." The more mobile nature of animal populations has tended to focus the attention of animal ecologists on animal populations and individuals rather than systems.

The history of succession as a scientific concept has been ably documented by McIntosh (1980a, 1985); therefore it is necessary in the present context to emphasize only those aspects that have a bearing on the interpretation of the process in thermodynamic terms.

The history of modern ecology has been confounded by a difference in philosophy that has persisted to the present day. Briefly this division may be categorized as the difference between the superorganismic views of Clements (1905, 1916, 1920, 1936) and his supporters, who regarded ecosystems as integrated wholes, and the individualistic views of Gleason (1917, 1926, 1939) and his followers. A comparable split developed within the ranks of animal ecologists, a holistic approach being taken by Nicholson (1933), Nicholson and Bailey (1935), Varley (1970), Margalef (1963, 1968), and Odum (1969, 1971), following the Chicago school of animal ecologists led by Allee et al. (1949). In opposition was the more individualistic stance adopted by Andrewartha and Birch (1954) and Erlich and Birch (1967).

According to McIntosh (1980a), Clements and Odum converge in their description of succession as an orderly predictable, unidirectional process which results in modification and control over the physical environment culminating in a mature (or climax) ecosystem, largely governed by climate. Gleason, on the contrary, describes succession as,

> ... an extraordinarily mobile phenomenon whose processes are not to be stated as fixed laws, but only as general principles of exceedingly broad nature and those results need not, and frequently do not ensue in any definitely predictable way.

The important feature of Clements' concept of the ecosystem was that the species populations formed well-integrated entities and that these entities interacted on a relatively fixed basis, forming, under constant natural conditions, a coherent whole of relatively fixed composition. This fixed composition held good over a considerable georgraphic area, forming the climatic climax; this was regarded as the basic unit of vegetation. The interactions between the various parts of the system Clements equated with organs, the whole forming a unity with certain characteristics of an individual organism. Gleason, on the other hand, believed that systems develop through the response of individual species to local conditions.

In a detailed review of the climax concept, Whittaker (1953) comes down fairly strongly on the side of Gleason, stating,

> Succession may thus be thought to occur, not as a series of distinct steps, but a highly variable and irregular change of populations through time, lacking orderliness or uniformity in detail though marked by certain fairly uniform overall tendencies.

The work of Curtis and McIntosh (1951), Bray and Curtis (1957), Curtis (1959), McIntosh (1967), and Whittaker (1967) on the ordination of plant associations along environmental gradients tended to support Gleason's individualistic approach, an approach supported by many ecologists (Connell and Slayter 1977; Drury and Nisbet 1973; Harper 1969, 1977a, 1977b). Gradient analysis thus functions as a "display" in which each species reveals its response to a particular signal array, similar to the way in which the various wavelengths in white light are displayed by passing it through a prism. The general conclusion reached by Whittaker (1967) was that

> ... species populations are gnerally distributed along complex gradients in the form of binomial curves, with their densities tapering on each side of the optima.

Further, Whittaker and Niering (1965) stated,

> Species tend to evolve also toward habitat differentiation, toward scattering of centers of maximum population density in relation to the environmental gradients, so that few species are competing with one another in their population centers.

The sum of the individualistic responses nevertheless leads to a particular steady state at the climax. Whittaker and Woodwell (1972) and Horn (1975a, 1975b) recognize also that systems starting from very different points may reach the same condition at the climax. However, no fixed associations could be discerned except in narrow environments (Curtis and McIntosh 1951).

Existing between the two fairly well-defined positions described as "holistic" and "individualistic" has been the middle ground occupied by Sir Arthur Tansley the eminent British ecologist. Tansley (1935) eventually rejected Clements' extreme organismic view as well as Gleason's individualistic concept. He conceded, however, that the community might be compared with a "quasi-organism." Tansley rather ingeniously compared the species of a plant community with the genes of an organism "both aggregates owing their 'phenotypic expression' to development in the presence of all the others of the aggregate and within a certain range of environmental conditions."

Gleason's position was criticized by Watt (1947) (also a founding father of British ecology) in a cogent manner,

> In short Gleason has minimized the significance of the relations between the components of the community in horizontal space and time. These relations constitute a primary bond in the maintenance of the integrity of the plant community: they give to it the unity of a co-ordinated system.

Langford and Buell (1969) maintained a similar point of view.

At least two features of the successional process appear to be generally agreed upon: one is that species diversity during succession increases up to the penultimate stage when a point of maximum diversity is reached; diversity then declines in the terminal stages with the emergence of one or more dominant species. Clements (1905) stated categorically that,

> ... the number of species is small in the initial stages [of succession], it attains its maximum in the intermediate stages, and again decreases in the ultimate formation, on account of the dominance of a few species.

This general pattern has subsequently been confirmed by many observers (Golley 1965; Loucks 1970; Margalef 1963, 1967; Odum 1960, 1969; Patrick 1967; Reiners et al. 1971; Tansley 1923; Whittaker 1953, 1969), although the generality of the process has been questioned by McNaughton and Wolf (1970). Dominant species generally have broader niches than subordinate species and the intensity to dominance is least on the most mesic sites (McNaughton and Wolf 1970). The trend toward increasing diversity is not a monotonic progression; periods of increasing diversity may be interspersed with stable situations in which certain species dominate before giving way to the next phase of increasing diversity. Clements stated:

> However faint their limits, real stages do exist as a consequence of the fact that each dominant or group of dominants holds its place and gives character to the habitat and community, until replaced by the next dominant.

Harper (1969) and Mellinger and McNaughton (1975) recognize similar stages during succession.

The other point of agreement is that fluctuation in the component species declines during succession, leading to increasing stability in face of the environment to which the ecosystem is exposed. Clements regarded stabilization as essentially synonymous with succession, although he recognized also that stability and diversity were not strictly associated,

"To his eternal credit he asserted in his first treatise on ecological research published in America (Clements 1905) the recently discovered maxim that stability in succession is not directly associated with species diversity." (McIntosh 1980b).

Gleason (1939) similarly stated that the dominant species, "... by their persistence always tend to reduce the variation in the physical environment" and Langford and Buell (1969) recognized the necessity for the existence of

> ... homostatic mechanisms which are operative in various species with a short life-history, buffering against environmental fluctuations and thus providing a *varied but steady state*, and mechanisms that lead to the stability of the association.

Odum (1969) expresses the view that a mature (climax) association able to buffer the physical environment to a greater extent than a young immature community,

> In a word, the 'strategy' of succession as a short-term process is basically the same as the 'strategy' of long-term evolutionary development of the biosphere — namely, increased control of, or homeostasis with, the physical environment in the sense of achieving maximum protection from its perturbations.

Pickett (1976) suggests that succession "is a complex gradient of decreasing physical stress" and Usher (1979) concludes that succession is a non-random process but neither is it an intrinsically regular or linear process.

Various attempts have been made to characterize the climax but in spite of assertions to the contrary, neither biomass, nutrient retention, nor primary productivity is maximized (Bormann and Likens 1979); nor do species

> ... evolve toward the formation of coherent groups, bound by co-adaptations resulting in similarity of distributions. They evolve instead to diversity of distributions: along environmental gradients, they form the kind of population continua that are observed in gradient analysis, with broadly overlapping, bell-shaped population curves with scattered centers.
>
> Whittaker 1951, 1969, 1970; Bormann and Likens 1979

Tansley (1935) discussing the climax makes the significant statement that:

> The climax represents the highest stage of integration and nearest approach to equilibrium that can be attained in a system under given conditions and with available components.... *The more relatively separate and autonomous the system, the more highly integrated it is and the greater the stability, of its dynamic equilibrium.* (Emphasis added by editor.)

In a later publication, Tansley (1949) states,

> ... a climax represents a well-marked position of vegetational stability which may persist in equilibrium with relatively stable environmental conditions ... [but] there are edaphic or biotic factors, not depending on the climate, permanently at work to stabilize the community.

On the other hand, in certain circumstances there are natural causes inducing cycling at the climax. Watt (1947) in his landmark paper "Pattern and Process in the Plant Community" describes the cycle of changes that occur in the flora of Tregaron Bog in North Wales. Hummocks are formed by various species of sphagnum moss. They are subsequently invaded by heather (*Calluna vulgaris*) that eventually fails to regenerate and decays. This creates a pool which is then invaded by sphagnum.

Over much of North America, long-term cycles involving fires at intervals of tens to hundreds of years appear to have been influential in affecting the current state of the vegetation (Heinselman 1973; E. A. Johnson, 1979, 1981; Loucks 1970; Rowe and Scotter 1973; Vogl 1974; Wright 1974). Wright concludes,

> A Clementsian climax is never reached, because the evolution of each forest stand is repeatedly interrupted by fire, but in the long-range the forest mosaic as a whole maintains an equilibrium.

Nevertheless, Langford and Buell (1969) state,

> The concept that shines through with greatest strength to this day is that great stretches of virgin country, dominated by one or several species of like growth form wherever the processes of succession have had time to reach a more or less permanent end point.

Apparent departures from the inherent tendency to dominance and a stable state have cast doubt on the value of the climax as a concept, emphasis being placed more on the dynamic properties of systems and the effect of disturbance (Curtis and McIntosh 1951; Curtis 1959; Whittaker 1953, 1957; Loucks 1970; Connell 1978; Drury 1973; Schugart and West 1970; Vogl 1980; Reiners 1983). However there is no real conflict in these views if the climax is regarded as the specific end point to which a system will proceed if all factors (and their associated variabilities) are held constant. The climax then becomes a perspective point on the horizon, analogous to the end point represented by thermodynamic equilibrium in an isolated system, in that it represents an effectively unattainable terminal state. The Arctic lake, of all systems in a pristine state, probably represents the closest attainable approach to a climax situation.

The decline in species diversity at the climax is essentially a phenomenon observed in plant dominated ecosystems. Grazing and predation, on the other hand, tend to stimulate diversity and reduce plant dominance. Paine (1966) showed that the experimental removal of the dominant starfish (*Pisaster*) from mussel beds (*Mytilus californianus*) off the coast of Washington resulted in a decline in diversity as mussels and barnacles (*Balanus glandula*) expanded and assumed dominance resulting in the elimination of a number of benthic invertebrates and algae. But an increase in diversity cannot proceed indefinitely; increase is eventually limited because each species population tends to maximize its *action*, with the assemblage conforming to the requirements of hierarchy formation within a homeorhetic system. This imposes a "top-down" control by the dominant species tending to increase the homogeneity on the system. Eventually a compromise has to be reached between the number of species in the assemblage and the needs of a limited number of species to survive. "Overgrazing" indeed can reduce species diversity and damage ecosystems, but as Slobodkin et al. (1967) believe, this does not occur in natural communities, but only where animals have been introduced or are controlled by man. This is supported by Harper's (1969) observation that after myxomatosis had eliminated the rabbit (*Oryctolagus cuniculus*) (introduced into England in the 12th century) from grasslands on the English Weald, many new species appeared, some of which had not previously been recorded from that region.

The Action Hypothesis accounts for all the divergent featues in the above brief description of the successional process. The constant interaction of the two countervailing force fields creates a homeorhetic system resulting in the continual search for a stable state. Succession is thus the result of interaction between the two nearly equal opposing forces each of which mainfest alternating periods of ascendancy. The climax thus represents both a maximum and a minimum. From the synecological, or community, perspective it is the maximum number of species, "selected" from the available set as provided by evolution and immigration, that can exist together. From the autecological, or individual, perspective in which each species pursues the "goal" of most action, it is minimum attainable diversity, resulting from the success of certain species and the failure of others to survive within the constraints imposed at the steady state. This is most clearly seen in the reduction in species diversity during the terminal stages of the successional process. This can be expressed as the "struggle" for maximum heterogeneity vs. minimum heterogeneity, or maximum energy flux (least action) vs. minimum energy flux (most action).

An interesting feature of the climax is that insofar as it represents both a maximum or a minimum, a tiny movement away from this position will make no difference (Feynman 1963). Hence, one dominant species will be readily replaceable by another of similar life-form and ecological characteristics. This accounts for the ready replacement of one species by another following a small environmental change and the multiplicity of dominants in many forests of broad-leafed trees, particularly the large number of species in the canopy of a tropical forest (Richards 1952).

Over evolutionary time, viable changes of a random nature are opportunistically employed to further either the short-term "goal" of most action and maximum homogeneity, or the long-term goal of *least action* and maximum diversity. In this the Action Hypothesis leads to a different conclusion from that of Odum (1969) who regards the goals of succession and evolution as the same. I believe that the successional goal is a stable state of most action, attained by the individual species proceeding toward a state of most action. However, evolutionary processes and immigration generally favor greater diversity and faster energy flux, driven by the ultimate ascendancy of the principle of least action.

11.4.2 THE PRODUCTION/BIOMASS RATIO

The Production/Biomass ratio (*P/B*) provides a measure of the energy input necessary to support unit biomass; it may be viewed either from the autecological perspective of an individual species population or from the synecological perspective. The P/B is dimensionally equivalent to the reciprocal of the mean age of a population, or the reciprocal of the mean energy residence time (MacArthur, in Leigh 1965).

Whittaker and Woodwell (1972) consider that the reciprocal of the P/B ratio, the Biomass Accumulation Ratio (B/P), is a more useful measure, as it represents the "turn-overtime." It thus becomes a relative measure of energy flux (the energy input per unit time necessary to support unit energy in the biomass): the smaller the B/P the greater the energy flux. Thus the B/P is essentially the time factor in the computation of *action* (energy *times* time).

Two general trends have been recognized with respect to the B/P ratio. In the first place it tends to increase along the food chain (Winberg et al. 1972; Winberg 1972; Saunders et al. 1980). In terms of the Action Hypothesis this implies an increasing generation time at each stage at each ascending hierarchical level with, concomitantly, a decline in the total energy at each level. Energy within an ecosystem is transferred from populations with a high degree of fluctuation (low B/P) and a rapid response time, to those with a low level of fluctuation (high B/P), each stage increasingly contributing to the stability of the whole and thus forming a hierarchy of coupled stabilizers. Thus it is the dominant species that imposes the main stability characteristics on the community (Frank 1968).

Basin (quoted by Waddington 1968) states,

> the only logical way in which it is possible to discriminate a number of activities into a hierarchy is by considering their reaction times, a higher level in the hierarchy always having a much longer reaction time than a level classified as lower.

Thus the B/P as a measure of the reaction time determines the hierarchical structure of the ecosystem. It is the time component of the action that is of importance, not the action (energy $\times$ time) itself, although individual species pursue the thermodynamic goal of maximum action. The increase in B/P of each species population along the food chain is attained through increasing size (Dickie et al. 1987), increasing life span, fewer juveniles, and greater uniformity.

At the world level, Mann and Brylinsky (1975) from their summary of worldwide observations on lakes made during the International Biological Program 1966–1974), concluded that the B/P decreases with decreasing latitude from the poles to the tropics. This is supported by Golley's (1972) conclusion that a very high level of energy input is required to repair a tropical forest compared with that required for the same purpose in a temperate forest. May (1981) concluded that tropical forests exhibit a turnover rate nearly double that of temperate forests such as occur at Hubbards Brook, New Hampshire in the northern United States.

The decrease in B/P ratio with decreasing latitude runs parallel with an increase in species diversity, an increase in mean temperature, and a decrease in climatic variability, between seasons and between years. These factors result in a gradient of increasing energy flux from the poles to the tropics.

11.4.3   *r*- AND *K*–SELECTION

The concept of *r*- and *K*-selection is an attempt to classify life-history characteristics (MacArthur and Wilson 1963, 1967); *r*-selection designates the tendency of certain species to evolve toward small size and rapid turnover time; *K*-selection denotes the opposite trend toward high biomass and stability. These characteristics have been set out by Southwood (1977) (Table 11.1). The notion has had considerable influence on ecological thinking (Gadgil and Solbrig 1972; Pianka 1970, 1978). The notion is based on the logistic equation (Kingsland 1981, 1982).

The logistic equation in its simplest from may be expressed,

$$\frac{dN}{dt} = rN \frac{(K - N)}{K}$$

where N is the number in the population at any given time, r is the intrinsic rate of reproduction, and K is the asymptotic value or carrying capacity, frequently referred to as the equilibrium value. As Hutchinson (1978) points out, the expression $(K - N)/K$ is a negative feedback term. This expression

TABLE 11.1
The Contrasting Characteristics of Extremes of the r–K Selection Spectrum

r-Selected Species	K-Selected Species
High Production/Biomass ratio	Low Production/Biomass ratio
Short generation time	Long generation time
Small size	Large size
High level of dispersal	Low level of dispersal
Much density independent mortality	High survival rate, especially of reproductive stages
Panmictic	Territorial
	Clonal
Intraspecific competition, often "scramble type"	Intraspecific competition, often "contest type"
Low investment in "defense" and other interspecific competition mechanisms	High investment in "defense" and other interspecific mechanisms
Time efficient	Food and space resource efficient
Populations often "overshoot"	Populations seldom "overshoot"
Population very variable "boom and bust"	Population relatively constant from generation to generation ≈ K

Source: Modified after Southwood, T.R.F. *J. Animal Ecol.*, 46, 337–365, 1977. With permission.

is usually interpreted in terms of density, implying that the rate of increase in a population decreases as density increases.

There are two objections to the application of this equation to natural populations (Kingsland 1981, 1982). The first, largely philosophical and in some respects not too significant, is that the infinitesimal calculus is used to derive the equation, although individuals are discrete entities; it assumes also that all individuals are equal. Second, all populations tend to be clumped rather than uniformly distributed, hence the meaning of density is equivocal. Density might function in two ways: (1) through social interaction or (2) through a shortage of some critical resource. In real populations the inflexion in the curve indicating the application of this constraint occurs long before there is any critical resource shortage, as is indicated by the fact that populations continue to increase, but at a decreased rate, after the point of inflexion has been passed. The density dependent model of population regulation has been described by Krebs (1979) as the ecologist's phlogiston theory, "useful years ago but now an obstruction to progress."

The concept has been further criticized by Schaffer (1979) on the grounds, that *r* is not justifiably equated with reproduction nor *K* with postbreeding survival. Green (1980) criticizes the concept on the grounds that *r* and *K* are not strictly comparable, *r* being a function of life-history parameters, where as *K* is not.

All species populations have a combination of *r*- and *K*-, *r* may be regarded as turnover time and *K* is essentially a descriptive term that indicates large size and great age. There is thus a gradient from species with high *r*-values and low *K* (species usually low in the trophic hierarchy) to species with high *K* and low *r*. The Action Hypothesis states that all species tend to maximize their action (the product of energy and time). Over evolutionary time, some species survive by opportunistically increasing the time component of their action, and this necessitates large size or other characteristics that inhibit their being consumed or overshadowed; these species occupy the "*K*-selected" pole of the dominance hierarchy, whether plant or animal. "*K*-selected" species are those which approach most closely the asymptotic value of their action and are usually dominant in the ecosystem. Other species emphasize the acquisition component of action accepting roles (or are "forced" into such roles to avoid elimination) near the "*r*-selected" pole. These roles accentuate rapid energy acquisition, frequently in specialized conditions, and short life span. Time is thus critical in determining hierarchical level, although without large size long life span cannot be attained. It is thus the special combination of large size and generation time, and the tendency to maximize both simultaneously, that ensures species populations form a dominance hierarchy.

11.4.4   Action, the Binding of Energy and Time, is the Inherent Stabilizing Property of Living Things

Thus both the *P/B* ratio and *r*- and *K*-selection are subsumed within the Action Hypothesis.

11.4.5   Stability

The relationship between diversity and stability has been a long-standing question, following Mac-Arthur's (1955) conclusion, based on information theory, that diverse systems are more stable than simple systems. He considered that the most diverse and stable system was the forest of the wet tropics, and this was contrasted with the fluctuating populations of small mammals on the norhtern tundra as reported by Elton (1927). It was also recognized that artificially simplified agricultural systems were highly unstable to the impact of so-called pests and diseases. A challenge to this view emerged with the finding that many Arctic lake systems are extremely stable (Johnson 1975, 1976; Rigler 1975b), showing very little change in configuration when sampled at an interval of several years (Figure 11.5). This stability applied not only to fish populations but also to zooplankton populations (Rigler 1974). These findings, thrown into stark relief by the fact that the stable lake ecosystems exist within the tundra itself, demanded explanation. It seemed that the primary reason for their stability in face of the environmental fluctuations experienced was their high degree of autonomy coupled with the great capacity of water to buffer environmental fluctuation.

The term stability has been used in many different contexts and with different meanings (Pimm 1991), but in the present context it simply implies continuity of the population characteristics of abundance and size frequency distribution in the face of the environmental fluctuation experienced over the recent history of the system.

The Action Hypothesis stipulates that stability is an inherent characteristic of the biological system, in that a time-delay on energy flux is fundamental of life. Thus any *closed* or autonomous ecosystem will approach a stable state, given a regular cycle of energy input. The lake ecosystem as a whole thus functions as a series of coupled shock absorbers, from the fluctuating primary producers to the long-lived, relatively constant fish populations, each stage imposing an increasing degree of control. Thus the dominant species endows the system with its main stability characteristics (Frank 1968; Langford and Buell 1969). The dominant species is the species that most closely approaches its maximum potential action over the longest period of time, and is thus limited by space and energy and its inherent physiology. In a forest ecosystem, the situation is different from that in a lake for the trees function both as main primary producer and dominant species. Food webs of considerable complexity and cross connections are thus established starting with the annual production of leaves, flowers, and fruit of the trees, the understory vegetation and decaying leaf litter.

Ecosystems generally accounted stable in the undisturbed natural state are diverse tropical systems (Elton 1958; MacArthur 1955; Odum 1953; Thienemann 1954), cave faunas (Heuts 1951; Poulson, 1963) as well as deep sea faunas (Hessler and Sanders 1967; Sanders 1969; Slobodkin and Sanders 1969). All these systems have the characteristics of energy sinks. The actual degree of stability attained depends on the genetics and physiology of the assembled species populations and ultimately the predictability of the energy input cycle.

Ecosystems are ultimately as stable as the environmental conditions allow; they cannot counter long-term changes in the environment but can follow trends after an appropriate time lag. Thus an ecosystem is only as stable as its long-term energy input cycle.

Thus ecosystems develop stability to the extent necessary to survive the environmental fluctuations to which they are normally exposed. A forest with the dominant trees up to 500 years old has, presumably, survived the vicissitudes of the environment over that period of time, whereas the nondominant species may have experienced cyclic changes over the same period as "good" and "bad" years succeeded each other.

Disturbance, beyond the normal range of environmental variation may be classified as external or internal. An external disturbance is a short-term stress producing only quantitative changes in the variables of state, but does not change the driving variables, whereas an internal perturbation implies a change in driving variables, such as the entry of a new species or an increase in nutrients (Hurd et al. 1971; Holling 1973).

Hurd et al. (1971) developed criteria of stability to external perturbation based on thermodynamic principles. They define stability as the ability of a system to maintain or return to its ground state after an external perturbation. The *degree* of stability is characterized by:

1. The amplitude of the deflection from the ground state,
2. The rapidity of response to the perturbation, and
3. The rate at which the deflection is damped.

In biological terms the first characteristic of stability can be envisaged in terms of the initial biomass of the species concerned: the greater the biomass the greater the deflection that can be tolerated without disruption of the system. The second and third characteristics are to some extent mutually antagonistic: a well-damped system returns slowly and monotonically to the ground state but a rapid response will tend to be less well-damped. Thus a stable system in the face of external perturbation will exhibit a well-balanced combination of rapid response time and strong dampening capacity. Together these two characteristics may be regarded as the "resilience" of a species population. The stability of a system to an internal perturbation, such as the addition of a new species or increased nutrient availability, is virtually unpredictable.

The actual stability of a small lake, in the face of severe external perturbation, is shown by an experiment carried out on Keyhole Lake, Victoria Island, in the Central Canadian Arctic. The lake was intensively fished between 1962 and 1967 during which time the population of Arctic charr was reduced to a remnant of its original state (Hunter 1970). It was then allowed to remain undisturbed until 1975 when it was again intensively sampled. After 8 years it had returned to the original configuration observed in 1962 (Figure 11.9) (Johnson 1983; Vanriel and Johnson 1995). Other experiments showed that these autonomous Arctic charr populations retain their original configuration even while the density is being greatly reduced (Figure 11.19).

Contrary to the Arctic lake, a tropical rain forest is highly susceptible to severe disturbance, taking years to return to its original configuration (Gomez-Pompa et al. 1972). Richards (1952), quoting Chevalier (1948) states that the forest on the site of ancient temples at Angkor Wat in Cambodia, destroyed some 5 to 6 centuries ago, now resembles the virgin forest of the district but still shows certain differences. Contrary to the rather "monolithic" stability of an Arctic lake fish population, the tropical forest is highly dynamic with continuous cycling of populations at all levels while maintaining an overall or *gestalt* stability

Stability is thus contingent on a closed or highly autonomous ecosystem, a relatively constant energy input and an environment free from long-term trends. In this respect, physical or biological boundaries as well as large area of relatively uniform conditions will provide a high degree of autonomy.

11.4.6 DIVERSITY

Species diversity may be quantified as a function of the number of species present in a specified area (species richness or species abundance) and the evenness with which species are distributed (species evenness or equitability). High diversity is indicated by a large number of species all relatively uniformly represented; low diversity is the equivalent of few species with disparate abundances. As diversity and species richness are generally correlated (Tramer 1967), the term diversity is used here for convenience in the general more colloquial sense of the number of species present.

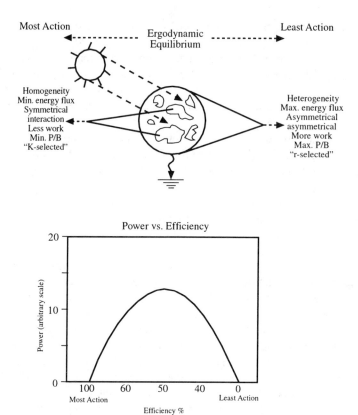

FIGURE 11.18 The upper panel illustrates dynamic interaction between most action and least action. Most and least action can be regarded as opposing "attractors." Over evolutionary time, the system gradually moves toward the right as energy flux increases. As both attractors are lethal, the system must remain in the vicinity of the mid-point between them. This is also in the vicinity of maximum power output.

High diversity develops in environments that are benign: where sunshine, moisture, and nutrients are abundant throughout the year, but where the relatively uniform energy flow is divided into a multiplicity of minor fluctuations. Low diversity develops in regions where energy input is limited to a short time interval or is low in total input. Diversity is an expression of the extent to which the incoming signal array has been dissected by the assemblage of organisms, which arrived either by *in situ* evolution or invasion from other regions. A signal is defined as any regular, identifiable fluctuation in the cycle of ambient energy.

Over the course of succession species diversity tends to increase, although this may occur in alternating stages in which one or a few species gains ascendancy for a time but is then followed by an increase in diversity (Harper 1969). In the penultimate stage leading to the climax, diversity declines as one or a few dominant species emerge. Similarly, Connell (1978) has shown that diversity increases on coral reefs and in tropical forest when the dominant species is removed.

Perhaps the world's most diverse fish fauna occurs in the River Amazon with about 1200 ± 200 species (Fitkau et al. 1975; Géry 1984; Sioli 1984; Lowe-McConnell 1987). The great length of the river and the divergent origins of its various tributaries give rise to a water mass that exhibits a complex series of fluctuations. At Manaus, 1500 km from the ocean, there is a fluctuation in water level of about 15 m. This causes alternate drying out and flooding of large areas of forest (*igapo*) along the banks and the creation of subsidiary channels (*fuños*) and *vareza* lakes that are connected with the main channel only at high water. Many species enter the *vareza* lakes at high water to reproduce, then return to the river as the water level falls.

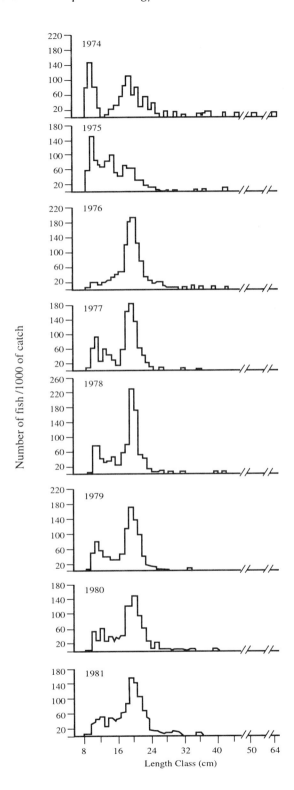

FIGURE 11.19 Little Nauyuk Lake, Arctic charr (*Salvelinus alpinus*) showing retention of the modal configuration despite severe reduction of the total population. The data were collected as soon as the ice cleared in late July.

Primary production in the main stem is very low, although high in the *vareza* lakes (Fitkau et al. 1975; Sioli 1975, 1984). The main organic input to the river is in the form of trees and branches, as well as vegetation mats with their associated fauna that fall from the banks; similarly, there is a high accidental input of terrestrial insects and other animals.

Few of the fishes appear to specialize on any particular food item. Despite the specialized teeth of many species, foods from the surface and the bottom are often found in the same individual (Lowe-McConnell 1975). In addition, "the stomach contents of one species collected at different times of the year were generally similar." This is in contrast to the work of Mckaye and Marsh (1983) who found a similar apparent anomaly in the specialized feeding adaptations in the Cichlid fish of Lake Malawi compared with the similarity of their actual stomach contents. That is, despite their highly specific feeding adaptations, many species were utlizing the same resource for much of the year. The specializations, he showed, reflect adaptations to specific habitats and feeding regimes at times of low abundance of the staple items.

High diversity in fish is associated with a relatively continuous energy input on which is superimposed a large number of small variations. As Paine (1966) remarks, "The animal diversity of a given system will probably be higher if the production is apportioned more uniformly through the year than occurring as a single major bloom."

The various factors contributing to the high number of species Lowe-McConnell attributes to (1) the age of the area, (2) freedom from major disturbance, (3) the large area of the basin, (4) the succession of habitats offered by the river, (5) the diverse niches in the lowland rivers and adjacent lakes, (6) the high proportion of the basin at low levels with comparatively stable conditions capable of supporting a large number of individuals, and (7) river capture and faunal exchange.

However, it is evident that fluctuation is not itself conducive to high diversity. There is a crossover point in the conditions favoring diversity development, in that small fluctuations foster diversity, but as fluctuations increase in intensity, diversity in inhibited. This is apparent in the marine system off the west coast of Peru which, although at much the same latitude as the River Amazon and with a very high energy input, exhibits low diversity. This region once supported the world's most productive, although somewhat erratic, fishery for anchoveta (*Engraulis ringens*) (Schaefer 1970).

At unpredictable intervals, on average about 5 to 10 years apart, the system is disrupted by a change in the pattern of oceanic currents, locally known as El Niño (Cushing 1982; Barber and Chavez 1983; Cane 1983; Rasmussen and Wallace 1983), resulting from major climatic and oceanographic phenomena. In "normal" years between El Niños the coastal upwelling continues throughout the year, driven by the Pacific oceanic circulation patterns. Wind-driven currents flowing northward toward the equator strike the South American coastline, with the result that the surface waters are driven offshore under the influence of the Coriolis force. Cold southward flowing water from the deeper regions of the oceans is thus brought to the surface. Although these events continue, to some extent, throughout the year, the winds are strongest and the effects most pronounced during the period of the southern winter, June through August. The area of upwelling is a relatively narrow band 10 to 20 km wide along the coast, terminated by a dynamic boundary, beyond which lie waters characteristic of subtropical oceans. Primary production in this upwelling of cold, nutrient-rich water starts before the waters reach the surface, so that by the time they do reach the surface, they are able to take maximum advantage of the intense sunshine. Primary productivity is among the highest recorded in the world reaching 3 to 10 $g.C.m^{-2}.day^{-1}$. (The comparative figure for Char Lake, Resolute Bay, 74°N is 20 $g.C.m^{-2}.y^{-1}$). The winds being strongest in June through August ensures an annual cycle of production reaching a peak at this time.

The food chain between the algae and the anchoveta is very short and direct ensuring a very high production of anchoveta. The main predators on the anchoveta are three species of fishing birds: a pelican, the alcatraz (*Pelecanus occidentalis thagus*), a booby, the piquero (*Sala variegata*), and a cormorant, the guanay (*Phalocorax bougainvilli*). With the insertion of the warm water band the system is disrupted and anchoveta production is reduced to a low level. This results in a heavy die-off among the fishing birds, but a relatively rapid return of the population once their food source has been reestablished. The system is evidently highly resilient in the long term, having evolved over a long time period.

Three factors appear responsible for maintaining the system at low diversity and a very high level of productivity: (1) the enormous annual spike of primary production, (2) the rapid removal of much of the biomass by fishing birds, and (3) it is an extremely "open" system functioning within boundaries that are far removed from the main production site. The erratic occurrence of the disruption of the system in El Niño years and the rapid transfer of energy out of the aquatic component of the system prevent the evolution of a stable hierarchy within the ocean. Presumably it is the capacity of the birds to recover more rapidly than marine species that establishes them as the main terminal predator.

This system is interpreted as being the result of a massive, highly coherent signal in non–El Niño years, stimulating a very favorable substrate to high annual production, within very weak boundary conditions. Combined with the occasional collapse of the system, it ensures lack of continuity with little possibility of the evolution of greater stability.

Similarly, when compared with that of South America, the species richness of the flora of the African tropical rain forest is relatively poor. This Richards (1973) attributes to relatively severe climatic fluctuations in Africa, due to drought, that do not occur in the Amazon Basin.

At the other diversity extreme is Lake Hazen (81° 55′ N; 81° 40′ W) at the northern end of Ellesmere Island in the High Arctic that has but a single fish species: Arctic charr. Energy input is limited despite almost 6 months of nearly continuous sunshine, because much of the incoming solar radiation is dissipated in melting ice; only in the warmest years does the ice clear completely from the lake. Lake Hazen is one of the most faunistically impoverished lakes in existence: apart from Arctic charr, the only fish species present, the fauna consists of one copepod *Cyclops scutifer*, two species of rotifer (*Keratella*), the water mite (*Libertia* sp.), the water beetle (*Hydroporus polaris*), three species of caddis fly, two species of crane fly, 22 species of midge, three of biting midge, and two species of syrphid flies (Oliver 1963; McLaren 1964; Johnson 1992b). Evidently a highly intermittent, low energy input gives rise to low system diversity. Nevertheless, the high autonomy of the lake imparts stability to the system as witnessed by samples taken 26 years apart (Figure 11.20).

If the base of a triangular envelope encompassing the extremes of fish species diversity is formed by the River Amazon and the Peruvian Coast ecosystem, the apex is at Lake Hazen. All other aquatic systems will fit within these limits.

Diversity is self-augmenting and greatly stimulated by physical structure as evident in tropical forests or on coral reefs. As the duration of the annual energy input decreases, irrespective of total energy input, diversity tends to increase. The effect of a sharp energy spike is apparent in many "pollutional" situations where high levels of nutrients in the spring create an unpleasant abundance of a single or a small number of species. As total energy input declines so does diversity. As diversity increases, energy flux increases, resulting in lower standing biomass per-unit-energy input (smaller B/P). As energy flux increases, so do energy density and system complexity, whether at the individual level or that of the ecosystem.

In summary, diversity increases in the absence of constraining factors, whether these are large fluctuations or a low level of energy input, or the effect of dominance or physical boundaries. Diversity is stimulated by a multiplicity of small fluctuations in an otherwise uniform and abundant energy environment. Further diversity is greatly enhanced by physical structure, whether inherent in the physical environment or developed by organism themselves. Two general trends emerge: (1) diversity increases from the poles to the tropics and (2) there has been a gradual, although not monotonic, increase in diversity over evolutionary time.

In the absence of change, the biosphere, functioning as a homeorhetic system, comes to a relatively steady state. Nevertheless, movements and migrations occurring continuously at various locations will impart a constant dynamic to the biosphere as a whole, but such changes will be gradually absorbed without disruption of the overall system *gestalt*. In the event of a disturbance of great magnitude such as the arrival of a bolide from space, the system is temporarily disorganized, and many species may be eliminated, but it then eventually settles down and is reordered in such

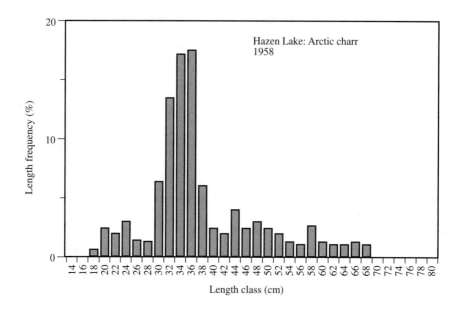

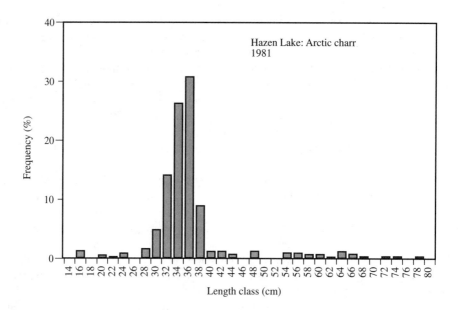

FIGURE 11.20 Lake Hazen, Artic charr (*Salvelinus alpinus*), showing little change in length-frequency distribution over a period of 23 years.

a manner as to fulfill the homeorhetic requirement that energy inflow and outflow must equilibrate. This may involve a further loss of species from among those that actually survived the catastrophe itself. From this point on, the old regime of force fields is resumed but with a different set of species.

11.4.7 STABILITY AND DIVERSITY

Stability develops within an energy sink, provided that the cycle of input remains relatively constant. In an autonomous system of low diversity and low energy input, high biomass develops as in an Arctic lake. This high biomass is stable to external perturbation, as the recovery trajectory is short and direct owing to short and direct energy pathways. The time to recovery is relatively short despite low annual energy input. Lake Hazen and other northern lakes may thus be seen to have stability comparable with that of a military tank designed to withstand a high degree of perturbation. *Per contra*, highly diverse systems whether terrestrial or aquatic can experience *small* perturbations and recover relatively quickly, as energy input is high and fluctuations are distributed rapidly along a large number of channels. Severe disturbance, however, causes the destruction of these channels resulting in a semipermanent change because of the long time necessary to rebuild and reestablish system complexity. Diverse systems exist only in benign environments. The highly diverse system, whether terrestrial or aquatic, is finely tuned to function in an environment of high energy input but of low variability. This is analogous to the stability of a jet airliner designed to operate in conditions that are relatively uniform but with many small variations. Innumerable feedback stabilizing mechanisms adjust the trim as necessary to maintain constant course and speed, but also like a diverse system the airliner is highly unstable when subjected to massive perturbation. Such a system is dependent on high energy flux.

11.5   GLOBAL ENERGY FLOW

Energy, under the influence of the Second Law inducing uniform distribution, in the form of heat tends to flow from regions of higher input at lower latitudes toward the poles. This energy distribution process, assisted by the Coriolis force, creates ocean and atmospheric currents, which, because of the disposition of the landmasses, results in uneven distribution. Owing to the tilt of the Earth's axis with respect to the plane of its orbit around the sun, the local climate becomes more variable with increasing latitude, within and between years. This increase in seasonality with latitude establishes an annual cycle of biological productivity: the greater the variability and the lower the energy input, the lower the diversity but the greater the biomass per unit energy input. Within this climatic gradient, ecosystems adapt to the local environment.

No ecosystem is completely autonomous so that from all ecosystems there is a transfer of biological energy or "leakage," from one subsystem to another. This global energy flow is from systems with high B/P (systems with a relatively slow turnover rate) to systems of lower B/P (systems with relatively rapid turnover); this is in the direction opposite to energy transfer within ecosystems, which is from populations of rapid turnover (low B/P) to populations of relatively slow turnover (low B/P). In the case of subsystems the principle of most action provides the dominant force field, whereas in the case of transfer between subsystems the principle of least action is predominant. However, with respect to the biosphere as a whole, the principle of most action must dominate system behavior; otherwise the world system could not continue to exist.

The unifying factor appears to be that energy transfer is from high fluctuation to low fluctuation; thus the various subsystems in the biosphere function as do populations within an ecosystem, as a series of coupled shock absorbers. The annual spike of production encourages the development of migration patterns and tends to increase the leakage. The extent of the transfer is modulated by the degree of autonomy of the system and the defenses against predation or grazing adopted by the populations concerned.

This transfer may be observed in the general movement of biological material during the annual migration patterns that have evolved, in many cases since the last glaciation: migration may be regarded as the biological equivalent of transhumance in human agriculture. Each year there is a massive migration of ducks, geese, raptors, waders, and many other types of birds to the Arctic, where they enjoy the brief bounty of the lakes and tundra, and the enormous migration of some 12 to 20 billion songbirds that winter in the subtropics moving to summer feeding and reproductive areas in higher latitudes where food is abundant and there are few year-round inhabitants (Cox, 1985). Of necessity, more individuals move south in the fall than arrive on the feeding and rearing grounds in spring.

As species diversity tends to increase from the poles to the wet tropics, there is, overall, a gradient of increasing energy flux. This is augmented by increasing average ambient temperatures from north to south stimulating increased respiration rates. This increase in energy flux with decreasing latitude is confirmed by the decrease in B/P over the same latitudes. The equatorial flow proceeds as far as the subtropics but mostly ceases there. The small seasonal climatic variation in the wet tropics induces more or less continuous year-round activity and more uniform energy flux over the course of the year, resulting in very little bird migration into or out of the region (MacArthur 1972), for an individual relinquishing its place would have great difficulty in regaining it in such a fast moving system. To this extent, wet tropical systems form an essentially closed system, where any straying organism from less rapidly turning over systems is rapidly incorporated into the general energy flow. This dynamically exclusive system thus attains a stable structure overall even though the component species populations may experience considerable fluctuation locally. The greater energy input and the higher and more uniform energy flux in the wet tropics promotes the development of ecosystem complexity.

11.6   EVOLUTION AND NATURAL SELECTION

11.6.1   EVOLUTION

If, as Darwin believed, life on Earth began only once and that "primordial organisms were all of the same or of very few kinds" (Dobzhansky 1968), then the great diversity of the present day biota implies a continuous, although not monotonic, increase. This is illustrated by the increase in the families of marine invertebrate and vertebrate species (Raup and Sepkoski 1982) (Figure 11.21).

Evolution, I postulate, results from the interaction of the two force fields over time. Most action ensures the ascendancy of acquisition and conservation of energy over ecological time, whereas least

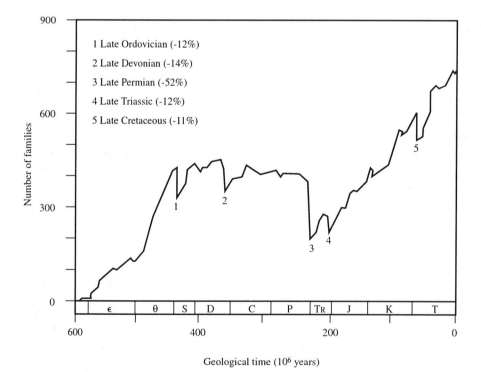

FIGURE 11.21   The gradual increase in species diversity in marine vertebrates and invertebrates over evolionary time. The numbers indicate times of severe reduction in the number of families present. (From Raup, D. M. and J. J. Sepkoski, *Science*, 215, 1501–1503, 1982. With permission.)

action inducing more rapid energy flux is ascendant over evolutionary time. Interaction between the two force fields ensures, in the first instance, an accumulation of energy followed by a gradual "streamlining" of the energy flow under the drive to increasing flux. With the emergence onto land, the third force field, that of gravity, becomes of great evolutionary significance. As the gradient of increasing diversity from poles to the tropics is contingent on an increase in energy flux, so the gradient of increasing diversity over evolutionary time is contingent on a similar increase in energy flux.

Organisms, being only partially stable, are liable to change. This endows them with a characteristic malleability allowing them to adapt to the prevailing environment, either by broadening the spectrum of signals utilized or by specialization to specific signals. In the absence of change, a system settles down in a state of ergodynamic equilibrium. Thus, if a species population in a system at equilibrium experiences a viable change, "evolution [being] opportunistic and immediate advantage ... more potent than eventual gain" (Dobzhansky 1968) that population will undergo either an increase or decrease in action, which in turn will either increase or decrease the flux through the system. An increase in action will be reflected in an increase in size, an increase in mean life span, an increase in uniformity, or an increase in acquisitive capacity. A decrease in action implies increased energy flux; for example, the advent of a new species.

Many evolutionary events, such as an increase in motility, demand increased energy expenditure, but this must be offset by some other advantage such as an increase in acquisitive ability.

The question remains: how can two countervailing processes occur without creating a chaotic situation? In that the overall evolutionary tendency has been toward increasing diversity, it must be assumed that the probability of any viable change resulting in an increase in energy flux is somewhat greater than the probability that it will result in a decrease in flux. This imperfect symmetry is the result of the self-augmenting characteristic of diversity (Hutchinson 1959; Golley 1965). The expression of these countervailing trends is evident in different time frames: over ecological time the system must assume a state of action; over evolutionary time, there is gradual "slippage" toward greater diversity and increased energy flux (shades of the Second Law: entropy in the shape of the maximum number of complexions or its sibling, the Principle of Least Action, always wins)!

The countervailing trends are clearly seen over evolutionary time. Opportunities for faster energy flow appear to have been extremely limited for the first 2.5 to 3.0 billion years of life on Earth for the diversity of species was low and changed little over the years, ecosystems being largely represented by stromatolite communities of algae and bacteria cemented together with sand grains (Golubic 1976).

With the advent of heterotrophy in the late Precambrian, stimulated by the Principle of Least Action, a great flowering of species occurred. This Stanley (1973a) attributed to a "grazing principle" which he derived from the work of Paine (1966). Hence, Stanley concluded that grazing stimulates diversity. According to Garrett (1970) intensive grazing of the stromatolites caused their disappearance in all regions except areas inhospitable to animals such as certain hypersaline lagoons in coastal Australia.

As heterotrophs evolved in the Cambrian, the increase in energy flux had a self-augmenting effect as the previously static energy accumulations in the abundant stromatolites were incorporated into a dynamic system. From this point on there has been a general increase in diversity, although this has not been a monotonic progression as, at various periods, diversity greatly decreased. Nevertheless, following each catastrophic decline, diversity once more began to increase.

A reenactment of evolutionary events during the Late Precambrian and early Cambrian eras occurred in Lake Nakuru in Kenya, East Africa. Lake Nakuru is a saline lake originally with a very low species diversity, comprised of two species of algae, one copepod, one rotifer, corixids, notonectids, and some 500,000 flamingoes belonging to, virtually, one species (Jacobs 1975). In 1962 the cichild fish, *Tilapia grahami*, was introduced, ostensibly for mosquito control, but soon the fish turned to feeding on algae. The *Tilapia* thrived reaching the high biomass of 1.5 g.m^{-2}, although remaining small in size. The flourishing *Tilapia* attracted many fish-eating birds not previously abundant in the region: pelicans, anhingas, cormorants, herons, egrets, grebes, terns, and fish eagles. Thus, with the addition of a single species, the ecological log-jam was broken and energy flux greatly increased.

Conversely, within certain evolutionary lines, there are recognizable trends toward increased energy acquisition and conservation fostered by the Principle of Most Action. This tendency for animal groups to evolve toward larger physical size is generally recognized in "Cope's Rule" (Newell 1949; Rensch 1959; Stanley 1973b). The advantages of an increase in size are many; according to Stanley the more salient are "improved ability to capture or ward off predators, greater reproductive success, increased intelligence (with increased brain size), better stamina, expanded size range of acceptable food, decreased annual mortality, extended individual longevity, and increased heat retention per unit volume." There are also disadvantages, such as a higher food requirement and the need for increasing structural support to overcome the effects of gravity, for, as Stanley points out, at a certain level further size increase becomes inadaptive.

One of the best documented is exemplified by the horse lineage; starting life as a five-toed mammal about the size of a dog (Marsh 1874), the original horses gradually improved their capacity for escape from predators through specialization for fast running on the world's grasslands. This entailed the loss of four functional toes and improved capacity for utilization of the coarse herbage of the plains, necessitating the evolution of grinding molars. The advantage of these attributes allowed size to increase until the wild horses of recent times evolved.

Other evolutionary trends toward reduced energy flux and increased efficiency of energy usage are those manifested in the decreased production of ova, internal fertilization, vivipary, and a high degree of parental investment in offspring resulting in a lesser need for juveniles and an increase in mean age.

However, it is important to note also that evolutionary lines resulting in a decrease in size are quite common, although less frequent than those showing an increase. A decrease in size implies emphasis on the energy-acquisition component, the Action, rather than the time component. This, it may be assumed, has been necessary for the organism to retain a place in the ecosystem at all costs.

Evolution is thus not the effect of one trend but takes place at the point of intersection of three force fields. As Simpson (1953) concludes, a major feature of evolution has been an increase in the *average* complexity of species. Greater complexity is a function of increasing energy density and increasing energy flux. Average complexity over evolutionary time has increased both with respect to individuals and ecosystems. That the increase in complexity of individuals is contingent on an increase in energy flux is supported by the work of Zotin (1984) who has shown that there has been a progressive increase in the value of the constant a, with increasing complexity, in the equation for respiration intensity* (Figure 11.22).

Not only has respiration intensity index increased over the broad spectrum of evolution, but within classes as well (Table 11.2).

"Respiratory intensity increases markedly from protozoa to mammals and birds in the animal kingdom and from monotremata to primates in the class of mammals. This points to obvious energetic trends in the evolutionary progress of these organisms" (Zotin 1984).

Thus evolution is a stepwise process, first energy is accumulated and then followed by more rapid dissipation. By this means the energy flow in ecosystems is streamlined or maintained at the highest possible level, commensurate with the constraints applied by the ambient energy input cycle and interspecific and intraspecific interactions. Evolution is thus irreversible, as the majority of organisms must continuously accommodate life in the conditions prevailing, if they are to maintain a place in an increasingly dynamic ecosystem. Nevertheless, some species, such as the horseshoe crab (*Xiphosura* (*Limulus*)) or the Coelacanth *Latimeria*, remain essentially unchanged over long time periods. In the event of worldwide catastrophic change, the remnants of the previous system,

---

* Because size has such an effect on basal metabolic rate, Zotin uses the value of the constant a in the equation for "respiration intensity" to compare the energy transfer rates of organisms of different sizes and different classes. Respiration intensity (q) is defined by the equation, $q_{o2=a}W - b$, where W is the mass and a and b are constants.

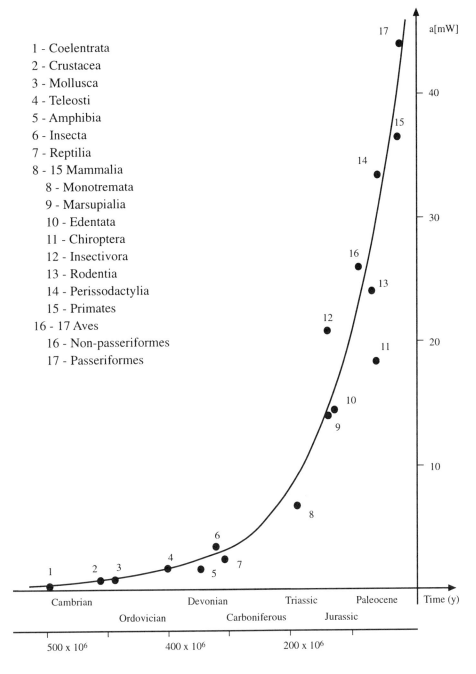

1 - Coelentrata
2 - Crustacea
3 - Mollusca
4 - Teleosti
5 - Amphibia
6 - Insecta
7 - Reptilia
8 - 15 Mammalia
    8 - Monotremata
    9 - Marsupialia
    10 - Edentata
    11 - Chiroptera
    12 - Insectivora
    13 - Rodentia
    14 - Perissodactylia
    15 - Primates
16 - 17 Aves
    16 - Non-passeriformes
    17 - Passeriformes

FIGURE 11.22 Increase in the "respiration index" with increase in complexity, as reflected in various animal groups over evolutionary time. (From Zotin, A. I., *Thermodynamics and Regulation of Biological Processes*, Berlin, New York: Walter de Gruyter, 1984. With permission.)

have to reorganize to meet the overriding demands of the most action principle, but these species will be adapted to more rapid energy flux, so that a new evolutionary symphony begins with a different set of players. Initially, these specialized players will be less efficient in energy acquisition in the new relatively unspecialized or destreamlined environment; hence there will be unexploited local energy accumulations and many opportunities for new evolutionary approaches.

TABLE 11.2
Values of Parameter "*a*" for Various Animal Groups (Poikilothermal at 20°C, Reptilia at 30°C, Homeothermal in the Thermoneutral Zone)

Order	*a*, mW
Monotremata	6.7
Marsupialia	13.9
Edentata	14.2
Chiroptera	18.3
Insectivora	20.8
Lagomorpha	23.0
Rodentia	24.2
Carnivora	29.2
Perissodactyla	33.6
Artiodactyla	33.6
Primates (Pongidae)	

The value of *a* increases with increasing complexity of the organism. (From Zotin, A. I., *Thermodynamics and Regulation of Biological Processes*, Berlin, New York: Walter de Gruyter, 1984. With permission.)

Coevolution may develop where the interaction between individuals of two species is not asymmetrical in nature, that is, where there is no established hierarchical relationship between the species. Coevolution will be fostered, initially, by some extrinsic or fortuitous mutual benefit. It has probably reached greatest development in the angiosperms and their associated insect pollinators, and more specifically in the orchidaceae. As Burger (1981) remarks,

> A unique pollination system allows orchids to persist at population densities that are much lower than those of most other plants of similar size.

Evolution results from the gradually changing conditions as the pace quickens and one must run ever harder to maintain a place in the system, not merely to stay in place, for the "place" and the pace are continually increasing. As Burger (1981) comments, "… survival in evolutionary time requires adaptation to an environment of increasingly sophisticated competitors and neighbours."

This increase in sophistication results from the continual acceleration of the energy flux, under the Principle of Least Action, attained through exploration of the ambient signal array. According to Burger (1981),

> Not only have the angiosperms [flowering plants] invented new ways of exploiting the environment, but they seem to have been driven into ever more refined ways of dividing the habitat among themselves.

To increase its action (energy *times* time) is the driving force of the coherent group, species population, or ecodeme. Some organisms trade large size and individual biomass for specialized acquisitive capacities, small size, and rapid turnover time ("*r*-selected" species), thus accepting a place in the lower reaches of the food hierarchy. Others with energy acquisition secured, gradually move up the food chain as size and lifespan increase, eventually reaching a position of dominance largely free from the constraints imposed by predator or grazing animal ("*K*-selected" species).

Thus, through the alternating ascendancy of the opposing forces, the system and its individual components are gradually ratcheted up to greater complexity and greater energy flux. Time's Arrow indicates the ultimate ascendancy of the Principle of Least Action: the inexorable trend toward greater energy flux. Evolution may therefore be regarded as a by-product of the "attempt" to reach a stable state in constantly accelerating energy flux within a changing series of constraints. Ergo-dynamic equilibrium is a temporary stable point in the continuing interactive process between time delay and acceleration of the energy flow, between symmetry and asymmetry, between homogeneity and heterogeneity, and between uniformity and hierarchical organization (Figure 11.18).

Complexity, either in individuals or in ecosystems, increases as energy is increasingly con-densed with a corresponding increase in energy flux. By analogy, the engine and pump of an air compressor must work harder as the pressure in the reservoir increases. Occasionally, under the stimulus of the Principle of Most Action, new, more energy efficient methods are explored, such as the introduction of homeothermy or insect pollination within the angiosperms. Such changes lead to a radical readjustment of ecosystem structure, ensuring the rise to dominance of mammals on land and in the sea and birds, and the dominance of angiosperms over much of the Earth's surface.

11.6.2 NATURAL SELECTION

The great developmental biologist, Karl Ernst von Baer, while accepting the concept of evolution, could not agree with Darwin's (1859) explanation of the evolutionary process through natural selection, for natural selection of itself, he maintained, contained no mechanism (von Baer 1873, in Hull 1973).

"Darwinism has triumphed," he states, "precisely because it denies purposes in nature and because it insists on explaining the appearances of purposiveness in nature by blind forces producing a host of life forms and elimination of the less fit by natural selection. We even see philosophy bowing before these conclusions."

The probable truth of this has lingered for a long time in the minds of many, and to date has never been properly answered, despite Ernst Mayr's vigorous defense of the concept (Mayr 1976). However, I believe that von Baer's words are as true today as when written over a century ago.

> I am convinced that everything that exists and continues to exist in nature arose and will continue to arise through natural forces and material substances. But these forces must be coordinated and directed. Forces which are not directed — so called blind forces — can never, as far as I can see, produce order.

An ecosystem develops through the interaction of the two force fields functioning within a set of environmental and biological constraints. The physical environmental constraints are the nature of the ambient energy cycle and the temperature cycle, exposure, or the land/water interface; the biological constraints are the nature of the receptor organisms, competitive effects, the top-down constraints of predators and grazers or, in forests, the attenuation of light by the dominant tree species. Together these constraints make up the partial boundaries of an ecosystem, but as there is no such thing as an open system; the ultimate boundaries are those of the biosphere.

Island biotas proceed toward increasing energy flux and increasing complexity at a relatively slow pace compared with the biotas of a large mainland because of their relatively small area and limited range of habitat and, in many cases, a relatively short history. Hence their biota cannot compete effectively when invaded by species raised in a more dynamic environment (Quammen 1997).

Ultimately, all species must fit within the worldwide mosaic of subsystems as the world system approaches ergodynamic equilibrium. Local change or disturbance, such as a volcanic eruption or El Niño event of uncertain periodicity, will inevitably maintain the world system in a "preergody-namic equilibrium state, but for the most part the general absence of change will necessitate that species fit within a whole that is governed by the Most Action imperative. This implies an overall degree of symmetry in species interactions. Natural selection is the adaptation, probably better described as the accommodation of species to the thermodynamic and physical constraints of the ecosystem in which they exist.

11.7 CONCLUSION

> The unity of knowledge may be a doctrine that has never been explicitly explained. To me, at any rate, it says that 'intellectually close' to any fact or argument of importance there are other facts or arguments which are also important and that these intellectually neighbouring facts or arguments are illuminated and rendered more easily understandable by the original fact or argument with which we started. It asserts furthermore that these relationships are reciprocal, so that the original fact or argument gains in stature and importance, from these intellectually neighbouring facts of arguments.
> (From Landsberg, P.T., *J. Non-Equilib. Thermodyn.,* 12, 45–60, 1987. With permission.)

Mankind can no longer be regarded as an organism set apart from his biological heritage. Like any other biological entity, we are subject to the same laws and constraints; indeed, many of the concepts developed in the Action Hypothesis are clearly expressed in our recent history. Perhaps the greatest change in the ecology of the Earth in recent millennia has been effected by mankind's change from a hunting and gathering existence to a more settled state made possible by the introduction of agricultue. This has allowed us to assume dominance and thus become limited only by space and energy. Space is gradually imposing limits on further expansion, but energy supply, in the form of "fossil" fuels, seems to have no immediate horizon and has thus allowed the development of great complexity culminating in the rise of great cities with their enormous entropy production. Contingent on our emergence as the dominant species, there has been a significant decrease in the species richness of the world, particularly among former dominants, whether plant or animal. In addition there has been a loss diversity, for many of the organisms that have escaped oblivion have been greatly reduced in abundance, relegated to isolated reserves a fraction the size of their former range.

In agriculture energy flow is canalized and food chains shortened on grasslands by controlled grazing through continually moving the herds to new pasture and the elimination of or protection from predators. Arable agriculture was probably initially confined to sites suitable for irrigation but then spread to more mesic, generally forested sites when technology for the removal of trees was sufficiently advanced. Arable agriculture in many ways mimics the natural situation evident in the anchoveta system off the Coast of Peru: a relatively short pulse of energy input contingent on the addition of nutrients and water is followed by removal of the product. The periodic collapse of the system (plowing) ensures simplicity. Local crop failures have had a similar effect on humans to failure of the anchoveta crop on Peruvian fishing birds. When a system is grossly simplified, there are no reserves; there is no possible return to hunting and gathering following crop failure; no turning to alternative resources in times of need as do the cichlids of Lake Malawi, bringing into play their highly specialized feeding mechanism.

Simplification of agricultural systems has greatly increased the instability of the whole and in the process has greatly increased the homogeneity of the Earth system.

As agriculturalists we have become, like the bowhead whale, the consummate K-strategists, reaching down to the lowest level in the trophic hierarchy and canalizing as much of the assimilated energy as possible, stimulating the production of young and nutritious food stuffs, and eliminating the "losses" contingent on transfer of the energy to higher trophic levels and more complex food webs. Energy flux in all parts of the ecosystem is thus accelerated except at the terminal level. In this way we have achieved great biomass, although in the human population this has been reached through abundance of individuals as opposed to the enormous individual size of the bowhead. Nevertheless, our species has undergone a gradual increase in size and life span since the time of our Australopithecine ancestors.

We have been able to adapt and accommodate to all climates and conditions in the world, becoming the consummate generalist utilizing a vast array of species to satisfy our every desire. These changes have relied greatly on the expenditure of capital accumulated by so-called "renewable resources," from forests, soils, and fresh waters to "fertilizers" (potash, guano, etc.) and fossil fuels as well as the introduction of prostheses to augment human labor, initially draft animals and then engines driven by "fossil" energy. Elimination of former ecological boundaries and the relatively unconstrained transfer of biological material has also resulted in the destabilization of many if not most local ecosystems.

11.8 DARWIN'S LEGACY

The notion of evolution permeated the collective consciousness of biologists from the beginning of the 19th century but it was the concept of natural selection, the major contribution of Darwin and Wallace, that created the storm of discussion and argument. Natural selection entailed the notion of competition for scarce resources in an overcrowded world as envisaged by Malthus when he regarded the human outcome of the advancing industrial revolution. These ideas were taken up by Herbert Spencer and his following of Social Darwinists who applied the notions of competition and "survival of the fittest" and "nature red in tooth and claw" to human activities. The term "Darwinism" thus assumed a perjorative connotation in the public mind expressed as the "law of the jungle" and is still frequently used in this context in the daily newspaper. But as Goodwin (1994) says, "Darwinism short changes our biological natures" for it omits the other vital aspect, the need to interact symmetrically and function as a cohesive unit, and, at the global level, to ensure the functioning of the whole as a cohesive unit.

The one person to contest Darwinism as propounded by Spencer and his followers was Prince Kopotkin (1890), sometimes described as the first sociobiologist, who sought a new social philosophy based on the grounds that there is as much cooperation and mutual assistance in the animal world as there is antagonism and competition, backing up his thesis with many examples. Kropotkin was never taken seriously but his philosophy, which implies the need for symmetrical interaction, homogeneity, and energy conservation, seems at last to be enjoying a modest revival of interest (Tuchman 1967; Ridley 1996). However, recognition of the existence of important cohesive forces will be difficult in face of the current emphasis on individualism.

Recognition of a dynamic interrelationship between symmetry and asymmetry, and between homogeneity and hierarchy, was expressed by Charles Darwin (1839) who observed of the Fuegians during his voyage on *H.M.S. Beagle* that,

> The perfect equality of all the inhabitants will for many years prevent their civilization. Until a chief arises who by his power can heap up possessions for himself there must be an end to all hopes of bettering their condition.

As Darwin recognized, without hierarchical organization, energy flow (in terms of wealth) remains at its lowest level. The problem is to maintain the balance between homogeneity and heterogeneity particularly when essentially autonomous systems are converted to more "open" ones through the elimination of barriers.

The Action Hypothesis provides a mechanism for natural selection in terms of interacting physical force fields, providing for limited competition and mutualy beneficial interaction; unlimited competition reduces the system to a less complex state from which position, over evolutionary time, it will gradually reform but in an unpredictible manner.

Not only does the Action Hypothesis provide a mechanism for natural selection or, as I believe, preferably, natural accommodation, it provides also the necessary understanding for solving age-old problems and apparent paradoxes such as "progress" and "directionality" in evolution. It answers, too, what Mayr (1962) refers to as the paradox of evolution, "the apparent contradiction between, on the one hand, the seeming purposefulness of organic nature, and on the other hand, the haphazardness of the evolutionary process." The biological process has both short- and long-term goals: the short-term goal is ergodynamic equilibrium, the long-term goal is instantaneous dissipation in a flash of photons, both are thermodynamic states rather than tangible conditions. Both are approached within a set of rules, although the route is dependent on viable random change and the opportunistic nature of the process.

These are the rules of the game: we live within a closed system and that system will tend to approach ergodynamic equilibrium in which the rates of energy input and output are equal, if necessary, impartially eliminating those species that cannot find a suitable niche. The words of the German economist, Wilhelm Ropke (1988), in his book *Germany in Transition* bear repeating in this context: German institutions in the post-war world, he maintains, have imparted to competition the framework rules and machinery of impartial supervision "which a competitive system needs as much as in any game or match, if it is not to degenerate into a vulgar brawl...."

If the Principle of Least Action is the dynamic equivalent of the Second Law, then the living world expresses the dominance of the countervailing force codified in the Principle of Most Action. The outcome, dependent on work done by time-delaying the energy flux, expresses all the characteristics that might be expected of "negentropy" or "ectropy."

The Action Hypothesis, like the origins of thermodynamics itself, developed out of observations made with essentially practical motives: the development and improvement of fisheries in the Canadian Northwest Territories. These observations revealed certain apparent anomalies which demanded explanation. Having attempted to provide an explanation, I believe that it is appropriate to conclude on a practical note. How does all this rather abstruse theoretical work contribute to fisheries science. I believe with Gauldie et al. (1989) that thermodynamic models demand a new approach to the management of populations of K-selected species such as the orange roughy, including greater understanding of the ecology, life history aspects, and the underlying mechanisms which control changes in abundance, "in order to flesh out the bare bones of essentially time-series statistical analyses of length and age frequencies which have such a high profile in modern fisheries science."

Finally, as Charles Darwin stated of his own work,

It can hardly be supposed that a false theory would explain in so satisfactory a manner as does the theory of natural selection [Action], the several classes of facts above specified.

ACKNOWLEDGMENTS

This work is dedicated to the memory of my wife Cecile, who for many years withstood the rigors of my annual absence in the Arctic; without her help and patience this work would never have been attempted and certainly never concluded. Unfortunately she did not live to see its completion. I thank Peter Vanriel, Bonnie Burns, and Eric Gyselman for their hard work and unfailing support as well as all members of the field parties, who, over the years, provided the raw data and much discussion. I am also much indebted to Charlie Kyiok and Peter Aggeloktok and their families for their help and for making our stay at Nauyuk Lake so rewarding. I am very much indebted to Dr. Harold Mundie for a very thorough review and I greatly appreciate Dr. Eric Schneider's constant stimulus since the publication of my first paper on the subject in 1981. I am also grateful to Dr. Harold Hamilton for this assistance and Commander Henry H. Parker, R.N. who tutored me in thermodynamics, but did not always agree with my thermodynamic interpretation of biological events. Any errors or omissions in this field are entirely my responsibility.

REFERENCES

Allee, W. C., Emerson, A. E., Park, O., Park, T., and Schmidt, K. P., *Principles of Animal Ecology.* Philadelphia: Saunders, 1949.

Anderson, P. W., More is different, *Science,* 177 (4047):393–396, 1972.

Andrewartha, H. G. and Birch, L. C., *The Distribution and Abundance of Animals.* Chicago: University of Chicago Press, 1954.

Barber, R. T. and Chavez, F. P., Biological consequences of El Niño. *Science,* 222:1203–1210, 1983.

Beverton, R. J. H. and Holt, S. J., *On the Dynamics of Exploited Fish Populations, Fisheries Investigations (Series 2),* 19. London: U.K. Min. of Ag, 1957.

Bormann, F. H. and Likens, G. E., *Pattern and Process in a Forest Ecosystem*. New York: Springer-Verlag, 1979.

Bormann, F. H., Likens, G. E., and Eaton, J. S., 1969. Biotic regulation of particulate and solution losses from a forest ecosystem. *BioScience,* 19:600–610, 1979.

Bourn, D. and Coe, M., The size, structure and distribution of the giant tortoise population of Aldabra. *Phil. Trans. Roy. Soc. Lond (B),* 282:139–175, 1978.

Bray, J. R. and Curtis, J. T., An ordination of upland forest communities in southern Wisconsin. *Ecol. Monogr.,* 27:325–349, 1957.

Burger, C., Why are there so many kinds of flowering plants? *Bioscience,* 31:572–591, 1981.

Cane, M. A., Oceanographic events during El Nino. *Science,* 222:1189–1195, 1983.

Chevalier, A., Biogéographie et écologie de la forêt dense ombrophile de la Côte d'Ivoire. *Rév. Bot. Appl.* 28:101–115, 1948.

Clements, F. E., *Research Methods in Ecology*: Lincoln, Nebraska: Univ. Nebraska Publ., 1905.

Clements, F. E., Plant Succession: An Analysis of the Development of Vegetation. Vol. Publ. 242. Washington: Carnegie Institute, 1916.

Clements, F. E., *Plant Indicators: The Relation of Plant Communities to Process and Practice*. Vol. Publ. 290: Carnegie Institution, Washington, 1920.

Clements, F. E., Nature and structure of the climax, *J. Ecol.,* 24:252–284, 1936.

Connell, J. H., Diversity in tropical rain forests and coral reefs. *Am. Nat.,* 98:399–414, 1978.

Connell, J. H. and Slayter, R. O., Mechanisms of succession in natural communities and their role in community stability and organization. *Am. Nat.,* 111:1119–1144, 1977.

Coulter, G. W., "Population changes within a group of fish species in Lake Tanganyika following their exploitation." *J. Fish. Biol.,* 2:235–259, 1970.

Cox, G. W., The evolution of avian migration systems between temperate and tropical regions of the New World. *Am. Nat.,* 126(4):451–474, 1985.

Curtis, J. T., *The Vegetation of Wisconsin*. Madison: University of Wisconsin Press, 1959.

Curtis, J. T. and McIntosh, R. P., An upland forest continuum in the prarie-border region of Wisconsin. *Ecology,* 32:476–496, 1951.

Cushing, D. W., *Climate and Fisheries.,* London: Academic Press, 1982.

Darwin, C., *The Voyage of the Beagle,* New York: Anchor Books (1962), 1839.

Darwin, C., *The Origin of Species*. London, John Murray, 1859.

Dickie, L. M., Kerr, S. R., and Boudreau, P. R., Size-dependent processes underlying regularities in ecosystem structure. *Ecol. Monogr.,* 57(3):233–250, 1987.

Dobzhansky, T., On some fundamental concepts of Darwinian biology. In *Evolutionary Biology,* edited by T. Dobzhansky, M. K. Hecht and W. C. Steere. New York: Appleton-Century-Crofts, 1968.

Drury, W. H. and Nisbet, I. C. T., Succession *J. Arnold Arboretum,* 54:331–368, 1973.

Eldredge, N. and Gould, S. J., Punctuated equilibria: an alternative to phyletic gradualism. In *Models in Paleobiology,* edited by T. J. M. Schopf. San Francisco: Freeman, 1972.

Elton, C. S., *Animal Ecology*. London: Sidgwick Jackson, 1972.

Elton, C. S., *The Ecology of Invasions by Plants and Animals*. London: Methuen, 1958.

Elton, C. S., *The Pattern of Animal Communities*. London: Methuen, 1966.

Erlich, P. R. and Birch, L. C., The "Balance of Nature" and "population control." *Am. Nat.* 101: 97–107, 1967.

Feynman, R. P., *The Feynman Lectures on Physics, Vol. 2*. Edited by F. R. P., L. R. B. and M. Sands. Vol. 2. Reading, MA: Addison-Wesley, 1963.

Finley, K. J., Isabella Bay, Baffin Island, an important historical and present-day concentration, area for the endangered bowhead whale (*Balaena mysticetus*) of the Eastern Canadian Arctic. Arctic 43:137–152, 1990.

Finley, K. J. and Darling, L. M., Historical data sources on the morphometry and oil yield of the bowhead whale. *Arctic,* 40(2):153–156, 1990.

Fitkau, E. J., Irmler, U., Junk, W. J., Reiss, F., and Schmidt, G. W., Productivity, biomass and population dynamics in Amazonian water bodies. In *Tropical Ecological Systems,* edited by F. B. Golley and E. Medina. New York: Springer, 1975.

Forbes S. A., On some interactions of organisms. In *Bulletin I*: Illinois State Laboratory of Natural History, 1880.

Forbes, S. A., The lake as a microcosm. *Bull. Sci. Assn. Peoria, Ill,* 1887.

Frank, P. W., Life histories and community stability. *Ecology,* 49:355–357, 1968.

Gadgil, M. and Solbrig, O., The concept of r- and K-selection: evidence from wild flowers and some theoretical considerations. *Am. Nat.,* 104:14–31, 1972.

Garrett, P., Phanerozoic stromatolites: noncompetitive ecologic restriction by grazing and burrowing animals. *Science,* 169:171–173, 1970.

Gauldie, R. W., West, I. F., and Davies, N. M., K-selection characteristics of orange roughy (*Hoplostethus atlanticus*) stocks in New Zealand waters. *J. Appl. Ichthyol.,* 5:127–140, 1989.

Gaymer, R., The Indian Ocean giant tortoise *Testudo gigantea* on Aldabra. *J. Zool. Lond.,* 154:341–363, 1968.

George, J. C., Bada, J., Zeh, J., Scott, L., Brown, S. E. and O'Hara, T., Preliminary age estimates of bowhead whales via aspartic acid racemization, Scientific Committee of the International Whaling Commission, 1998.

Géry, J., The fishes of Amazonia. In *The Amazon: Limnology and Landscape Ecology of a Might Tropical River and Its Basin,* edited by H. Sioli. Dortretch: Dr. J. W. Junk, 1984.

Gleason, H. A., The structure and development of the plant association. *Torrey Bot. Club Bull.,* 44:463–481, 1917.

Gleason, H. A., The individualistic concept of the plant association. *Bull. Torrey Bot. Club,* 53:1–20, 1926.

Gleason, H. A., The individualistic concept of the plant association. *Am. Mid. Nat.,* 21:92–101, 1939.

Goldman, C. R., Antarctic freshwater ecosystems. In *Antarctic Ecology,* edited by M. W. Holdgate. New York: Academic Press, 1970.

Golley, F. B., Structure and function of an old-field broomsedge community. *Ecol. Monogr.,* 35:113–117, 1965.

Golley, F. B., Energy flux in ecosystems. Paper read at Ecosystem structure and function, at Corvallis, Oregon, 1972.

Golubic, S., Organisms that build stromatolites. In *Stromatolites,* edited by M. R. Walter. Amsterdam: Elsevier, 1976.

Gomez-Pompa, A., Vasquez-Yanez, C., and Guavera, S., The tropical rain forest: a non-renewable resource. *Science,* 177:762–765, 1972.

Goodwin, B., *How the Leopard Changed Its Spots: The Evolution of Complexity.* Rockefeller Center. Touchstone-Simon and Schuster, 1994.

Gould, S. J. and Eldredge, N., Punctuated equilibria. *Paleobiology,* 3:115–151, 1977.

Graham, S. A., Climax forests of the Upper Peninsula of Michigan. *Ecology,* 22:355–362, 1941.

Green, R. F., A note on K-selection. *Am. Nat.,* 116:291–296, 1980.

Grubb, P. J., The growth ecology and population structure of giant tortoises on Aldabra. *Phil. Trans. Roy. Soc. Lond.,* (*B*) 260:327–372, 1971.

Gunther, A. C. L. G., The giant land tortoises (living and extinct) in the Collection of the British Museum, 20. London: Taylor and Francis, 1877.

Harper, J. L., A Darwinian approach to plant ecology. *J. Ecol.,* 55:247–270, 1967.

Harper, J. L., The role of production in vegetational diversity. In *Diversity and Stability in Ecological Systems,* edited by G. M. Woodwell and H. H. Smith: Brookhaven Symposium on Biology, 22, 1969.

Harper, J. L., *Population Biology of Plants.* New York: Academic Press, 1977a.

Harper, J. L., The contributions of terrestrial plant ecology studies for the development of the theory of ecology. In *Changing Scenes in the Natural Sciences.* edited by C. E. Goulden. Philadelphia: Philadelphia Academy of Natural Sciences, 1977b.

Heinselman, H. L., Fire in the virgin forests of the Boundary Waters Canoe area, Minnesota. *J. Quart. Res.,* 3:329–382, 1973.

Hessler, R. R. and Sanders, H. L., Faunal diversity in the deep seas. *Deep-Sea Res.,* 14:65–78, 1967.

Heuts, J. J., Ecology, variation and adaptation in the blind cave fish Caecobarbus geertsii Blgr. *Ann. Soc. Roy. Zool. Belg.,* 82:155–230, 1951.

Holling, C. S., Resilience and stability in ecological systems. *Ann. Rev. Ecol. Syst.,* 4:1–23, 1973.

Horn, H. S., Forest succession. *Sci. Am.,* 232:90–98, 1975a.

Horn, H. S., Markovian properties of forest succession, community structure and evolution. In *Ecology and Evolution of Communities,* edited by M. L. Cody and J. M. Diamond. Harvard: Bellknap Press, 1975b.

Hull, D. L., *Darwin and His Critics: The Reception of Darwin's Theory of Evolution by the Scientific Community.* Chicago: University of Chicago Press, 1973.

Hunter, J. G., The production of Arctic charr (*Salvelinus alpinus* Linneus) in a small arctic lake: Fisheries Research Board Canada, 1970.

Hurd, L. E., Mellinger, M. V., Wolf, L. L., and McNaughton, S. J., Stability and diversity at three trophic levels in terrestrial successional ecosystems. *Science,* 173:1134–1136, 1971.

Hutchinson, G. E., Homage to Santa Rosalia, or why are there so many kinds of animals? *Am. Nat.,* 93:145–159, 1959.

Hutchinson, G. E., *An Introduction to Population Biology.* New Haven and London: Yale Univ. Press, 1978.

Jacobs, J., Diversity, stability and maturity in ecosystems influenced by human activities. In *Unifying Concepts in Ecology,* edited by W. H. Van Dobben and Lowe-McConnell. The Hague: W. Junk, 1975.

Johnson, E. A., Fire recurrence in the sub-arctic and its implications for vegetation composition. *Can. J. Bot.,* 57:1374–1379, 1979.

Johnson, E. A., Vegetation organization and dynamics of lichen woodland communities in the Northwest Territories. *Ecology,* 62:200–215, 1981.

Johnson, L., Keller Lake: characteristics of a culturally unstressed salmonid community. *J. Fish. Res. Board Can.,* 29:731–740, 1972.

Johnson, L., The dynamics of Arctic fish populations. Paper read at Circumpolar Conference on Northern Ecology, at Ottawa, 1975.

Johnson, L., The ecology of arctic populations of take trout, *Salvelinus namaycush,* lake whitefish, *Coregonus clupeaformis,* and Arctic char, *S. alpinus,* and associated species in unexploited lakes of the Canadian Northwest Territories. *J. Fish. Res. Board Can.,* 33:2459–2488, 1976.

Johnson, L., The Arctic charr. Pages 15–98 in *Charrs, Salmonid Fishes of the Genus Salvelinus,* edited by E. K. Balon. The Hague: Dr. W. Junk, 1980.

Johnson, L., The thermodynamic origin of ecosystems. *Can. J. Fish. Aquat. Sci.,* 38:571–590, 1981.

Johnson, L., Homeostatic characteristics in single species fish stocks in article lakes. *Can. J. Fish. Aquat. Sci.,* 40:987–1024, 1983.

Johnson, L., Hypothesis testing: Arctic charr, giant land tortoises, marine and freshwater molluscs and tawny owls. Paper read at Third workshop on Arctic charr, at Tromso, 1985.

Johnson, L., The thermodynamic origin of ecosystems: a tale of broken symmetry. In *Entropy, Information and Evolution,* edited by B. H. Weber, D. J. Depew and J. D Smith. Cambridge, MA: M.I.T. Press, 1988.

Johnson, L., The thermodynamics of ecosystems. Pages 1–47 in *The Natural Environment and Biogeo-Chemical Cycles,* edited by O. Hutzinger. Heidelberg: Springer-Verlag, 1990.

Johnson, L., An ecological aproach to biosystem thermodynamics. *Biol. Philos.,* 7:35–60, 1992a.

Johnson, L., Hazen Lake. Pages 549–559 in *The Book of Canadian Lakes,* edited by R. J. Allen, M. Dickman, C. B. Gray and V. Cromie. Burlington, Ontario: The Canadian Asassociation on Water Quality, 1992b.

Johnson, L., Pattern and process in ecological systems: a step in the development of an ecological theory. *Can. J. Fish. Aquat. Sci.,* 51:226–246, 1994a.

Johnson, L., The far-from-equilibrium ecological hinterlands. Pages 51–103 in *Complex Ecology: The Part Whole Relationship in Ecosystems,* edited by B. Patten and S. E. Jørgensen, 1994b.

Johnson, L., Long-term experiments on the stability of two fish populations in previously unexploited arctic lakes. *Can. J. Fish. Aquat., Sci.,* 51(1):209–225, 1994c.

Jones, E. W., The structure and reproduction of forest of the north temperate zone. *New. Phytol.,* 44:130–148, 1945.

Kingsland, S. E., Modelling nature: theoretical and experimental approaches to population ecology 1929–1950. Ph.D. dissertation, Toronto, 1981.

Kingsland, S. E., The refractory model: the logistic mcurve and the history of population ecology. *Q. Rev. Biol.,* 57:29–52, 1982.

Kleiber, M., *The Fire of Life,* New York: Wiley, 1961.

Kolata, G. B., Theoretical ecology: beginnings of a predictive science. *Science,* 183:400–401, 1974.

Krebs, C. J., Small mammal ecology. *Science,* 203:350–351, 1979.

Kropotkin, P., Mutual aid among animals. *Nineteenth Century,* 28:337–354, 1890.

Landsberg, P. T., Entropy and the unity of knowledge. *J. Non-Equilib. Thermodyn.,* 12:45–60, 1987.

Langford, A. N. and Buell, M. F., Integration, identity and stability in the plant association. *Adv. Ecol. Res.,* 6:83–135, 1969.

Leigh, E. G., On the relationship between productivity, biomass, diversity, and stability of a community. *Proc. Natl. Acad. Sci. U.S.A.,* 53:777–783, 1965.

Likens, G. E., Bormann, F. H., Johnson, N. M., Fisher, D. W., and Pierce, R. S., Effects of forest cutting and herbicide treatment on nutrient budgets in Hubbard's brook watershed-ecosystem. *Ecol. Monogr.* 40:23–47, 1970.

Loucks, O. L., Evolution of diversity, efficiency and community stability. *Am. Zool.* 10:17–25, 1970.

Lowe-McConnell, R. H., *Fish Communities in Tropical Freshwaters*. London: Longman, 1975.

Lowe-McConnell, R. H., Ecological studies in tropical fish communities. Cambridge: Cambridge University Press, 1987.

MacArthur, R. H., Fluctuations of animal populations and measure of community stability. *Ecology*, 36:510–533, 1955.

MacArthur, R. H., *Geographical Ecology*. New York, Harper and Row, 1972.

MacArthur, R. H. and Wilson, E. O., An equilibrium theory of insular zoogeography. *Evolution*, 17:373–387, 1963.

MacArthur, R. H. and Wilson, E. O., *The Theory of Island Biogeography*. Vol. 1, Princeton Monogr. Popn. Biol., Princeton: Princeton University Press, 1967.

Mann, K. H. and Brylinsky, M., Estimating productivity of lakes and reservoirs. In *Energy Flow: Its Biological Dimension*, edited by T. W. M. Cameron and L. W. Billingsley. Ottawa: Royal Society of Canada, 1975.

Margalef, R., On certain unifying principles in ecology. *Am. Nat.*, 97:357–374, 1963.

Margalef, R., Some concepts relative to the organization of plankton. *Oceanogr. Mar. Biol. Annu. Rev.*, 5:257–289, 1967.

Margalef, R., *Perspectives in Ecological Theory*. Chicago, Univ. Chicago Press, 1968.

Marsh, O. C., Notice of new equine mammals from the Tertiary formation. *Am. J. Sci.*, 7:247–258, 1874.

May, R. M., A cycling index for ecosystems. *Nature (Lond.)*, 292:105–106, 1981.

Mayr, E., Accident or design: the paradox of evolution. In *The Evolution of Living Organisms*, edited by G. W. Leeper. Melbourne: Melbourne University Press, 1962.

Mayr, E., *Evolution and the Diversity of Life*. Harvard: Bellknap Press, 1976.

McCallum, W. R. and Regier, H. A., The biology and bioenergetics of Arctic charr in Char Lake N.W.T. Canada. In *Biology of the Arctic Charr*, edited by L. Johnson and B. Burns. Winnipeg, Manitoba: University of Manitoba Press, 1984.

McIntosh, R. P., The continuum concept of vegetation. *Bot. Rev.*, 33:130–187, 1967.

McIntosh, R. P., The background and some current problems in ecology. *Synthese*, 43:195–255, 1980a.

McIntosh, R. P., The relationship between succession and the recovery process in ecosystems. In *The Recovery Process in Ecosystems*, edited by J. Cairns. Ann Arbor: Ann Arbor Science, 1980b.

McIntosh, R. P., *The Background of Ecology: Concept and Theory*. London: Cambridge University Press, 1985.

McKaye, K. R. and Marsh, A., Food switching by two algae-scraping Cichild fishes in Lake Malawi, Africa. *Oecologia*, 56:245–284, 1983.

McLaren, I. A., Zooplankton of Lake Hazen, Ellesmere Island and a nearby pond, with special reference to Cyclops scutifer. *Can. J. Zool.*, 42:613–629, 1964.

McNaughton, S. J. and Wolf, L. L. Dominance and the niche in ecological systems. *Science*, 176:131–139, 1970.

Mellinger, M. V. and McNaughton, S. J., Structure and function of the vascular plant communities in Central New York. *Ecol. Monogr.*, 45:161–182, 1975.

Moore, W. J., *Schrödinger, Life and Thought*. Cambridge, England: Cambridge University Press, 1989.

Motz, L. and Weaver, J. H. *The Story of Physics.*, New York: Avon Books, 1989.

Nerini, M. K., Braham, H. W., Marquette, W. M., and Rugh, D. J., Life history of the bowhead whale *Balaena mysticeuts* (Mammalia: Cetacea). *J. Zool.*, 204:443–468, 1984.

Newell, N. D., Phyletic size increase — an important trend illustrated by fossil invertebrates. *Evolution*, 3:103–124, 1949.

Nicholson, A. J., The balance of animal populations. *J. Animal Ecol.*, 2:132–178, 1933.

Nicholson, A. J. and Bailey, V. A., The balance of animal populations. *Proc. Zoo. Soc. Lond. Part*, 3:551–598, 1935.

Odum, E. P., *Fundamentals of Ecology*. Second ed. Philadelphia: Saunders, 1953.

Odum, E. P., Organic production and turnover in old field succession. *Ecology*, 41:34–49, 1960.

Odum, E. P., The strategy of ecosystem development. *Science*, 164:262–270, 1969.

Odum, E. P., *Fundamentals of Ecology*. Philadelphia: W. B. Saunders, 1971.

Oliver, D. R., Entomological studies in the Lake Hazen area, Ellesmere Island including lists of arachnida, collembola and insecta. *Arctic*, 16:175–180, 1963.

Onsager, L., Reciprocal relations in irreversible processes. *Phys. Rev.*, 37:405–562, 1931.

Paine, R. T., Food web complexity and species diversity. *Am. Nat.*, 100:65–75, 1966.

Patrick, R., The effect of invasion rate species pool and size of area on the structure of the diatom community. *Proc. Natl. Acad. Sci. U.S.A.*, 58:1335–1342, 1967.

Philo, L. M., Shotts, E. B., et al. Morbidity and mortality. *The Bowhead Whale.* Edited by J. Burns, J. Montague and C. Cowles. Kansas, Allen Press. 275–312 in Special Publication #2, Society for Marine Mammalogy: 275–312, 1993.

Pianka, E. R., On r-selection and K-selection. *Am. Nat.,* 104:592–597, 1970.

Pianka, E. R., *Evolutionary Ecology.* New York: John Wiley, 1978.

Pickett, S. T. A., Succession: an evolutionary interpretation. *Am. Nat.,* 110 (971):107–119, 1976.

Pimm, S. L., *The Balance of Nature?* Chicago: University of Chicago Press, 1991.

Platt, J. R., Strong inference. *Science,* 146:347–353, 1964.

Polanyi, M., Life's irreducible structure. *Science,* 160:1308–1312, 1968.

Poulson, T. L., Cave adaptation in Amblyopsid fishes. *Am. Mid. Nat.,* 70:257–290, 1963.

Prigogine, I., Time, structure and fluctuations. *Science,* 201:777–785, 1978.

Prigogine, I. and Stengers, I., *Order Out of Chaos.* New York: Bantam, 1984.

Quamen, D., *The Song of the Dodo.* New York: Simon and Schuster, 1997.

Rasmussen, E. M. and Wallace, J. M., Meterological aspects of El Nino/Southern oscillation. *Science,* 222:1195, 1983.

Raup, D. M. and Sepkoski, J. J., Mass extinctions in the marine fossil record. *Science.* 215:1501–1503, 1982.

Reiners, W. A., Disturbance and basic properties of ecosystem energetics. In *Disturbance and Ecosystems: Components and Response,* edited by H. A. Mooney and M. Godron. New York: Springer-Verlag, 1983.

Reiners, W. A., Worley, I. A., and Lawrence, D. B., Plant diversity in a chronosequence at Glacier Bay, Alaska, *Ecology,* 52:51–70, 1971.

Rensch, B., *Evolution Above the Species Level.* New York: Columbia University Press, 1959.

Richards, P. W., *The Tropical Rain Forest.* Cambridge: Cambridge University Press, 1952.

Richards, P. W., Africa, the "odd-man out." In *Tropical Forest Ecosystems and Africa and South America: A Comparative Review,* edited by B. J. Meggers, E. S. Ayensu and W. D. Duckworth. Washington, D.C.: Smithsonian Institute Press, 1973.

Ricker, W. E., *Computation and Interpretation of Biological Statistics of Fish Populations.* Vol. Bull. 191: Fish. and Mar. Serv. Ottawa, Environment Canada, 1975.

Rigler, F. H., The concept of energy flow and nutrient flow between trophic levels. In *Unifying Concepts in Biology,* edited by W. H. van Dobben and R. H. Lowe-McConnell. The Hague: W. Junk, 1975a.

Rigler, F. H., The Char Lake Project. In *Energy Flow — Its Biological Dimension,* edited by T. W. M. Cameron and L. W. Billingsley. Ottawa: Royal Society of Canada, 1975b.

Rigler, F. H., MacCallum, M. E., and Roff, J. C., Production of zooplankton in char Lake. *J. Fish. Res. Board Can.,* 31:637–646, 1974.

Ropke, W., *Germany in Transition,* 1988.

Ross, W. G., The annual catch of Greenland (bowhead) whales in waters north of Canada. 1717–1915: a preliminary compilation. *Arctic,* 32:91–121, 1979.

Rowe, J. S. and Scotter, G. W., Fire in the boreal forest. *Quaternary Res.* 3:444–464, 1973.

Sanders, H. L., Benthic marine diversity and the stability-time hypothesis. In *Diversity and Stability in Ecological Systems,* edited by G. M. Woodwell and H. H. Smith: Brookhaven Symposium in Biology 22, 1969.

Saunders, G. W., Cummins, K. W., Gak, D. Z., Pieczynska, E., Straskrabova, V., and Wetzel, R. G., Organic matter and decomposers. In *The Functioning of Freshwater Ecosystems,* edited by E. D. LeCren and R. H. Lowe-McConnell. Cambridge: Cambridge University Press, 1980.

Schaefer, M. B., Men, birds and anchovy in the Peru Current: Dynamic interactions. *Trans. Am. Fish. Soc.,* 99:461–467, 1970.

Schaffer, W. M., The theory of life-history evolution and its application to Atlantic salmon. In *Fish Phenology: Anabolic Adaptiveness in Teleosts; Symposium of the Zoological Society, 44,* edited by P. J. Miller, London: Academic Press, 1979.

Scheer, B. T., A universal definition of work. *Bioscience,* 26:505–506, 1970.

Schrödinger, E., *What Is Life?* Cambridge: Cambridge University Press, 1944.

Schugart, H. H., Jr., and West, D. C., Forest succession models. *Bioscience,* 30:308–313, 1970.

Semper, K., *Animal Life as Affected by the Natural Conditions of Existence.* New York: Appleton, 1881.

Simpson, G. G., *The Major Features of Evolution.* Vol. 434. New York: Columbia Univ. Press, 1953.

Sioli, H., Tropical rivers as expressions of their terrestrial environments. In *Tropical Ecological Systems,* edited by F. B. Golley and E. Medina. New York: Springer, 1975.

Sioli, H. (Ed.), *The Amazon: Limnology and Landscape Ecology of a Mighty Tropical River and Its Basin.* Dortrecht: Dr. J. W. Junk, 1984.

Slobodkin, L. B., The strategy of evolution. *Am. Sci.,* 52:342–357, 1964.

Slobodkin, L. B., On the inconstancy of ecological efficiency and the form of ecological theories. *Trans. Connecticut Acad. Arts Sci.,* 44:293–305, 1972.

Slobodkin, L. B. and Sanders, H. L., On the contribution of environmental predictability to species diversity. In *Diversity and Stability in Ecological Systems,* edited by G. M. Woodwell and H. H. Smith. Brookhaven: Brookhaven Symp. Biol., 22, 1969.

Slobodkin, L. B., Smith, F. E., and Hairston, N. G., Regulation in terrestrial ecosystems and the implied balance of nature. *Am. Nat.,* 101 (918):109–124, 1967.

Soodak, H. and Iberall, A., Homeokinetics: a physical science for complex systems. *Science,* 201:579–582, 1978.

Southwood, T. R. E., Habitat, the templet for ecological strategies. *J. Animal Ecol.,* 46:337–365, 1977.

Stanley, S. M., An ecological theory for the sudden origin of multicellular life in the Late Precambrian. *Proc. Natl. Acad. Sci. U.S.A.,* 70(5):1486–1489, 1973a.

Stanley, S. M., An explanation for Cope's Rule. *Evolution,* 27:1–26, 1973b.

Stoddart, D. R. and Wright C. A., Ecology of Aldabra Atoll. *Nature,* 213:1174–1177, 1967.

Swingland, I. R., Reproductive, effort and life history strategy of the Aldabran giant tortoise. *Nature,* 269:402–404, 1977.

Swingland, I. R. and Coe, M., The natural regulation of giant tortoise populations on aldabra Atoll. Reproduction. *J. Zool. Lond.,* 186:285–309, 1978.

Swingland, I. R. and Lessels, C. M., The natural regulation of giant tortoise populations on Aldabra Atoll, movement, polymorphism, reproductive success and mortality. *J. Animal Ecol.,* 48:639–654, 1979.

Szent-Gyorgi, A., *An Introduction to Molecular Biology.* New York: Academic Press, 1960.

Szent-Gyorgi, A., Introductory remarks. In *Light Is Life,* edited by W. D. McElroy and B. Glass. Baltimore: Johns Hopkins Press, 1961.

Tansley, A. G., *Practical Plant Ecology.* New York: Dodd, Mead, 1923.

Tansley, A. W., The use and abuse of vegetational concepts and terms. *Ecology,* 16:284–307, 1935.

Tansley, A., *The British Isles and Their Vegetation.* Cambridge: Cambridge Univ. Press, 1949.

Thienemann, A., Ein drittes bioznötiches Grund prinzip. *Arch. Hydrobiol.,* 49:421–422, 1954.

Tramer, E. J., Bird species diversity: components of Shannon's formula. *Ecology,* 50:927–929, 1967.

Tuchman, B., *The Proud Tower.* New York: Bantam Books, 1967.

Usher, M. B., Markovian approaches to ecological succession. *J. Animal Ecol.,* 48:413–426, 1979.

Vanriel, P. and Johnson, L., Action principles as determinants of ecosystem structure: The autonomous lake as a reference system. *Ecology,* 76(6):1741–1757, 1995.

Varley, G. C., The concept of energy flow applied to a woodland community. Paper read at Brit. Ecol. Soc. Symp., 1970.

Vogl, R. J., Effects of fire on grasslands. In *Fire in Ecosystems,* edited by T. T. Kozlowski and C. E. Ahlgren. New York: Academic Press, 1974.

Vogl, R. J., The ecological factors that produce purturbation-dependent ecosystems. In *The Recovery Process in Damaged Ecosystems,* edited by J. Cairns. Ann Arbor, Michigan: Ann Arbor Science, 1980.

von Baer, K. E., *The Controversy over Darwinism.* Translated by D. L. Hull abd published in "Darwin and his Critics" by D. L. Hull (1973). Vol. 130, Augsberger Allgemeine Zeitung, 1873.

Waddington, C. H., Towards a theoretical biology. *Nature (Lond.),* 218:525–527, 1968.

Watson, A., Physics — where the action is. *New. Sci.* (Jan. 1986): 42–44, 1986.

Watt, A. S., Pattern and process in the plant community. *J. Ecol.,* 35:1–22, 1947.

Westbroek, P., *Life as a Geological Force.* New York: W. M. Norton, 1991.

Whittaker, R. H., A criticism of the plant asociation and climax concepts. *Northwest Sci.,* 25:17–31, 1951.

Whittaker, R. H., A consideration of climax theory: a climax as a population and pattern. *Ecol. Monogr.,* 23:41–78, 1953.

Whittaker, R. H., Recent evolution of ecological concepts in relation to the eastern forests of North America. *Am. J. Bot.,* 44:197–206, 1957.

Whittaker, R. H., Vegetation of the Siskiyou Mountains, Oregon and California. *Ecol. Monogr.,* 30:279–338, 1960.

Whittaker, R. H., Dominance and diversity in land plant associations. *Science,* 147:250–260, 1965.

Whittaker, R. H., Gradient analysis in vegetation. *Biol. Rev.,* 49:207–264, 1967.

Whittaker, R. H., Evolution of diversity in land plant communities. Pp. 178–196. In: *Diversity and Stability in Ecological Systems,* edited by G. M. Woodwell and H. H. Smith. Brookhaven: Brookhaven Symp. Biol., 22, 1969.

Whittaker, R. H., *Communities and Ecosystems.* London and Toronto: Macmillan, 1970.

Whittaker, R. H. and Niering, W. A., Vegetation of the Santa Catalina Mountains, Arizona: a gradient analysis of the south slope. *Ecology,* 46:429–452, 1965.

Whittaker, R. H. and Woodwell, G. M., Evolution of natural communities. Pp. 137–159. In: *Ecosystem Structure and Function,* edited by J. A. Wiens. Corvallis: Oregon State University Press, 1972.

Winberg, G. G. and Babitsky, V. A. et al. Biological productivity of different lake types. In: *Productivity Problems in Fresh Waters,* edited by Z. Kajak and A. Hillbrich-Ilkowska. Warszawa: Polish Scientific Publishers, 1972.

Winberg, G. G., Etudes sur le bilan biologique energetique et la productivité des lacs en Union Soviétque. *Verh. Internat. Verein. Limnol.,* 18:39–64, 1972a.

Wright, H. E. Jr., Landscape development, forest fires and wilderness management. *Science,* 186:487–495, 1974.

Zotin, A. I., Bioenergetic trends of evolutionary progress of organisms. In: *Thermodynamics and Regulation of Biological Processes.* Berlin-New York: Walter de Gruyter, 1984.

What you gain in precision,
You lose in plurality

CHAPTER 12

This chapter takes its starting point from the Third Law of Thermodynamics, which states that entropies and entropy productions at absolute temperature, zero, and 0 K, are all zero. This implies that the order at 0 K is absolute — there is no entropy. This also implies that the activity (creativity) is zero, because all activities involve production of entropy. However, this is impossible at 0 K.

Obviously, life, characterised by a high and specific activity, will require a higher temperature than 0 K, namely, a temperature which allows transport processes of a certain minimum rate. Life also requires the presence of the right elements (oxygen, hydrogen, nitrogen, phosphorus, and so on) to construct the biochemical compounds which characterise our carbon-based life. The chapter formulates eight prerequisites for life as we know it, based on thermodynamic considerations. They are applied on the question: has there been life on the planet Mars or somewhere else in our solar system? Seven of the eight prerequisites are or have at least been fulfilled on Mars. Whether or not the eighth has been fulfilled is hard to say.

This chapter also discusses the concept of openness. Ecosystems must be open, or at least nonisolated, to be able to survive or even to exist. Openness implies that opportunities for development (and evolution) are created, but also that the ecosystem becomes nondeterministic, mainly due to its enormous complexity.

12 The Third Law of Thermodynamics Applied in Ecosystem Theory

Sven E. Jørgensen

CONTENTS

12.1 Introduction of the Third Law of Thermodynamics and Its Relationship
to the Second Law of Thermodynamics..289
12.2 The Temperature Range Needed for Life Processes...292
12.3 Physical Openness...293
12.4 Conditions for Creation of Life...294
12.5 Ontic Openness 296
12.6 Exergy as Limiting Factor 300
References 301

12.1 INTRODUCTION OF THE THIRD LAW OF THERMODYNAMICS AND ITS RELATIONSHIP TO THE SECOND LAW OF THERMODYNAMICS

The First Law of Thermodynamics is often applied to an ecosystem when energy balances of an ecosystem are first made. The Second Law of Thermodynamics is applied when we consider the entropy production of ecosystems as a consequence of the maintenance of the system far from thermodynamic equilibrium. This chapter is concerned with the application of the Third Law of Thermodynamics on ecosystems.

The lesser-known Third Law of Thermodynamics states that the entropies, S_0, of pure chemical compounds are zero, and that entropy production, ΔS_0, by chemical reactions between pure crystalline compounds is zero at absolute temperature, 0 K. The third law implies, since both $S_0 = 0$ (absolute order) and $\Delta S_0 = 0$ (no disorder generation), that disorder does not exist and cannot be created at absolute zero temperature. But at temperatures > 0 K disorder can exist ($S_{\text{system}} > 0$) and be generated ($\Delta S_{\text{system}} > 0$). The third law defines the relation between entropy production, ΔS_{system}, and the Kelvin temperature, T,

$$\Delta S_{\text{system}} = \int_0^T \Delta c_p \, d \ln \text{T} + \Delta S_0 \quad [\text{ML}^2\text{T}^{-2}\text{TEMP}^{-1}] \tag{1}$$

where Δc_p is the increase in heat capacity by the chemical reaction. Since order is absolute at absolute zero, its further creation is precluded there. At higher temperatures, however, order can be created.

The consequences of the Third Law of Thermodynamics are more easily expressed in terms of exergy. At 0 K, from Equation (1), the exergy of a system is always 0 ($Ex = kT_0I = 0$ when $T_0 = 0$ K); no useful work can be performed and no further order produced (as this is already absolute). Entropy production implies that degradation of energy from a state of high utility (large T) to a state of low utility (small T) occurs. Therefore, a system can only create an internal state of high exergy through energy dissipation. Maintenance of an internal state of high exergy implies, inevitably, that energy or exergy is dissipated — it is the "cost" of maintaining a system far from thermodynamic equilibrium. An internal state of high exergy, i.e., a system state maintained far from thermodynamic equilibrium, and exergy dissipation or destruction are two sides of the same coin. The system is inexorably drawn toward the thermodynamic equilibrium. An input of exergy is therefore needed to maintain a certain distance from thermodynamic equilibrium, and the farther the system is from thermodynamic equilibrium, i.e., the more complex the system is or the more gradients the system has built up, the more exergy is needed to ensure the maintenance of the system with its present complexity.

Ecosystems have therefore a global attractor state, the thermodynamic equilibrium, but will never reach this state as long as they are not isolated and receive exergy from outside to combat the decomposition of their compounds. As ecosystems *have* an energy through-flow, the attractor becomes the steady state, where the formation of new biological compounds is in balance with the decomposition processes. As seen from these perspectives of the Second Law of Thermodynamics for open (nonisolated) systems, it is vital to an ecosystem to be nonisolated. The consequences of this openness may be described by use of the concept of exergy, as follows:

Exergy, the work potential inherent in solar radiation and certain geochemical compounds, is built by photosynthesis and chemosynthesis into biological structure as organic compounds (charge phase). In photosynthesis, for example, if the power of solar radiation is W/t (work/time) and the average temperature of the system is $T1$, then the exergy gain per unit of time from radiant energy fixation is,

$$dEx/dt = T1W\,(1/T0 - 1/T2)\,/t, \quad [ML^2T^{-3}] \tag{2}$$

where $T0$ is the temperature of the environment and $T2$ is the surface temperature of the sun (Eriksson et al., 1976). The spectral differences between the incoming and outgoing electromagnetic waves may also be used to express exergy gain per unit of time; see Ulanowicz, 1986. This exergy flow generated by the photosynthetic capture of photons is used to synthesise and maintain structure (the gradients toward thermodynamic equilibrium). The more structure on ecosystem has, the more exergy it can capture and utilise, but the more it also needs for maintenance. The structural compounds are progressively decomposed in catabolism (discharge phase, see Patten et al., 1997), involving the consumption of exergy in work performed, and discharge of the residue to the environment as low exergy heat (only heat transferred from a higher to a lower temperature can produce work). The energy is conserved in accordance with the First Law of Thermodynamics, but it is converted from work (exergy) to heat (with zero or near-zero exergy). In the process, due to the differences between sun and earth surface temperatures (5780 vs. 287 K) 20 infrared photons (heat) are produced for every solar photon (Patten et al., 1997). How much work is done depends on the processes of the different pathways utilised. Hence more or less exergy may be consumed for a given quantity of radiant and chemical energy degraded to heat, more or less exergy may be stored in biomass with different levels of information — exergy is not conserved. The energy is conserved but not the work capacity due to the irreversibility of all real processes. The relationship can be captured with implicit function notation, writing exergy, Ex, as a function of energy, E, through work, W: $Ex(W(E))$. The work, W, is context specific; hence exergy is not conserved though its basis, energy, is.

It can be stated that it is *necessary* for an ecosystem to transfer the generated heat (entropy) to the environment and to receive exergy (solar radiation) from the environment for formation of dissipative structure. The next obvious question would be: will energy source and sink also be *sufficient* to initiate formation of dissipative structure, which can be used as a source for entropy combating processes?

The answer to this question is "yes." It can be shown by the use of simple model systems and basic thermodynamics; see Morowitz (1968). He shows that a flow of energy from sources to sinks leads to an internal organisation of the system, and to the establishment of element cycles. The type of organisation is, of course, dependent on a number of factors: the temperature, the elements present, the initial conditions of the system and the time available for the development of organisation. It is characteristic for the system, as pointed out above, that the steady state of an open system does *not* involve chemical equilibrium.

An interesting illustration of the creation of organisation (dissipative structure) as a result of an energy flow through ecosystems concerns the possibilities to form organic matter from the inorganic components which were present in the primeval atmosphere. Since 1897 many simulation experiments have been performed to explain how the first organic matter was formed on earth from inorganic matter. All of them point to the conclusion that energy interacts with a mixture of gases to form a large set of randomly synthesised organic compounds. Most interesting is the experiment performed by Stanley Miller and Harold Urey at the University of Chicago in 1953, because it showed that amino acids can be formed by sparking a mixture of CH_4, H_2O, NH_3, and H_2; corresponding approximately to the composition of the primeval atmosphere; see Figure 12.1.

Prigogine and his colleagues have shown that open systems that are exposed to an energy through-flow exhibit coherent self-organisation behaviour and are known as dissipative structures. Formations of complex organic compounds from inorganic matter as mentioned above are typical examples of self-organisation. Such systems can remain in their organised state by exporting entropy outside the system, but are dependent on outside energy fluxes to maintain their organisation, as already mentioned and emphasised above.

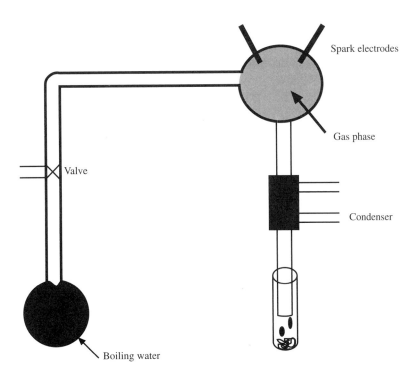

FIGURE 12.1   The apparatus used by Stanley Miller and Harold Urey to simulate reactions in the primeval atmosphere.

Glansdorff and Prigogine (1971) have shown that the thermodynamic relationship of far-from-equilibrium dissipative structures is best represented by coupled nonlinear relationships, i.e., auto-catalytic positive feedback cycles.

A system at 0 K on the other hand is without any creative potential, because no dissipation of energy can take place at this temperature. A temperature greater than 2.726 ± 0.01 K, where 2.726 K is the temperature of deep space, is therefore required before order can be created. Given this necessary condition, simple energy flow through a system, which would require that the system is open or at least nonisolated, provides a sufficient condition. Creation of order is inevitable. On earth, the surface temperature difference between sun and planet guarantees this. Morowitz (1968) showed, as mentioned above, that energy through-flow is sufficient to produce cycling, a prerequisite for the ordering processes characteristic of living systems.

12.2   THE TEMPERATURE RANGE NEEDED FOR LIFE PROCESSES

The input of exergy for ecosystems is in the form of the solar photon flux. This comes as small portions (quanta) of exergy ($= hv$, where h is Planck's constant and v is the frequency), which implies that the exergy at first can only be utilised at molecular (lowest) levels in the hierarchy. The appropriate atoms or molecules must be transported to the place where order is created. Diffusion processes through a solid are extremely slow, even at room temperature. The diffusion of molecules through a liquid is about three orders of magnitude faster than in a solid at the same temperature. Diffusion coefficients for gases are ordinarily four orders of magnitude greater than for liquids. This implies that the creation of order (and also the inverse process, disordering) is much more rapid in liquid and gaseous phases than in solids. The temperature required for a sufficiently rapid creation of order is consequently considerably above the lower limit mentioned above, 2.726 K. As far as diffusion processes in solids, liquids, and gases are concerned, gaseous diffusion allows the most rapid mass transport. However, many molecules on earth that are necessary for ordinary carbon-based life do not occur in a gaseous phase, and liquid diffusion, even though it occurs at a much slower rate, is of particular importance for biological ordering processes.

The diffusion coefficient increases significantly with temperature. For gases, the diffusion coefficient varies with temperature approximately as $T^{3/2}$ (Hirschfelder et al., 1954), where T is the absolute temperature. Thus, we should look for systems with the high order characteristic of life, at temperatures considerably higher than 2.726 K. The reaction rates for biochemical anabolic processes on the molecular level are highly temperature dependent (see Straskraba et al., 1999). The influence of temperature may be reduced by the presence of reaction-specific enzymes, which are proteins formed by anabolic processes. The relationship between the absolute temperature, T, and the reaction rate coefficient, k, for a number of biochemical processes can be expressed by the following general equation (see any textbook in physical chemistry):

$$\ln k = b - A/R*T, \qquad (\mathrm{ML^{-3}T^{-1}}) \qquad (3)$$

where A is the so-called activation energy, b is a constant, and R is the gas constant. Enzymes are able to reduce the activation energy (the energy that the molecules require to perform the biochemical reaction). Similar dependence of the temperature is known for a wide spectrum of biological processes, for instance, growth and respiration. Biochemical and biological kinetics point, therefore, toward ecosystem temperatures considerably higher than 2.726 K.

The high efficiency in the use of low-entropy energy at the present "room temperature" on earth works hand in hand with the chemical stability of the chemical species characteristic of life on earth. Macromolecules are subject to thermal denaturation. Among the macromolecules, proteins are most sensitive to thermal effects, and the constant breakdown of proteins leads to a substantial turnover of amino acids in organisms. According to biochemistry, an adult man synthesises and degrades approximately 1 g of protein nitrogen per kilogram of body weight per day. This corresponds to a protein turnover of about 7.7% per day for a man with a normal body temperature. A too-high temperature of the ecosystem (more than about 340 K) will therefore enhance the breakdown processes too much. A temperature range between 260 and 340 K seems from these considerations the most appropriate to create the carbon-based life that we know on earth. An enzymatic reduction of the activation energy makes it possible to realise basic biochemical reactions in this temperature range, without a too high decomposition rate, which would be the case at a higher temperature. In this temperature range anabolic and catabolic processes can, in other words, be in a proper balance.

12.3   PHYSICAL OPENNESS

An energy balance equation for ecosystems might be written as follows in accordance with the principle of energy conservation:

$$Ecap = Qevap + Qresp + \cdots + \Delta E_{bio}. \quad [ML^2T^{-2}] \tag{4}$$

Here Ecap is external energy captured per unit of time. A part of the incoming energy, solar radiation being the main source for the ecosystems on earth, is captured and a part is reflected unused, determining the albedo of the globe. The more developed structure an ecosystem possesses, the more of the incoming energy it is able to capture, i.e., the lower the albedo.

In ecosystem steady states, the formation of biological compounds (anabolism) is in approximate balance with their decomposition (catabolism). That is, in energy terms,

$$\Delta E_{bio} \approx 0, \quad \text{and} \quad Ecap \approx Qevap + Qresp + \cdots \quad [ML^2T^{-2}] \tag{5}$$

The energy captured can in principle be any form of exergy (electromagnetic, electrical, magnetic, chemical, kinetic, etc.) but for the ecosystems on earth, shortwave exergy containing solar radiation (electromagnetic energy) plays the major role. The exergy captured per unit of time is, however, not conserved, but is in this case used to pay the cost of maintenance per unit of time = Qevap + Qresp.... Thereby, exergy able to perform work is converted to heat, not able to perform work. The exergy is lost as heat while the energy is conserved as energy captured per unit of time is equal to heat lost per unit of time.

The overall results of these processes requires that Ecap be > 0, which entails openness (or at least nonisolation).

The following reaction chain summarises the consequences of energy openness: *source*: solar radiation → *anabolism* (charge phase): incorporation of high quality energy, with entrained exergy (information), into complex biomolecular structures, entailing antientropic system movement away from equilibrium → *catabolism* (discharge phase): deterioration of structure involving release of chemical bond energy and its degradation (exergy consumption) to lower states of usefulness for work → *sink*: dissipation of degraded (low exergy and negentropy) energy as heat to the environment (and, from earth, to deep space), involving entropy generation and return toward thermodynamic equilibrium. This same chain can also be expressed in terms of matter: *source*: geochemical substrates relatively close to thermodynamic equilibrium → *anabolism*: inorganic chemicals are molded into complex organic molecules (exergy, negentropy, and distance from equilibrium all increase) → *catabolism*: synthesised organic matter is ultimately decomposed into simple inorganic molecules again; exergy and distance from equilibrium decrease, and entropy increases → *cycling*: the inorganic molecules, returned to near-equilibrium states, become available in the nearly closed material ecosphere of earth for repetition of the matter charge–discharge cycle.

Input environments of ecosystems serve as sources of high quality energy whose high contents of exergy and low entropy raise the organisational states of matter far from equilibrium. Output environments, in contrast, are sinks for energy and matter lower in exergy, higher in entropy, and closer to equilibrium. Since, in the organisation of ecosystems, output environments feed back to become portions of input environments, living systems operating in the ecosphere, which is energetically nonisolated but materially nearly closed, must seek an adaptive balance between these two aspects of their environmental relations in order to sustain their continued existence.

The same concepts may be used for single chemical processes. $\Delta G°$ the difference in the standard free energy for biogeochemical processes in ecosystems can be calculated, as $G°$ for various compounds may be found in physical-chemical handbooks. Since $\Delta G° = -RT \ln K$, it is also possible to find the equilibrium constant K. R is the gas constant and T the absolute temperature. If the concentrations of reactants are known, it is possible to determine the corresponding concentrations at thermodynamic equilibrium. For instance, the process,

$$\text{protein} + \text{oxygen} \leftrightarrow \text{carbon dioxide} + \text{water} + \text{ammonium} + \text{energy as heat}$$

has a very high equilibrium constant, corresponding to relatively high concentrations of the product compounds and a low concentration of protein, as oxygen today is present in the atmosphere at an almost constant concentration. The process is inevitably driven toward the right side of the chemical equation, which implies that the free energy (exergy) is consumed and converted to the low-exergy form of energy as heat. This does, however, not imply that it is impossible to form proteins against this decomposition tendency, but rather that the formation of proteins does not happen spontaneously (except in an extremely low concentration corresponding to thermodynamic equilibrium). Formation of protein will require transactional coupling to another process or processes that is or are able to deliver the free energy (exergy) needed for this process. This process is photosynthesis (the mechanism for energy charge in the charge–discharge cycle of the molecular cascade), which has physical openness as a prerequisite.

12.4   CONDITIONS FOR CREATION OF LIFE

The conditions for creation of life ordering processes out of disorder (or more specifically chemical order by formation of complex organic molecules and organisms from inorganic matter) can now be deduced from the First, Second, and Third Laws of Thermodynamics.

1. It is necessary that the system be open (or at least nonisolated) to exchange energy (as well as mass) with its environment;
2. An influx of exergy is compulsory;
3. An outflow of high-entropy energy (heat originated from exergy consumption) is necessary (this means that the temperature of the system inevitably must be greater than 2.726 K);
4. Entropy production accompanying the transformation of energy (work) to heat in the system is a necessary cost of maintaining the order; and
5. Mass transport processes at a rate that is not too low are necessary (a prerequisite). This implies that the liquid or gaseous phase must be anticipated. A higher temperature will imply a better mass transfer, but also a higher reaction rate. An increased temperature also means a faster breakdown of macromolecules, and therefore a shift toward catabolism. A temperature in the range of approximately 260 to 340 K must therefore be anticipated for carbon-based life.

The rates of biochemical reactions on the molecular level are determined by the temperature of the system and the exergy supply to the system. Hierarchical organisation ensures that the reactions and the exergy available on the molecular level can be utilised on the next level, the cell level, and so on throughout the entire hierarchy: molecules → cells → organs → organisms → populations → ecosystems. The maintenance of each level is dependent on its openness to exchange of energy and matter. The rates in the higher levels are dependent on the sum of many processes on the molecular level. They are furthermore dependent on the slowest processes in the chain: supply of energy and matter to the unit → the metabolic processes → excretion of waste heat and waste material. The first and last of these three steps limit the rates and are determined by the extent of openness, measured by the area available for exchange between the unit and its environment relative to the volume. These considerations are based on allometric principles (Peters, 1983; Straskraba et al., 1999).

In addition to the five conditions given above, it is necessary to add a few biochemically determined conditions. The carbon-based life on earth requires first of all an abundant presence of water to deliver the two important elements hydrogen and oxygen, as solvent for compounds containing the other needed elements (see below), as a compound which is liquid at a suitable temperature with a suitable diffusion coefficient, a specific heat capacity to buffer temperature fluctuations, and proper vapour pressure to ensure an appropriate cycling (purification) rate of this crucial chemical compound.

Life on earth is characterised by about 25 elements. Some of these elements are used by life processes in micro amounts, and it cannot be excluded that other elements could have replaced these elements on other planets somewhere else in the universe. Several metal ions are, for instance, used as coenzymes and are often important parts of high molecular organic complexes. Other ions may be able to play similar roles for biochemical processes and complexes. It is on the other hand difficult to imagine carbon-based life without at least most of the elements used in macro amounts, such as nitrogen for amino acids (proteins — the enzymes) and amino bases, phosphorus for ATP and phosphorus esters in general, and sulphur for formation of some of the essential amino acids.

The biochemically determined conditions can therefore be summarised in the following points:

6. Abundant presence of the unique solvent, water, is a prerequisite for the formation of life forms similar to the life forms as we know from the earth.
7. The presence of nitrogen, phosphorus, and sulphur and some metal ions seems absolutely necessary for the formation of carbon-based life.

A last and eighth condition should be mentioned in this context: the seven other conditions should be maintained within reasonable ranges for a long period of time. The genes may ensure that if an advantageous property of an organism has been developed, the property can be heritaged and the following generations will be able to maintain the advantageous property. All progress made can to a certain extent be preserved, provided that organisms with genes have been formed. Many mechanisms are probably involved in emergence of a progressive property in the first hand, but indisputable random processes based on trial and error are important in the emergence of progressive properties. This implies that carbon life is not formed overnight. The history of evolution on earth shows that a suitable temperature was achieved and water was abundant. Probably 10^8 years or more (Nielsen, 1999) were needed to form, from inorganic components dissolved in water, the first living cells with some type of primitive genes to ensure a continuous development (evolution). Numerous theories have been published to explain how this development may have happened: inorganic matter formed organic molecules by a through-flow of exergy (compare with Figure 12.1), organic molecules formed high molecular organic compounds, self-catalytic processes occurred, complex organic molecules were brought randomly in contact by adsorption on clay particles, and many other processes are mentioned in the presented theories. Which of the theories is right is not important in this context. The focal point is that the seven above-mentioned conditions must be fulfilled for a sufficient long period of time, which leads to the eighth condition:

8. As the formation of life from inorganic matter requires a long time, probably in the order of 10^8 years or more, the seven conditions have to be maintained in the right ranges for a long time, which probably exceeds about 10^8 years.

After the Mars pathfinder mission it was discussed whether Mars hosts or has hosted life. Clearly, the conditions 1 to 7 are not met on Mars today. The climate is too harsh and water is far from being present in the amount needed for the planet to bear life. There are, however, many signs of a warmer and wetter climate at an earlier stage. It looks, therefore, as if the seven conditions may have been valid and the question is: have they also prevailed for a sufficient period of time? If later missions to Mars show that it is the case, the next obvious question is: will life inevitably be the result of self-organising processes, if the eight conditions (the eighth condition about a sufficient time, should of course be included) are fulfilled? According to the tentative Fourth Law of Thermodynamics, named ELT, see Chapter 9, it should be expected that primitive life has been present on Mars at an early stage, provided that the warmer and wetter conditions have prevailed for a sufficient time. The further evolution from (maybe prokaryotic) unicellular organisms to more and more complex organisms as we know from the earth could not be realised on Mars, because the climate changed and the water disappeared.

Another possibility for life in our solar system exists on Europa, one of the moons of Jupiter (Sveinsdottir, 1997). Europa is characterised by a coverage of ice. It implies that there is plenty of water on Europa, which means that one of the important conditions for life is fulfilled. Some researchers (Sveinsdottir, 1997) believe that the chance to find life on Europa is higher than on Mars. Europa has of course much less sunlight and the surface temperature is probably too low, but volcanic activity in the deeper parts of the oceans on Europa is very probable, and it could provide the needed energy (exergy) for the formation and maintenance of life.

12.5 ONTIC OPENNESS

It can be deduced from the First, Second, and Third Laws of Thermodynamics that (physical) nonisolation (open to an energy through flow, but not necessary to exchange of mass) is necessary to maintain an ecosystem at its present stage far from thermodynamic equilibrium. It can also be shown that through-flow of exergy is sufficient to create some kind of structure/organisation/order as cycling processes inevitably will emerge that may be considered prerequisite for order, structure, and organisation. The next chapter will discuss whether it is possible to assess which structures, organisations, orders, and cycling processes will prevail. This leads to the question: is the world deterministic? — in the sense that if we would know the initial conditions in all details, could we also predict in all details how a system would develop — or is the world ontic open?

We cannot and will probably never be able to answer these two questions, but the world is, under all circumstances, too complex to enable us to determine the initial conditions. The uncertainty relations similar to Heisenberg's uncertainty relations in quantum mechanics are also valid in ecology. This idea has been presented in Jørgensen (1992, 1997), but will be summarised below, because the discussion in the next chapter is dependent on this uncertainty in our description of nature. The world may be ontic open = nondeterministic, because the universe has been created that way, or it may be ontic open = nondeterministic, because nature is too complex to allow us to know a reasonable fraction of the initial conditions even for a subsystem of an ecosystem. We shall probably never be able to determine which of the two possibilities will prevail, but it is not of importance, because we have to accept ontic openness in our description of nature.

If we take two components and want to know all the relations between them, we would need at least three observations to show whether the relations were linear or nonlinear. Correspondingly, the relations among three components will require 3*3 observations for the shape of the plane. If we have 18 components we would correspondingly need 3^{17} or approximately 10^8 observations. At present this is probably an approximate, practical upper limit to the number of observations which can be invested in one project aimed for one ecosystem. This could be used to formulate a practical uncertainty relation in ecology:

$$10^5 * \Delta x / \sqrt{3^{n-1}} \leq 1 \qquad (6)$$

where Δx is the relative accuracy of one relation, and n is the number of components examined or included in the model.

The 100 million observations could, of course, also be used to give a very exact picture of one relation. Costanza and Sklar (1985) talk about the choice between the two extremes: knowing "everything" about "nothing" or "nothing" about "everything." The first refers to the use of all the observations on one relation to obtain a high accuracy and certainty, while the latter refers to the use of all observations on as many relations as possible in an ecosystem.

Equation (4) formulates a practical uncertainty relation, but, of course, the possibility that the practical number of observations may be increased in the future cannot be excluded. Ever more automatic analytical equipment is emerging on the market. This means that the number of observations that can be invested in one project may be one, two, three, or even several magnitudes larger in one or more decades. However, a theoretical uncertainty relation can be developed. If we go to the limits given by quantum mechanics, the number of variables will still be low, compared to the number of components in an ecosystem.

One of Heisenberg's uncertainty relations is formulated as follows:

$$\Delta s * \Delta p \geq h/2\pi \tag{7}$$

where Δs is the uncertainty in determination of the place, and Δp is the uncertainty of the momentum. According to this relation, Δx of Equation (4) should be in the order of 10^{-17} if Δs and Δp are about the same. Another of Heisenberg's uncertainty relations may now be used to give the upper limit of the number of observations,

$$\Delta t * \Delta E \geq h/2\pi \tag{8}$$

where Δt is the uncertainty in time and ΔE in energy.

If we use all the energy that earth has received during its lifetime of 4.5 billion years, we get:

$$173 * 10^{15} * 4.5 * 10^9 * 365.3 * 24 * 3600 = 2.5 * 10^{34} \text{ J}, \tag{9}$$

where $173 * 10^{15}$ W is the energy flow of solar radiation. Δt would, therefore, be in the order of 1^{-69} sec. Consequently, an observation will take 10^{-69} sec, even if we use all the energy that has been available on earth as ΔE, which must be considered the most extreme case. The hypothetical number of observations possible during the lifetime of the earth would therefore be:

$$4.5 * 10^9 * 365.3 * 3600/10^{-69} = \simeq \text{ of } 10^{85} \tag{10}$$

This implies that we can replace 10^5 in Equation (4) with 10^{60} since

$$1^{-17}/\sqrt{10^{85}} = \sim 10^{-6}$$

If we use $\Delta x = 1$ in Equation (4), we get:

$$\sqrt{3}^{n-1} \leq 10^{60} \tag{11}$$

or

$$n \leq 253$$

From these very theoretical considerations, we can clearly conclude that we shall never be able to obtain a sufficient number of observations to describe even one ecosystem in all details. These results are completely in harmony with the Niels Bohr complementarity theory. He expressed it as follows, "It is not possible to make one unambiguous picture (model or map) of reality, as uncertainty limits our knowledge." The uncertainty in nuclear physics is caused by the inevitable influence of the observer on the nuclear particles; in ecology the uncertainty is caused by the enormous complexity and variability.

No map of reality is completely correct. There are many maps (models) of the same piece of nature, and the various maps or models reflect different viewpoints. Accordingly, one model (map) does not give all the information and far from all the details of an ecosystem. In other words, the theory of complementarity is also valid in ecology.

The use of maps in geography is a good parallel to the use of models in ecology. As we have road maps, airplane maps, geological maps, maps in different scales for different purposes, we have in ecology many models of the same ecosystems and we need them all, if we want to get a comprehensive view of ecosystems. A map, furthermore, cannot give a complete picture. We can always increase the size of the scale and include more details, but we cannot get all the details — for instance, where all the cars of an area are situated just now — and if we could, the picture would be invalid a few seconds later because we want to map too many dynamic details at the same time. An ecosystem also consists of too many dynamic components to enable us to prepare a model of all the components simultaneously and, even if we could, the model would be invalid a few seconds later, where the dynamics of the system have changed the "picture."

In nuclear physics we need to use many different pictures of the same phenomena to be able to describe our observations. We say that we need a pluralistic view to cover our observations completely. Our observations of light, for instance, require that we consider light as waves as well as particles. The situation in ecology is similar. Because of the immense complexity we need a pluralistic view to cover a description of the ecosystems according to our observations. We need many models covering different viewpoints.

In addition to physical openness, there is also an epistemological openness inherent in the formal lenses through which humans view reality. Gödel's theorem, which was published in January 1931, strongly introduces an epistemic openness. The theorem requires that mathematical and logical systems (i.e., purely epistemic, as opposed to ontic) cannot be shown to be self-consistent within their own frameworks but only from outside. A logical system cannot itself (from inside) decide on whether it is false or true. This requires an observer from outside the system, and this means that even epistemic systems must be open.

We can distinguish between ordered and random systems. The latter type can only be described by all the details of the system, while an ordered system can be described by application of an algorithm. The information *needed* to describe the random system corresponds (Boltzmann, 1905) to the entropy of the system, S

$$S = k \ln N, \qquad [\text{ML}^2\text{T}^{-2}\text{TEMP}^{-1}] \qquad (12)$$

where N is the number of possible (micro)states, and k is the so-called Boltzmann's constant, 1.3803 * 10-23 J/(molecule*degree). The information needed to describe the ordered system is less as it can be described through the algorithm, a rule of order which requires less information than a detailed description of all the components of the system. Many ordered systems have emergent properties defined as properties that a system possesses in addition to the sum of properties of the components — the system is more than the sum of its components. Wolfram (1984a, b) calls these *irreducible systems* because their properties cannot be revealed by a reduction to some observations of the behaviour of the components. It is necessary to observe the entire system to capture its behaviour, becuse everything in the system is dependent on everything else due to direct and indirect linkages. The presence of irreducible systems is consistent with Gödel's theorem, which states it will never be possible to give a detailed, comprehensive, complete, and comprehensible description of the world. Most natural systems are irreducible, which places profound restrictions on the inherent reductionism of science.

In accordance with Gödel's theorem, properties of order and emergence cannot be observed and acknowledged from within the system, but only by an outside observer. It is consistent with the proverb: "you cannot see the forest for the trees," meaning that if you only see the trees as independent details inside the forest, you are unable to observe the system, the forest as a cooperative unit of trees. This implies that the natural sciences, aiming toward a description or ordering of the systems of nature, have meaning only for open systems. A scientific description of an isolated system, i.e., the presentation of an algorithm describing the observed, ordering principles valid for the system, is impossible. In addition, sooner or later an isolated ontic system will reach thermodynamic equilibrium, implying that there are no ordering principles, but only randomness. We can infer from this that an isolated epistemic system will always ultimately collapse inward on itself if it is not opened to cross fertilisation from outside. Thomas Kuhn's account of the structure of scientific revolutions would seem to proceed from such an epistemological analogy of the second law.

This does not imply that we can describe *all* open systems in *all* details. On the contrary, the only complete detailed and consistent description of a system is the system itself. We can furthermore never know if a random system or subsystem is ordered or random, because we have not found the algorithm describing the order. We can never know if it exists or we may find it later by additional effort. This is what modelling and model-making in accordance with our definition of life (Patten et al., 1997) are all about. A model is always a simplified or homomorphic description of some features of a system, but no model can give a complete or isopmorphic description. Therefore, one might conclude that it will always require an infinite number of different models to realise a complete, detailed, comprehensive, and consistent description of any entire system. In addition, it is also not possible to compute or totally explain our thoughts and conceptions of our limited, but useful description of open natural systems. Our perception of nature goes, in other words, beyond what can be explained and computed, which makes it possible for us to conceive irreducible (open) systems, though we cannot explain all the details of the system. This explains the applicability and usefulness of models in the adaptations of living things ("subjects," Patten et al., 1997) to their environment. It also underlines that the models in the best case will be able to cover only one or a few out of many views of considered systems. If we apply the definition of life proposed in Patten et al., 1997: Life is things that make models — this implies that all organisms and species must make their way in the world based on only partial representations, limited by the perceptual and cognitive apparatus of each, and the special epistemologies or models that arise therefrom. The models are always incomplete, but sufficient to guarantee survival and continuance, or else extinction is the price of a failed model.

Following from Gödel's theorem, a scientific description can only be given from outside for open systems. Natural science cannot be applied to isolated systems (the universe is considered open due to the expansion) at all. A complete, detailed, comprehensive, and consistent description of an open system can never be obtained. Only a partial, though useful description (model) covering one or a few out of many views can be achieved.

Ecosystems are nondeterministic. Due to the enormous complexity of ecosystems we cannot, as already stressed, know all the details of ecosystems. When we cannot know all the details, we are not able to describe fully the initial stage and processes that determine the development of the ecosystems — as expressed above ecosystems are therefore irreducible. This implies that our description of ecosystem developments must be open to a wide spectrum of possibilities. It is consistent with the application of chaos and catastrophe theory; see, for instance, Jørgensen (1992, 1994, 1997). Ulanowicz (1997) makes a major issue of the necessity for systems to be causally open in order to be living — the open possibilities may create new pathways for development that may be crucial for survival and further evolution in a nondeterministic world. He goes so far as to contend that a mature insight into the evolutionary process is impossible without a revision of our contemporary notions on causality. Ulanowicz (1997) uses the concept of propensity to come around the problem of causality. On the one side we are able to relate the development with the changing internal and external factors of ecosystems. On the other side, due to the uncertainty in our predictions of development caused by our lack of knowledge about all details, we are not able to give deterministic descriptions of the development, but we can only indicate which propensities will be governing.

Ecosystems have ontic openness. They are irreducible and due to their enormous complexity, which prohibits us from knowing all details, we will only be able to indicate the propensities of their development. Ecosystems are not deterministic systems.

12.6   EXERGY AS LIMITING FACTOR

When the first life emerged from the inorganic soup on earth about 3.8 billion years ago, the major challenge to primeval life was to maintain what was already achieved. This problem was eventually solved in the present form by introduction of DNA or DNA-like molecules that were able to store information about the already obtained results on how to convert inorganic matter to life-bearing organic matter. It requires, however, both matter and energy to continue the development, i.e., to continuously build more ordered structure of life-bearing organic compounds on the shoulders of the previous development or expressed by use of the term gradients: to increase the size of the gradients. The amounts of elements on earth are almost constant, i.e., the earth is today a nonisolated (closed) system (energy is exchanged with space, but (almost) no exchange of matter takes place, excluding minor input from meteorites and the minor output of hydrogen), while energy is currently supplied by the solar radiation. The concept of limiting elements (factors) is embodied in the conservation laws; see also Patten et al. (1997). Further development (evolution), when one or more elements are limiting, is consequently not possible by formation of more organic life-bearing matter just by uptake of more inorganic matter, but only by a better use of the elements, i.e., by a reallocation of the elements through a better organisation of the structure. It requires that more information (exergy) is embodied in the structure. More ecological niches are thereby utilised, and more life forms will be able to cope better with the variability of the forcing functions. Exergy is not limiting in this phase of development, because each new unit of time brings more exergy to combat the catabolic processes and to continue the energy-consuming and exergy-building anabolic processes. Exergy stored in the biomass is steadily increased under these circumstance, because the structure able to capture exergy is increased. The in-flowing exergy, the solar radiation, is either captured by the structure of ecosystems or reflected (unused). Kay and Schneider (1992) have shown that mature ecosystems (in the sense of E. P. Odum, 1968) capture about 80% of the incoming exergy, which is close to what is physically feasible. Most of the captured exergy is used for evapotranspiration and only a minor part is used for gross production (2% on average), which again has to cover respiration and exergy built into a new biological structure. Exergy captured will therefore become limiting sooner or later, when a sufficiently large structure has been formed. Under these circumstances, the additional constraints on the processes of the ecosystem are a better utilisation of the incoming exergy, meaning that less of the exergy captured is used for maintenance and more is therefore available for further growth of the ordered structure, i.e., to increase the exergy stored in the system = ΔEx_{bio}. The above mentioned reallocation of the elements in this phase will also be used to obtain organisms which together — as a whole — can more economically use the incoming flow of exergy through the system, corresponding to less exergy demand for maintenance and more available for further growth in its broadest sense. Mauersberger (1983, 1995) has used these considerations to determine the nonlinear relations between rates and chemical affinities of biological processes in an ecosystem. Mauersberger postulates that the development of bioceonosis is controlled during the finite time interval of length Δt by the chemical affinities, X, so that the time integral,

$$I = \int_{t+\Delta t}^{t} E(B(t'), X(t')) \, dt' \tag{13}$$

over the generalised excess entropy production, E, within $(t - \Delta t) \le t' \le t$ becomes an extreme value (minimum) subject to the initial values $B_0 = B (t - \Delta t)$ and to the mass equations,

$$dBk(t)/dt = fk(B(t), Y(X(t))), \tag{14}$$

which connect B and X through the rates of the biological processes, Y, for the kth species. This optimisation principle postulates that locally, and within a finite time interval, the deviation of the bioprocesses from a stable stationary state tends to a minimum. This is similar to, and a more formal version of, Lionel Johnson's least dissipation principle (Johnson, 1995) which goes back to Lord Rayleigh. Mauersberger has used this optimisation principle on several process rates as functions of state variables. As a result, he finds well known and well accepted relations for uptake of nutrients of phytoplankton, primary production, temperature dependence of primary production, and respiration. It supports the use of the optimisation principle locally and within a finite time interval. This leads inevitably to the need for a Fourth Law of Thermodynamics, which is tentatively presented in the next chapter: how can we describe the results of the opposition between anabolism and catabolism or between the above-mentioned global and local optimisation principles?

REFERENCES

Boltzmann, L., The second Law of Thermodynamics. (Populare Schriften Essay no. 3, address to Imperial Academy of Science in 1886). Reprinted in English. in *Theoretical Physics and Philosophical Problems, Selected Writings of L. Boltzmann*. D. Reidel, Dordrecht, 1905.

Eriksson, B., Eriksson, K. E., and Wall, G., *Basic Thermodynamics of Energy Conversions and Energy Use*. Institute of Theoretical Physics, Göteborg, Sweden, 1976.

Glansdorff, P. and Prigogine, I., *Thermodynamic Theory of Structure, Stability, and Fluctuations*. Wiley-Interscience, New York, 1971.

Hirschfelder, J. O., Curtiss, C. F., and Bird, R. B., *Molecular Theory of Gases and Liquids*. John Wiley & Sons, New York, 631 pp., 1954.

Johnson, L., The far-from-equilibrium ecological hinterlands. In: B. C. Patten, S. E. Jørgensen, and S. I. Auerbach (editors). *Complex Ecology. The Part-Whole Relation in Ecosystems*. Prentice-Hall PTR, Englewood Cliffs, NJ, pp 51–104, 1995.

Jørgensen, S. E., Parameters, ecological constraints and exergy. *Ecol. Modelling,* 62: 163–170, 1992.

Jørgensen, S. E., *Fundamentals of Ecological Modelling* (second edition) (Developments in Environmental Modelling, 19). Elsevier, Amsterdam, 628 pp., 1994.

Jørgensen, S. E., The growth rate of zooplankton at the edge of chaos: ecological models. *J. Theor. Biol.* 175: 13–21, 1995.

Jørgensen, S. E., *Integration of Ecosystem Theories: A Pattern,* Second Edition. Kluwer Academic Publishers, Dordrecht, 400 pp., 1997.

Kay, J. and Schneider, E. D., Thermodynamics and measures of ecological integrity. In: *Proc. Ecological Indicators,* pp. 159–182. Elsevier, Amsterdam, 1992.

Mauersberger, P., General principles in deterministic water quality modeling. In: G. T. Orlob (editor). *Mathematical Modeling of Water Quality: Streams, Lakes and Reservoirs* (International Series on Applied Systems Analysis, 12). Wiley, New York, 42–115, 1983.

Mauersberger, P., Entropy control of complex ecological processes. In: B. C. Patten and S. E. Jørgensen (editors). *Complex Ecology: The Part-Whole Relation in Ecosystems*. Prentice-Hall, Englewood Clifs, NJ, 130–165, 1995.

Morowitz, H. J., *Energy Flow in Biology*. Academic Press, New York, 1968.

Nielsen, R. H., The Archaean Seas were teeming of life. *Ingeniøren. Naturvidenskab.* Number 6, February, 16–17, 1999.

Patten, B. C., Straskraba, M., and Jørgensen, S. E., Ecosystem Emerging: 1. Conservation. *Ecol. Modelling* 96: 221–284, 1997.

Peters, R. H., *The Ecological Implications of Body Size*. Cambridge University Press, Cambridge, 1983.

Straskraba, M., Jørgensen, S. E., and Patten, B. C., 1999. Ecosystem emerging: 3. Dissipation. *Ecol. Modelling,* 117: 1–41, 1999.

Sveinsdottir, S., Highest probability of life on the Moon of Jupiter, Europa. *Ingeniøren,* Number 21, May, 10–11, 1997.

Ulanowicz, R. E., *Growth and Development. Ecosystems Phenomenology*. Springer-Verlag, New York, 204 pp., 1986.

Ulanowicz, R. E., *Ecology, the Ascendent Perspective*. Columbia University Press, New York, 201 pp., 1997.

We cannot dispose the future,
but we can propose the trends

CHAPTER 13

This chapter proposes a Fourth Law of Thermodynamics, also called ELT in Chapter 9, to describe how an ecosystem will develop away from thermodynamic equilibrium as a consequence of an energy (exergy) flow through the system. At this stage it should be considered as a hypothesis. *If a system receives a through-flow of exergy (1) the system will utilise this exergy flow to move away from thermodynamic equilibrium, (2) more than one pathway is offered to move away from thermodynamic equilibrium, the pathway yielding most stored exergy (measured in J/m² or J/m³) by the prevailing conditions, i.e., with the most ordered structure and the longest distance to thermodynamic equilibrium, will have a propensity to be selected.*

Several observations support the hypothesis, which is also consistent with other theoretical approaches in systems ecology:

Ecosystems develop toward increasing ascendancy.
Ecosystems develop toward an increasing ratio indirect to direct effect.
Ecosystems develop toward an increasing exergy (energy) throughput.

Two approaches seem to be in contradiction, namely, that ecosystems develop toward increasing exergy dissipation and toward minimum (specific) exergy dissipation (see also Chapter 11). It is possible to unite these two different theories, namely, by use of the proposed hypothesis of the Fourth Law of Thermodynamics. Ecosystems apparently develop toward a maximum exergy dissipation when we observe the development from a first stage ecosystem to a mature ecosystem, but there is an upper limit for how much exergy can be dissipated. If we exclude seasonal variations, the amount of exergy dissipated cannot exceed the amount of exergy actually captured from the solar radiation, but an ecosystem will (rapidly) develop toward this maximum dissipation level. This is characteristic for a mature ecosystem. This implies, however, that a further development of a mature ecosystem cannot be described by use of maximum dissipation as an orientor. In the mature stage the system can still develop by increasing cycling and the information embodied in the biological structure. This type of growth does not imply increased dissipation, but the exergy stored in the ecosystem will still increase. It means that the amount of specific exergy (exergy stored relative to the biomass) increases and that the amount of exergy dissipated to maintain the system relative to the stored exergy decreases or remains constant.

The hypothesis seems well supported but a different question is: How can we determine the exergy stored in an ecosystem? This is an impossible task because an ecosystem is too complex. We can only calculate the exergy of an ecological model. As exergy is a relative measure anyhow, it is recommended to talk about an exergy index, not to pretend that we find the exergy of an ecosystem. It is shown how we can find a reasonably good estimation of an exergy index valid for a model of an ecosystem.

The exergy index has to be used to develop structural dynamic models, which are models that can simulate the changes in ecosystem properties as a consequence of changes in the forcing functions. These types of models, sometimes called next generation models, can also be developed using expert knowledge about the reactions of the ecosystem, but in this case, the exergy index is used as a goal function. The most crucial properties are currently changed in accordance with what would give the maximum exergy under the prevailing conditions. Ten model case studies have been successfully carried out by this application of the exergy index as a goal function.

The last section presents what we can derive from thermodynamics, particularly the proposed hypothesis, about the properties of ecosystems. The last section before the conclusion attempts to develop an evolution index based on exergy calculations. The idea is controversial, but it is presented here as an introduction of the idea.

13 A Tentative Fourth Law of Thermodynamics

Sven E. Jørgensen

CONTENTS

13.1 Introduction 305
13.2 Observations Applied to Test the Hypothesis 307
13.3 How to Calculate an Exergy Index 308
13.4 Other Orientors Describing Growth of an Organism or
      Ecosystem Development..316
13.5 Development of Ecosystems..320
13.6 Ecosystem Properties 331
13.7 Structural Dynamic Models of Ecosystems 334
13.8 A Very First Attempt to Express the Biological Evolution by
      a Relative Information Index 342
13.9 Conclusions and Summary 344
References 345

13.1 INTRODUCTION

Ecosystems are operating far from thermodynamic equilibrium. The Second Law of Thermodynamics is not violated, because the amount of exergy received from the solar radiation is less than or equal to the amount of exergy dissipated to the environment as heat, which corresponds to the exergy utilised for maintenance of the ecosystems. This chapter is concerned with the crucial question: what determines the composition of the ecosystem among the many possible ones? Or expressed differently: what determines which gradients toward thermodynamic equilibrium will prevail in an ecosystem? A hypothetical Fourth Law of Thermodynamics is proposed in this chapter as an answer to these questions.

The three laws of thermodynamics can be considered as constraints on the development of ecosystems: only processes that follow the conservation principle (first law) and that consume exergy (produce entropy or dissipate energy) (second law) are possible. A flow of energy (exergy) through the system, which means that the system must be open or at least nonisolated, is absolutely necessary for its existence (partly deduced from the third law; see also Chapter 12). A flow of exergy through the system is also *sufficient* to form an ordered structure (also named a dissipative structure, Prigogine, 1980).

Morowitz (1992) calls this latter formulation the Fourth Law of Thermodynamics, but it would be more appropriate to expand this law to encompass a statement about *which* ordered structure among the possible ones will be selected, or which factors determine how an ecosystem will grow and develop. This expanded version was formulated as a tentative Fourth Law of Thermodynamics in Jørgensen (1992a, 1997), but was expressed without the name of the Fourth Law of Thermodynamics in Jørgensen and Mejer (1977, 1979), Mejer and Jørgensen (1979), and in Jørgensen (1982).

1-56670-272-0/01/$0.00+$.50
© 2001 by CRC Press LLC

This chapter focuses on this tentative and expanded version of the Fourth Law of Thermodynamics and its implications for ecosystem properties and development. It is, to a great extent, based upon a recent presentation of the hypotheses (Jørgensen et al., 2000).

In thermodynamic terms, a growing system is moving away from thermodynamic equilibrium. At thermodynamic equilibrium, the system cannot do any work, the components are inorganic and have zero free energy, and all gradients are eliminated. Everywhere in the universe you will, however, find structures and gradients, resulting from growth and developmental processes. Dissipation attempts to tear down the structures and eliminate the gradients, but dissipation cannot operate unless the gradients were there in the first place. An obvious question is therefore: what determines the build up of gradients?

Biological systems in particular have many possibilities to move away from thermodynamic equilibrium. It is therefore crucial in ecology to know which pathways among many possible ones an ecosystem will select for development. That would be the key to describe the processes characteristic for ecosystem development. Considering the ontic openness presented in the former chapter focusing on the Third Law of Thermodynamics, it would, however, be more appropriate not to discuss the selection of components and processes for development of an ecosystem but rather to discuss the propensity of direction for development (Ulanowicz, 1997). The possible solution to that problem may be formulated as a hypothesis:

If a system receives a through-flow of exergy (1) the system will utilise this exergy flow to move away from thermodynamic equilibrium. (2) If more than one pathway is offered to move away from thermodynamic equilibrium, the one yielding most *stored* exergy (measured in J/m^2 or J/m^3) by the prevailing conditions, i.e., with the most ordered structure and the longest distance to thermodynamic equilibrium, will have a propensity to be selected.

The tentative Fourth Law of Thermodynamics may also be considered as an extended version of "Le Chatelier's Principle." Formation of biomass may be described as,

$$\text{Energy + nutrients = molecules with more free energy (exergy) and organisation} \qquad (1)$$

If we pump energy into a system, the equilibrium will, according to Le Chatelier's Principle, shift toward a utilisation of the energy. It means that molecules with more free energy and organisation are formed. If more pathways are offered, the pathways, that give most relief, i.e., use most energy and thereby form molecules with most embodied free energy (exergy) will win according to the proposed tentative Fourth Law of Thermodynamics.

As it is not possible to prove the first three laws of thermodynamics by deductive methods, the tentative fourth law can only at the best be proved by inductive methods. It implies that the tentative fourth law should be investigated (a falsification is attempted) in as many concrete cases as possible.

The next section of the chapter is devoted to a presentation of such concrete cases where the hypothesis has been tested in the sense that falsifications have been attempted. A method to calculate important contributions to exergy is presented and discussed in the third section. The following section examines the consistency of the hypothesis with other theories describing ecosystem development. The sixth section focuses on typical ecosystem properties and examines whether they are consistent with the proposed tentative Fourth Law of Thermodynamics. The seventh section presents structural dynamic modelling applying exergy as a goal function. It is a recent development in ecological modelling which strongly supports the hypothesis formulated as a tentative Fourth Law of Thermodynamics, as observations reconcile with the structural changes predicted by use of exergy as a goal function. The eighth section attempts to express the biological evolution by application of a relative information index. The last section will summarise the important results and conclusions of the chapter.

This presentation of the tentative Fourth Law of Thermodynamics is, to a great extent, based on Jørgensen (1997) and Jørgensen et al. (2000).

13.2 OBSERVATIONS APPLIED TO TEST THE HYPOTHESIS

It has been possible to find ecologically relevant case studies where several pathways are available to utilise the flow of exergy and where the exergy gained by the system can be compared (Jørgensen, 1997). These (few) case studies support the tentative Fourth Law of Thermodynamics, as it was attempted to falsify the hypothesis — but without success.

The sequence of oxidation of organic matter (see for instance Schlesinger, 1997) is as follows: by oxygen, nitrate, manganese dioxide, iron (III), sulphate, and carbon dioxide. It means that oxygen, if present, will always out-compete nitrate which will out-compete manganese dioxide and so on. The amount of exergy stored as a result of an oxidation process is measured by the number of ATPs formed. ATP represents storage of 42 kJ exergy per mole. The available exergy in form of ATPs decreases in the same sequence as indicated above, as should be expected if the tentative Fourth Law of Thermodynamics was valid; see Table 13.1.

Numerous experiments have been performed to imitate the formation of organic matter in the primeval atmosphere on earth $4 * 10^9$ years ago (see, for instance, Jørgensen, 1997). Various sources of energy have been sent through a gas mixture of carbon dioxide, ammonia, and methane. Analyses have shown that a wide spectrum of various compounds including several amino acids is formed under these circumstances, but generally only compounds with rather large free energy (i.e., high exergy storage) will form an appreciable part of the mixture (Morowitz, 1968).

The entire evolution has been toward organisms with an increasing number of information genes (genes actually utilised) and more types of cells, i.e., toward storage of more exergy due to the increased information content.

At the biochemical level, we find that different plants operate three different biochemical pathways for the process of photosynthesis: (1) the C3 or Calvin Benson cycle; (2) the C4 pathway, and (3) the crassulacean acid metabolism (CAM) pathway. The latter pathway is less efficient than the two other possible pathways, measured as g plant biomass formed per unit of energy received. Plants using CAM pathway can, however, survive in harsh, arid environment, in which plants following C3 and C4 pathways cannot. The photosynthesis will, however, switch to C3 as soon as the availability of water is sufficient; see Shugart (1998). The CAM pathways give the highest exergy storage under harsh, arid conditions, while the two other pathways give the highest exergy storage under other conditions. These observations are completely in accordance with the tentative Fourth Law of Thermodynamics.

Givnish and Vermelj (1976) made the assumption that leaves optimise the size by the pay-off of having leaves of a given size vs. maintaining leaves of a given size. They can by this assumption, which corresponds to optimisation of exergy storage, explain how the size of leaves in a given environment depends on solar radiation and humidity.

If should also be mentioned that the general relationship between animal body size, W, and population density, D, is (Peters, 1983) $D = A/W$, where A is a constant. The highest possible

---

TABLE 13.1
Available kJ/equiv. to Build ATP for Various Oxidation Processes of Organic Matter at pH = 7.0 and 25°C

Reaction	Available kJ/equiv.
$CH_2O + O_2 \rightarrow CO_2 + H_2O$	125
$CH_2O + 0.8NO_3^- + 0.8H^+ \rightarrow CO_2 + 0.4N_2 + 1.4H_2O$	119
$CH_2O + 2MnO_2^- + H^+ \rightarrow CO_2 + 2Mn^{2+} + 3H_2O$	85
$CH_2O + 4FeOOH + 8H^+ \rightarrow CO_2 + 7H_2O + Fe^{2+}$	27
$CH_2O + 0.5SO_4^{2-} + 0.5H^+ \rightarrow CO_2 + 0.5HS^- + H_2O$	26
$CH_2O + 0.5CO_2 \rightarrow CO_2 + 0.5CH_4$	23

packing of biomass is therefore independent of the size of the organisms, which supports that it is attempted to optimise exergy storage in an ecosystem. If, for instance, the exergy dissipation should be maximised, which is not the case, then the packing would be in accordance with the following relationship: $D = \text{constant}/W^{0.65-0.75}$ (see Peters, 1983), as the specific respiration is proportional to W in the exponent 0.65–0.75.

If a resource (for instance, one of the limiting elements for plant growth) is abundant, it will recycle faster. This is a little strange, because recycling is not needed when a resource is nonlimiting. Nevertheless, it has been shown that the exergy stored (Jørgensen, 1997) increases when an abundant resource recycles faster. Ulanowicz (1997) also claims that ascendency increases with faster recycling of an abundant resource, and modelling studies (see below) have shown that ascendency and storage of exergy are well correlated.

Brown et al. (1993) and Marquet and Taper (1998) have examined the patterns of animal body size. They can explain the frequency distributions for the number of species as a function of body size by optimisation of the fitness. The fitness may be defined as the rate that resources (Brown, 1995), in excess of those required for maintenance of the individual, can be used for reproduction. This definition of fitness is close to "ability to increase the stored exergy." Their results may therefore be considered a support of the presented hypothesis.

Ulanowicz (1997) is using ascendancy to formulate the following hypothesis for development of an ecosystem. In the absence of overwhelming external disturbances, living systems exhibit a natural propensity to increase in ascendancy. As ascendancy is correlated well, as shown by modelling studies (Jørgensen, 1994a), with the stored exergy of the system. This hypothesis is very close to the hypothesis presented above as the tentative Fourth Law of Thermodynamics. Ascendancy is, in the presented hypothesis, replaced by exergy stored and in the absence of overwhelming external disturbances is replaced with "under the prevailing conditions." In addition, it has been shown in Patten and Fath (1999) that increased cycling will increase the through flow and storage of the network components (nodes). As ascendancy is dominated by through flow, it is thereby shown that exergy and ascendancy are closely related.

Several modelling cases, mostly of aquatic ecosystems, have been examined and they have all supported the tentative law by calculating exergy indexes to be used for a description of the structural dynamic changes in the focal ecosystem. The exergy index is calculated as shown in the next section. Models with dynamic structure are models based on nonstationary, time-varying differential equations. Structural dynamic models have also been developed by use of knowledge about which species with which properties (used as parameters in the model) would be dominant under which circumstances; see, for instance, Recknagel et al. (1994), Reynolds (1996), and Patten (1997). The structural dynamic models referred here as support for the tentative Fourth Law of Thermodynamics, have, however, applied optimisation of an exergy index to describe the current changes of the parameters. The details of this type of model will be presented in the seventh section of this chapter.

13.3   HOW TO CALCULATE AN EXERGY INDEX

Exergy is (further details, see Chapters 7 and 10) defined as the work the system can perform when it is brought into equilibrium with the environment or another well-defined reference state (Figure 13.1; see also Chapter 7).

If we presume a reference environment that represents the system (ecosystem) at thermodynamic equilibrium, which means that all the components are inorganic at the highest possible oxidation state (all the free energy has been previously utilised to do work) and homogeneously distributed in the system (no gradients), the situation illustrated in Figure 13.2 is valid. As the chemical energy embodied in the organic components and the biological structure contributes far more to the exergy content of the system, there seems to be no reason to assume a (minor) temperature and pressure differences between the system and the reference environment. Under these circumstances we can calculate the

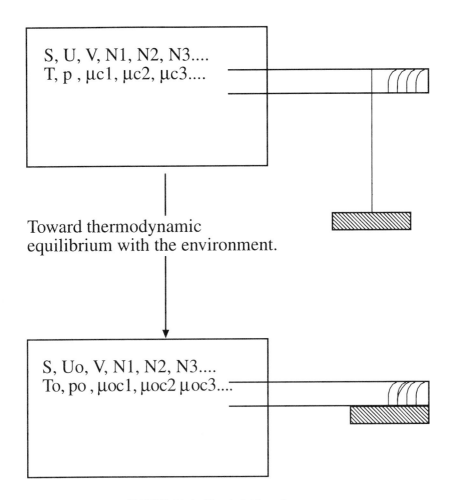

FIGURE 13.1 The definition of exergy.

exergy content of the system as coming entirely from the chemical energy: $\Sigma(\mu_c - \mu_{co})N_i$. Only what Szargut (1998) calls chemical exergy is included in the computation of exergy. The physical exergy (Szargut et al., 1988; Szargut, 1998) is omitted in these calculations as there is no temperature and pressure difference between the system and the reference system. We find by these calculations the exergy of the system compared with the same system at the same temperature and pressure but in form of an inorganic soup without any life, biological structure, information, or organic molecules. As $(\mu_c - \mu_{co})$ can be found from the definition of the chemical potential replacing activities by concentrations, we get the following expressions for the exergy:

$$Ex = RT \sum_{i=0}^{i=n} c_i \ln c_i / c_{ieq} \qquad (2)$$

where R is the gas constant, T is the temperature of the environment (and the system; see Figure 13.2), while c_i is the concentration of the ith component expressed in a suitable unit, e.g., for phytoplankton in a lake c_i could be expressed as mg/1 or as mg/1 of a focal nutrient. c_{ieq} is the concentration of the ith component at thermodynamic equilibrium and n is the number of components. c_{ieq} is, of course, a very small concentration (except for $i = 0$, which is considered to cover the inorganic compounds), but is not zero, corresponding to a very low probability of forming complex organic compounds spontaneously in an inorganic soup at thermodynamic equilibrium.

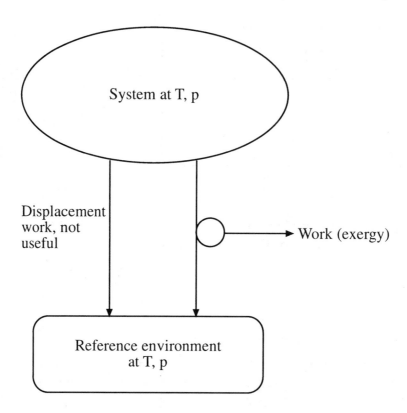

FIGURE 13.2   The definition of the concept exergy used to compute the exergy index for an ecological model is illustrated.

It can be shown (Mejer and Jørgensen, 1979) that the chemical exergy contributed by the components in an open system with a through flow is,

$$Ex = R * T * \sum_{i=0}^{n} [c_i * \ln(c_i / c_{ieq}) - (c_i - c_{ieq})], \qquad [ML^2 T^{-2}] \qquad (3)$$

where R is the gas constant, T is the absolute temperature, c_i is the concentration in the ecosystem of component i, index 0 indicates the inorganic components of the considered element, and c_{ieq} is the corresponding concentration of component i at thermodynamic equilibrium in the reference state, the inorganic soup of the system. The problem by application of these equations is related to the size of c_{ieq}. The contributions from the inorganic components are usually very low and can in most cases be neglected.

Shieh and Fen (1982) have suggested estimating the exergy of structurally complicated material on the basis of the elementary composition. This has, however, the disadvantage that a higher organism and a microorganism with the same elementary composition will get the same exergy, which is in complete disagreement with the exergy embodied in the information.

The problem related to the assessment of c_{ieq} has been discussed and a possible solution proposed in Jørgensen et al. (1995), but the most essential arguments should be repeated here. For dead organic matter, detritus, which is given the index 1, it can be found from classical thermodynamics (see, for instance, Russel and Adebiyi, 1993),

$$\mu_1 = \mu_{1eq} + RT \ln c_1 / c_{1eq} \qquad (4)$$

where μ indicates the chemical potential. The difference $\mu_1 - c_{1eq}$ is known for organic matter, e.g., detritus, which is a mixture of carbohydrates, fats, and proteins.

Generally, c_{ieq} can be calculated from the definition of the probability P_{ieq} to find component i at thermodynamic equilibrium,

$$P_{ieq} \equiv c_{ieq} \Big/ \sum_{i=0}^{N} c_{ieq} \tag{5}$$

If we can find the probability, P_i, to produce the considered component i at thermodynamic equilibrium, we have determined the ratio of c_{ieq} to the total concentration. As the inorganic component, c_0, is very dominant by the thermodynamic equilibrium, Equation (5) may be rewritten as,

$$P_{ieq} \approx c_{ieq} / c_{0eq} \tag{6}$$

By a combination of Equations (4) and (6), we get,

$$P_{1eq} = [c_1 / c_{0eq}] \exp[-(\mu_1 - \mu_{1eq}) / RT] \tag{7}$$

For the biological components, 2, 3, 4...N, the probability, P_{ieq}, consists of the probability of producing the organic matter (detritus), i.e., P_{1eq}, and the probability, $P_{i,a}$, to obtain the information embodied in the genes, which again determine the amino acid sequence. Living organisms use 20 different amino acids and each gene determines the sequence of about 700 amino acids (Li and Grauer, 1991). $P_{i,a}$ can be found from the number of permutations among which the characteristic amino acid sequence for the considered organism has been selected. It means that we have the following two equations available to calculate P_i:

$$P_{ieq} = P_{1eq} P_{i,a}$$

($i \geq 2$; 0 covers inorganic compounds and 1 detritus) and $\qquad$ (8)

$$P_{i,a} = 20^{-700g}$$

where g is the number of genes. Equation (6) is reformulated to,

$$c_{ieq} \approx P_{ieq} c_{0eq} \tag{9}$$

Equations (9) and (3) are combined:

$$Ex \approx RT \sum_{i=0}^{N} [c_i \cdot \ln (c_i / (P_{ieq} c_{0eq})) - (c_i - P_{ieq} c_{0eq})], \tag{10}$$

This equation may be simplified by the use of the following approximations (based upon $P_{ieq} \ll c_i$, $P_{ieq} \ll P_0$ and $1/P_{ieq} \gg c_i$, $1/P_{ieq} \gg c_{0eq}/c_i$: $c_i/c_{0eq} \approx 1$, $c_i \approx 0$, $P_i c_{0eq} \approx 0$ and the inorganic component can be omitted. The significant contribution is coming from $1/P_{ieq}$, see Equation (8). We obtain,

$$Ex \approx -RT \sum_{i=0}^{N} c_i \cdot \ln (P_{ieq}) \tag{11}$$

where the sum starts from 1, because $P_{0, eq} \approx 1$.

Expressing as P_{ieq} in Equation (8) and P_{leq} as in Equation (7), we obtain the following expression for the calculation of an exergy index:

$$Ex / RT = \sum_{i=0}^{N} [c_i \cdot \ln(c_1 / (c_{0eq})) - (\mu_1 - \mu_{1eq})] \sum_{i=1}^{N} c_i / RT - \sum_{i=2}^{N} c_i \ln P_{i,a}$$

As the first sum is minor compared with following two sums (use for instance $c_i / c_{0eq} \approx 1$), we can write,

$$Ex / RT = (\mu_1 - \mu_{leq}) \sum_{i=1}^{N} c_i / RT - \sum_{i=2}^{N} c_i \ln P_{i,a} \qquad (12)$$

This equation can now be applied to calculate contributions to the exergy index by important ecosystem components. If we consider only detritus, we know that the free energy released per g of organic matter is about 18.7 kJ/g. R is 8.4 J/mol and the average molecular weight of detritus is assumed to be 100,000. We get the following contribution of exergy by detritus per litre of water, when we use the unit g detritus exergy equivalent/l:

$$Ex_1 = 18.7 c_i \text{ kJ/l} \qquad \text{or} \qquad Ex_1 / RT = 7.34 * 10^5 c_i \text{ g/l} \qquad (13)$$

A typical unicell alga has on average 850 genes. We may use the number of genes, while recent development points toward the use of DNA per cell. We have previously purposely used the number of genes and not the amount of DNA per cell, which would include unstructured and nonsense DNA. In addition, a clear correlation between the number of genes and the complexity has been shown (Li and Grauer, 1991). However, recently it has been discussed that the nonsense genes *are* playing an important role, for instance, that they may be considered as parts which are able to repair genes when they are damaged. If it is assumed that an alga has a total of 850 genes, i.e., they determine the sequence of $850 \times 700 = 595,000$ amino acids, the contribution of exergy per litre of water, using g detritus equivalent/l as concentration unit would be,

$$Ex_{algae} / RT = 7.34 * 10^5 c_i - c_i \ln 20^{-595000} = 25.2 * 10^5 c_i \text{ g/l} \qquad (14)$$

The contribution to exergy from a simple prokaryotic cell can now be calculated similarly as,

$$Ex_{prokar.} / RT = 7.34 * 10^5 c_i + c_i \ln 20^{329000} = 17.2 * 10^5 c_i \text{ g/l} \qquad (15)$$

The above shown calculations are based on the number of genes and not on the amount of total DNA. The information that codes for proteins in the genes of higher organisms is contained in small packets, called exons. They are interrupted by stretches of DNA that apparently code for nothing, which came to be called introns (Lewin, 1994). Genes may also duplicate and form multigene families. Sometimes a gene may be copied via an RNA intermediate from which the introns and regulatory sequences are removed. Such pseudo genes are therefore nonfunctional (Lewin, 1994). It may be discussed whether it is correct to base the calculations of exergy originated from all the stored, applied, and non-applied information or only from the information used to code the proteins. The argument for the latter should be that information which cannot be interpreted has no value. Introns and pseudo genes may, however, have functions, namely, as "spare parts" for repair of damaged genes or for a faster emergence of new and better genes due to more rapid mutations of introns and pseudo genes. Inclusion of *all* DNA in a cell implies that the so-called c-value should be applied; see Fonseca (1999). The c-value has been found for more than 300,000 organisms and can be found on the Internet: www. rbgkew.org.uk/cva/database 1.html.

TABLE 13.2
Approximate Number of Nonrepetitive Genes

Organisms	Number of Nonrepetitive Genes Factor*	Conversion
Detritus	0	1
Minimal cell (Morowitz, 1992)	470	2.3
Bacteria	600	2.7
Algae	850	3.4
Yeast	2000	5.8
Fungus	3000	9.5
Sponges	9000	26.7
Moulds	9,500	28.0
Plants, trees	10,000–30,000	29.6–86.8
Worms	10,500	30.0
Insects	10,000–15,000	29.6–43.9
Jellyfish	10,000	29.6
Zooplankton	10,000–15,000	29.6–43.9
Fish	100,000–120,000	287–344
Birds	120,000	344
Amphibians	120,000	344
Reptiles	130,000	370
Mammals	140,000	402
Human	250,000	716

Sources: T. Cavalier-Smith (1985), Li and Grauer (1991), and Lewin (1994).
* Based on energy contained in detritus. 1 g detritus has in average 18.7 kJ exergy.

Organisms with more than one cell will have DNA in all cells determined by the first cell. The number of possible microstates becomes therefore proportional to the number of cells. Zooplankton has approximately 100,000 cells and (see Table 13.2) 15000, genes per cell, each determining the sequence of approximately 700 amino acids. ln P_{zoo} can, therefore, be found as,

$$-\ln P_{zoo} = -\ln(20^{-15000*700} * 10^{-5}) \approx 315 * 10^5 \qquad (16)$$

As seen the contribution from the numbers of cells is insignificant. Similarly, $P_{fish,a}$ and the P-values for other organisms can be found by use of the figures in Table 13.2 which is based on the number of genes. If the c-values are used to cover all the information embodied in DNA of a considered organisms, the weighting factor will obviously be larger.

The application of these values for a model consisting of inorganic material, IM, phytoplankton, P, zooplankton, Z, fish, F, and detritus, D, would yield,

$$Ex/RT = 0 \text{ IM} + P(1.79 * 10^6) + Z(31.5 * 10^6) + F(2.52 * 10^8)$$
$$+ (D + P + Z + F) * (7.34 * 10^5) \text{ [g / l]} \qquad (17)$$

The contributions from phytoplankton, zooplankton, and fish to the exergy of the entire ecosystem are significant and far more than corresponding to the biomass. Notice that the unit of Ex/RT is g/1. Exergy can always be found in joules per litre, provided that the right units for R and T are used. Equation (21) can be rewritten by converting g/l to g detritus/1 by dividing by (7.34 * 10⁵),

$$Ex / RT = P(3.4) + Z(44) + F(344) + (D) \text{ [exergy as g deteritus/1]} \qquad (18)$$

As can be seen from the last equations, the exergy index is dominated by the contributions coming from the information, originated from the genes of the organisms, because the concentrations of phytoplankton, zooplankton, and fish are multiplied by factors $>$ or $\gg 1$ and the factors are determined by the information embodied in the genes.

The exergy index is found as the concentrations of the various components, c_i, multiplied by weighting factors, β_i, reflecting the exergy that the various components possess due to their chemical energy and to the information embodied in the genes,

$$Ex = \sum_{i=n}^{i=0} \beta_i c_i \qquad (19)$$

The index 0 covers all the inorganic components which, of course, in principle should be included, but can be neglected in most cases. The contributions from detritus and from the biological components are higher due to an extremely low concentration of these components in the reference system (the ecosystem converted to an inorganic dead system). The inorganic components, including oxygen, may, of course, have different concentrations in the system and in the reference system, but as the ratios between the concentrations of inorganic components in the two comparable systems are much lower than the ratios between the concentrations of biological components, the contributions from inorganic components to exergy are negligible compared with the contributions of the biological components.

The calculation of exergy index accounts, by use of this equation, for the chemical energy in the organic matter as well as for the information embodied in the living organisms. It is measured by the extremely small probability to form the living components, for instance, algae, zooplankton, fish, mammals, and so on, spontaneously from inorganic matter. The weighting factors may also be considered quality factors reflecting how developed the various groups are and to what extent they contribute to the exergy due to their content of information which is reflected in the computation as shown in Equations 17 to 19. This is according to Boltzmann (1905), who gave the following relationship for the work, W, that is embodied in the thermodynamic information:

$$W = RT \ln N \qquad (ML^2T^{-2}) \qquad (20)$$

where N is the number of possible states, among which the information has been selected. N is as seen for species the inverse of the probability to obtain the valid amino acid sequence spontaneously.

It is furthermore consistent with the following reformulation of Reeves (1991): "information appears in nature when a source of energy (exergy) becomes available but the corresponding (entire) entropy production is not emitted immediately, but is held back for some time (as exergy)."

Svirezhev (1998) has shown that Equation (3) can be rewritten in the form (see Chapter 10).

$$Ex = R * T * \left(A \sum_{i=0}^{N} [P_i * \ln(P_i / P_{ieq}) + A \ln A / A_o - (A - A_o) \right) \qquad (21)$$

P_{ieq} is defined above (see Equations (5) and (6)), and

$$P_i \equiv c_i \bigg/ \sum_{i=0}^{N} c_i \qquad (22)$$

A is the total matter,

$$A = \sum_{i=0}^{N} c_i \qquad (23)$$

and A_o is the total matter at thermodynamic equilibrium. The vector $P = (P_o, P_1,...,P_n)$ describes the structure of the system. P_i are intensive variables and

$$K = \sum_{i=0}^{N} [P_i * \ln (P_i / P_{\text{ieq}})] \tag{24}$$

is the so-called Kullback measure that is the (additional) information when the distribution is changed from P_{ieq} to P_i. Note that K is a specific measure (per unit of matter). The product $Ex_{\text{inf}} = A * K$ may be considered the total amount of information for the entire system, which has been accumulated in the transition from some reference state corresponding to the thermodynamic equilibrium, i.e., some prevital state, to the current state of living matter.

A is an extensive variable and $Ex_{\text{mat}} = A \ln (A/A_o) - (A - A_o)$ represents the increase of exergy due to change in the total mass of the system.

Exergy is therefore the sum of two terms: Ex_{inf} resulting from structural changes inside the system and Ex_{mat} caused by a change of total mass of the system.

The total exergy of an ecosystem *cannot* be calculated exactly, as we cannot measure the concentrations of all the components or determine all possible contributions to exergy in an ecosystem. If we calculate the exergy of a fox, for instance, the above shown calculations will only give the contributions coming from the biomass and the information embodied in the genes, but what is the contribution from the blood pressure, the sexual hormones, and so on? These properties are at least partially covered by the genes but is that the entire story? We can calculate the contributions from the dominant components, for instance, by the use of a model or measurements that covers the most essential components for a focal problem.

Exergy calculated by Equation (19) has some short-comings; it is proposed to consider the exergy found by these calculations as a *relative exergy index*,

1. We account only for the contributions from the organisms' biomass and information in the genes. Although these contributions most probably are the most important ones, it cannot be completely excluded that other important contributions are omitted.
2. We do not account for the information embodied in the network — the relations between organisms. The information in the model network that we use to describe ecosystems is negligible compared with the information in the genes, but we cannot exclude that the real, much more complex network may contribute considerably to the total exergy of a natural ecosystem.
3. We have made approximations in our thermodynamic calculations. They are all indicated in the calculations and are, in most cases, negligible.
4. We can never know all the components in a natural (complex) ecosystem. Therefore, we will only be able to utilise these calculations to determine exergy indices of our simplified images of ecosystems, for instance, of models.

Exergy indices are, however, useful, as they have been successfully used as goal function (orientor) to develop structural dynamic models. The *difference* in exergy by *comparison* of two different possible structures (species composition) is decisive. Moreover, exergy computations always give only relative values, as the exergy is calculated relative to the reference system.

As already stressed, the presented calculations do not include the information embodied in the structure of the ecosystem, i.e., the relationships between the various components, which is represented by the network. The information of the network encompasses the information of the components and the relationships of the components. The latter contribution is calculated by Ulanowicz (1991) as a part of the concept of ascendancy. In principle, the information embodied in the network should be included in the calculation of the exergy of structural dynamic models, as the network is also dynamically changed (Pahl-Wostl, 1995). It may, however, often be omitted in most dynamic model calculations due to:

1. The contributions from the network relationships of models (not from the components of the network, of course) are minor, compared with the contributions from the components. This is due to the extreme simplifications made in the model compared with the real network, although they attempt to account for the major flows of energy or mass.
2. In most cases a relative value of the exergy in form of an exergy index is sufficient to describe the direction of ecosystem development/growth.
3. If the network is changing in addition to the components (the nodes) of the network, it should/could be considered what this change would contribute to the exergy index of the system.
4. The calculations of exergy indexes will always be an approximation focusing on the most important components with respect to the changes taking place. A model (e.g., a conceptual model) is often used as the basis for these calculations, and a model is always a simplification of the real system. The ecosystem is too complex to know all the components.

It has been shown that ascendancy and exergy indices of models are well correlated (see Jørgensen, 1994a). This is not surprising, as the network, which is the basis for the ascendancy calculations, is a result of the ability of the biological components in the network to cooperate and build interrelationships. So, the network, including the transfer of mass or energy from one component to another, indirectly reflects the size of and the information embodied in the components of the network.

Two quantities play a role in networks: the information embodied in the network and the transfer of energy or mass, which is highly dependent on the information embodied in the nodes of the network.

13.4 OTHER ORIENTORS DESCRIBING GROWTH OF AN ORGANISM OR ECOSYSTEM DEVELOPMENT

Boltzmann (1905) proposed that "life is a struggle for the ability to perform work," which is exergy. The ability of a single species to perform work is proportional to its biomass. Margalef (1968), Straskraba (1979, 1980), and Brown (1995) have proposed to use biomass as the goal function.

Lotka (1922) proposed to use maximum power as the goal function to describe ecosystem development. Maximum power is defined as the transformation of energy to perform work per unit of time (Odum, 1983). The transformation of energy to perform work is correlated to the amount of exergy available (stored) in the system. The more exergy stored, the more exergy can be transformed to perform work. This correlation was demonstrated by Salomonsen (1992). He showed that the ratio of exergy and maximum power of two lakes at significantly different levels of eutrophication was approximately the same. Is it a "what was first, the chicken or the egg" problem? Or, is Mauersberger (1983, 1995) "minimum entropy principle" right, and is this principle consistent with the maximum power principle and/or with the principle of maximum exergy storage? Are there situations where ecosystems attempt to minimise the energy dissipation (Johnson, 1995), and maximise the storage of exergy, in spite of the general relationship that more structure (more exergy stored) will require more exergy (energy spent on maintenance of already stored exergy)? This question will be further discussed in the next section.

Odum (1983) introduced another goal function named emergy; see Chapter 8. Figure 13.3 shows the idea behind this concept. The idea is that it is not the actual content of energy that counts, but the embodied energy. It is the energy (originated from the ultimate source of energy — the solar energy) that it costs to construct the considered component. If it costs 10 units of phytoplankton,

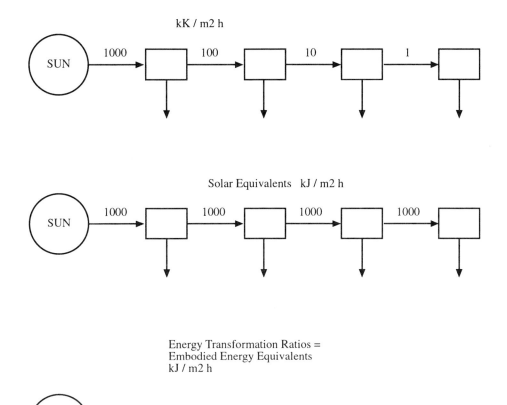

FIGURE 13.3 The definition of the concept emergy is illustrated.

for instance, to construct one unit of zooplankton, the energy of zooplankton should be multiplied by ten to obtain the embodied energy in zooplankton relative to the energy of phytoplankton. It has been shown that there is a good correlation between exergy index and emergy when the calculations are based on realistic ecosystem models derived from data and observations (Jørgensen, 1994). Different organisms have, however, developed different strategies to the prevailing conditions in different ecological niches. In accordance with the emergy calculations, a tree adapted to shade should have lower emergy than a tree adapted to full sunlight, if we presume that they have the same biomass and growth rate. As we know that they may have an equal role for ecosystem development and they have approximately equal exergy (which is dependent on the embodied information), this difference in emergy cannot be correct. Emergy calculates how much solar energy it costs to build the structure while exergy expresses the actual work capacity. If emergy is calculated for an entire ecosystem, the differences between exergy and emergy will be levelled out, but emergy for individual species/components may sometimes be a wrong measure of the potential to develop the ecosystem further away from thermodynamic equilibrium.

Aoki (1988, 1989, 1993) has compared the entropy production = utilisation of exergy for maintenance and exergy storage of different lake ecosystems; see Chapter 8. He finds that the more eutrophic a lake is, the more exergy is stored and the more exergy is also captured and used for maintenance. It is also consistent with Salomonsen (1992) and Jørgensen (1982): a more eutrophic lake has more biomass and more exergy than a mesotrophic or oligotrophic lake. The food web in a eutrophic lake may be simpler, and the top-carnivorous fish species may be missing due to low transparency (top-carnivorous fish are mainly finding the prey visually). This implies that the specific exergy (exergy per unit of biomass) is lower in a eutrophic lake than in mesotrophic and oligotrophic lakes, but the higher biomass usually more than compensates for the reduction in specific exergy.

Kay and Schneider (1992) propose that the development of an ecosystem is best described as follows: the system will use all possible avenues to capture per unit of time as much of the incoming exergy as possible and the captured exergy will be degraded (dissipated) to cover the exergy needed for maintenance. As the degradation of exergy is a focal point in Kay and Schneider's theory, they call their formulation an extended version of the Second Law of Thermodynamics. The relationship between the use of exergy per unit of time can be described by use of the following equation (see Jørgensen et al., 1999):

$$\Delta Ex_{cap} = \Delta Ex_{bio} + \Delta E_{resp+eva} \qquad (ML^2T^{-3}) \qquad (25)$$

where ΔEx_{cap} is the exergy captured by the system per unit of time, ΔEx_{bio} is the exergy stored (accumulated) by the structure per unit of time and $\Delta E_{resp+eva}$ is the exergy degraded by respiration and evapotranspiration processes to heat energy per unit of time. Kay and Schneider mean that the system will optimise $\Delta E_{resp+eva}$, while the previously presented formulation of the Fourth Law of Thermodynamics claims that $F(t) = \int \Delta Ex_{bio}(t)\, dt$ (measures the integrated growth over time from 0 to t) is optimised. As very little of the exergy captured per unit of time is utilised for building new structure (except for a system at the early stage), the exergy degradation per unit of time = energy converted to heat per unit of time = $\Delta E_{resp+eva}$ are for most (at least the most developed) ecosystems >> ΔEx_{bio}, it implies that $\Delta E_{resp+eva} \approx \Delta Ex_{cap} \cdot Ex_{bio}(t)$, resulting from an integration of ΔEx_{bio}, determines the size of the structure and how well organised the ecosystem is. It determines how much exergy the system has on stock for later consumption (included for dissipation). It determines also how much exergy the system can capture per unit of time in the future and how much exergy the system dissipates due to maintenance, because there is a close relationship between the size of the structure on the one side and the exergy captured per unit of time and the exergy used for maintenance per unit of time on the other side.

Note, however, that Equation (25) presents a relationship between rates which hardly can be used to optimise the development of an ecological system, because

1. Rates cannot continuously increase, as they have upper limits. It is for instance impossible to capture more than 100% of the solar radiation (or rather 85 to 90% because of physical constraints). A longer term development will therefore need a function derived from integration over time of rates to be able to describe a continuous development under prevailing conditions.
2. Rates in ecological systems are continuously changed as the environment determining the rates is varying over time. It is therefore significant to account the results (integration) of rates over time, as the system cannot be determined by an occasional high rate for a short, insignificant time.
3. The thermodynamic laws apply in their formulation variables derived from integration over time, for instance, "energy is conserved" and "entropy will increase for all real processes."

The hypotheses "to optimise exergy dissipation or exergy captured" and "to optimise the exergy of the system under the prevailing conditions" are just two sides of the same coin, when we describe the development of an ecosystem from the early stage to the mature stage; see also the next section of this chapter. Increased exergy stored in the structure of the system under development will also enable the system to capture more exergy, as already mentioned, but the difference between the two theories occurs when the system has attained the maximum rate of capturing exergy (as mentioned above, 85 to 90% of the incoming solar radiation). Kay and Schneider's hypothesis is still valid in this situation, because the system dissipates as much exergy as it can, but this amount of exergy cannot increase (seasonal variations are excluded) because the system cannot capture more exergy than the about 80% of the incoming solar radiation. Kay and Schneider's hypothesis is therefore consistent with the hypothetical Fourth Law of Thermodynamics, but exergy dissipation is not a good orientor for further development of a mature system, because it will not change (increase), or only seasonal changes will occur, while the exergy stored still can increase, as it will be touched on in the next section.

An illustration of the presented concepts and their relation to the conservation and dissipation principles and to the hypotheses of Kay and Schneider on the one side and Jørgensen on the other is shown in Figure 13.4.

Choi et al. (1999) have recently presented several results pointing toward a decrease in the ratio respiration/biomass (R/B) as an ecosystem develops. They claim that this ratio is a good measure of the distance from thermodynamic equilibrium. This is according to Odum (1969) and

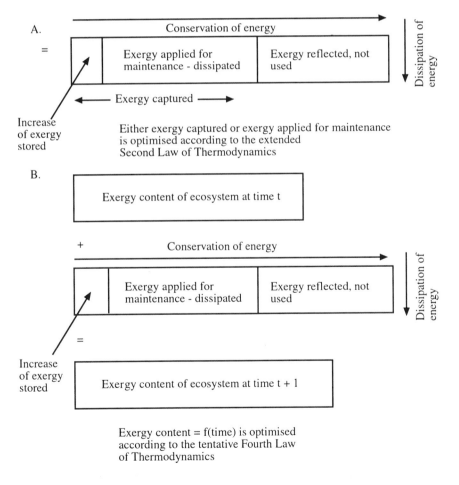

FIGURE 13.4 The difference between the expanded Second Law of Thermodynamics (A) and the tentative Fourth Law of Thermodynamics (B) is illustrated. They are both in accordance with the First and Second Laws of Thermodynamics.

also completely in accordance with the hypothesis presented in this chapter as the Fourth Law of Thermodynamics. Choi et al. 1999 show further that the R/B ratio for lakes decreases and the body size increases with increasing depth of the lake. In accordance with their results, perturbations imply increased R/B. They apply the R/B ratio as a direct measure of perturbations.

13.5   DEVELOPMENT OF ECOSYSTEMS

Ulanowicz (1986) uses growth and development as the extensive and intensive aspects of the same process. Growth implies increase or expansion, while development focuses on the increase in organisation which is considered independent of the size of the system. Ulanowicz considers growth and development as aspects of a unitary process and he applies the concept ascendancy (which is strongly correlated to exergy, as mentioned above) to cover both the changes in size and organisation.

There are three possible directions of development for an ecosystem.

1. The biomass increases — it is generally called growth. It implies of course increased exergy, as the exergy is increasing with increasing biomass.
2. The network is changed to give a better utilisation of the available resources, for instance, by a higher extent of recycling. Recycling will generally imply a better utilisation of the flow of exergy by solar radiation and an increase of the through-flow, which implies that more exergy will be available for the network and the organisms — the total exergy will increase. This is in accordance with Patten et al. (2000), where it is shown, by use of network computations, that increased cycling will imply increased through-flow and storage of exergy, *without* changes in the input of exergy. This is consistent with the interpretation in the last section: when the exergy captured has reached the possible level, the exergy storage can still increase, as it is clearly seen in mature ecosystems.
3. The information of the entire ecosystem, encompassing the information embodied in the organisms and in the network, increases. The increased information implies that the resources are utilised better and more ecological niches can be used. The biomass and the efficiency of the exergy inflow may increase as a consequence of the increased information, but the exergy will increase anyhow due to the increased information according to Equation (20).

All three directions of development imply increased exergy. Or, expressed differently, all three directions can be measured by use of the same concept, namely, exergy. All three directions of development also mean increased gradients, as a more developed network will also have a bigger gradient toward thermodynamic equilibrium than a simple network. More information also means a bigger gradient toward no information. Notice that the information gradient can increase almost unlimited and that it is not broken down even when information is transferred to other components. It is therefore not surprising that the further development of a mature ecosystem, which has used up all the available elements and is capturing as much of the inflowing exergy as possible, is playing on increased information for further development.

The successional development of ecosystems from an early to a mature stage (see, for instance, E. P. Odum, 1969), illustrates that the two concepts, exergy storage and exergy utilisation, are parallel. An ecosystem at an early stage of development, for instance, an agricultural field, has only a small exergy storage and exergy utilisation. The biomass per square meter is small compared with the mature system, i.e., the exergy storage is small. The structure is simple and only little energy (exergy) is needed for respiration or growth as they are both to certain extent proportional to the biomass. The total surface area of the plants is furthermore small, which implies that they are not able to catch and utilise much solar radiation.

As the system develops, the structure becomes more complicated, animals with more information per unit of biomass, i.e., with more genes, populate the ecosystem and the total biomass per square

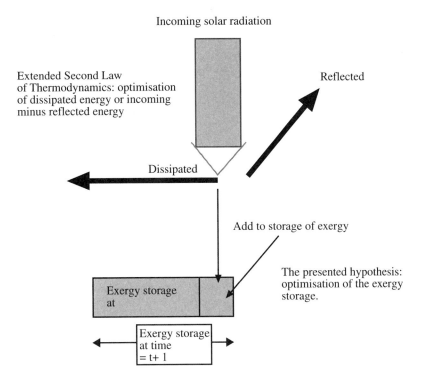

Incoming solar radiation

Extended Second Law
of Thermodynamics: optimisation
of dissipated energy or incoming
minus reflected energy

Reflected

Dissipated

Add to storage of exergy

Exergy storage
at

The presented hypothesis:
optimisation of the exergy
storage.

Exergy storage
at time
= t+ 1

FIGURE 13.5   The exergy utilisation of an ecosystem under development is shown vs. time. Notice that the consequence of the growth in exergy is increased utilisation of exergy for maintenance.

meter increases. It implies that both the exergy storage and the exergy needed for maintenance increase; see Figure 13.5. A very mature ecosystem, for instance a natural forest, has a very complex structure and well-organised food webs. It contains a high concentration of biomass per square meter and contains much information in a wide variety of organisms. The entire structure tries to utilise solar radiation either directly or indirectly, resulting in a high utilisation of the solar exergy flux.

The catabolic energy demand is related to the total biomass and the overall organisation. It represents the exergy needed for maintaining the ecosystem far from thermodynamic equilibrium. This is a parallel to what is experienced by man-made systems: A large town with many buildings of different types (skyscrapers, cathedrals, museums, scientific institutes, etc.) obviously needs much more maintenance than a small village consisting of a few almost identical farmhouses.

The development of ecosystems may also be described (Kay and Schneider, 1990) as a steady growth of a gradient between the ecosystem and thermodynamic equilibrium. The force to break down the gradient will increase with increasing gradient. This tendency to break down the gradient is represented by respiration and evapotranspiration, that spend exergy and produce entropy. As long as the exergy received from solar radiation can compensate for this need of exergy to maintain the gradient, it is possible for the system to stay far from thermodynamic equilibrium. If even more exergy can be captured than needed for maintenance of the gradient, the surplus exergy increases the stored exergy, which means that the system moves further away from thermodynamic equilibrium and thereby increases the gradient even more.

The amount of information (see also Chapter 9) stored in biomass may still increase in a mature ecosystem due for example to

1. Immigration of (slightly) better fitted species, and
2. Emergence of new genes or genetic combinations. This latter possibility is covered by the concept of "evolution."

The system stops growing in biomass when the most limiting inorganic component has been fully utilised for biomass construction. Then the mature stage of the ecosystem has been attained. Nutrients and water are often the limiting factors in growth of plants. These resource cycles give possibilities for formation of new biomass with perhaps more information, but the total biomass is not changed by this reallocation of resources. This constraint by the laws of conservation is essential for the development of more and more complex living structures. As living organisms compete for limited food supplies, they invent and develop thousands of new and ingenious strategies (Reeves, 1991). Some species invest in movement; speed can be a valuable asset both for capturing prey and for avoiding predators. Others use protective armour or chemical poisons. Each species thus defines the terms under which it engages in the harsh business of life. Better feedbacks to assure maintenance of a high biomass level to changed circumstances, better buffer capacities, better specialisation to populate all possible ecological niches, and better adapted organisms to meet the variability in forcing functions are all developed. Thus, the biomass is maintained at the highest level over a longer time and the information level will increase due to the steady development of better feedback and more self-organisation. Both contributions are reflected in a higher exergy. The exergy of the mature system can therefore still grow further, namely, by increase of the information. In other words, the system better utilises its resources and becomes more fitted to the prevailing conditions. Adaptation and specialisation require information, which implies that a better fitness to prevailing conditions is more probable by a system with more stored information in the genes.

Neither exergy nor information is conserved. Exergy is lost by all transfers of energy, but exergy and information are also lost by death of organism as ß in Equation (19) decreases from a value >>1 to 1 (detritus exergy equivalent is applied as unit). Exergy and information may, however, be gained when phytoplankton is converted to zooplankton by grazing. Exergy and information are therefore not cycling in the same manner as we know for mass and energy. Their distribution in ecosystems, as with energy and matter, follow a complicated pattern, determined by the life processes.

The two hypotheses of maximisation of exergy storage and of exergy capture are, as already pointed out, not consistent when we have to describe the further development of a mature system in which domain Mauersberger's minimum principle may be valid (see Mauersberger, 1983 and 1995).

Ecosystems locally decrease entropy production by transporting energy-matter from more probable to less probable spatial locations. Therefore less exergy is lost and more exergy stored.

A possible hypothetical formulation of the tentative Fourth Law of Thermodynamics trying to unite the hypotheses of Kay and Schneider with the hypotheses of Mauersberger may be:

If a system is moved away from thermodynamic equilibrium by application of a flow of exergy, it will utilise all avenues available to build up as much dissipative structure (store as much exergy) as possible. An ecosystem at an early stage will try to store more exergy by increasing the amount captured (decrease the reflection) while a mature ecosystem will gain more exergy by decreasing the exergy lost by the maintenance processes (mainly achieved by increase of the information content of the system). These two situations are illustrated in Figure 13.6.

The overall description of the development of ecosystems is illustrated in Figure 13.7. In the first phase the structure and its biomass increase rapidly mainly due to the rapid growth of r-strategists. The gradients, the amount of exergy captured by the system, and the exergy required for maintenance are all increasing. A transition phase is shown in Figure 13.8 between the first and third phase. In the third phase, the limiting elements are used up, which implies that a further increase in the physical structure measured by the biomass is not possible. A better use of the resources and more storage of information can, however, continue and will work hand in hand. More exergy cannot be captured or dissipated, but the resources can be reallocated to give more exergy stored and a development toward relatively less dissipation of exergy (Mauersberger's principle). A development toward K-strategists is therefore favoured as they will have bigger size and therefore less specific exergy needs. Microorganisms also shift from r-strategists with quick exploitation of the resources to K-strategists with a slow exploitation of resources (Gerson and

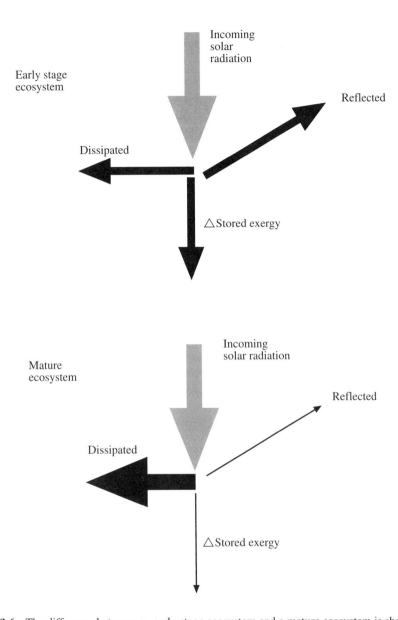

FIGURE 13.6 The difference between an early stage ecosystem and a mature ecosystem is shown. An early stage ecosystem has great advantage by reducing the reflected part of the solar radiation, which is made by increasing the structure of the system (could be measured by the stored exergy). The mature ecosystem requires more exergy for maintenance due to its more developed structure. The total storage of exergy can only be increased by reduction of the exergy used for maintenance of an increasing structure, as the reflected part cannot be reduced (much) further due to physical constraints.

Chet, 1981). Late successional plants with K-strategy produce litter which is poor in nutrients and simple sugar but high in lignine (Heal and Dighton, 1986), which change the physico-chemical environment of the top soils. These processes cause modifications in the composition as well as in the activity of microbial communities, favouring the K-selected organisms. The shift from r-strategists to K-strategists may therefore be considered a process with synergistic effects. Notice that the shift from r-strategists to K-strategists is consistent with the three directions of development presented above. The biomass does not change, but the amount of information and

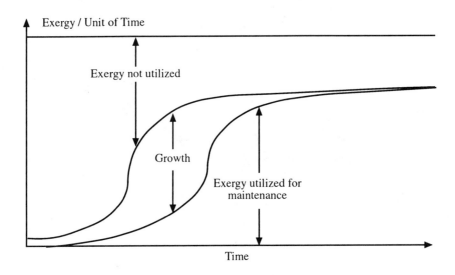

FIGURE 13.7 The development of an ecosystem according to Odum (1969) and to the general description in the text is described by plotting exergy storage vs. exergy captured. The amount of exergy which can be captured is limited to about 80 to 90% of the incoming solar radiating, but the exergy stored can continue by increase of information and a better utilisation of the resources. The graph is consistent with Figure 13.8, where measured or estimated values of the two variables are applied.

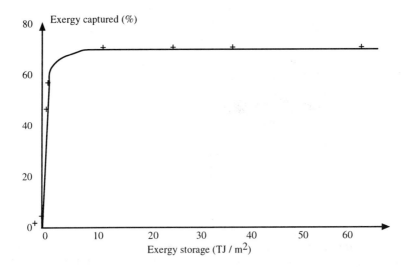

FIGURE 13.8 The exergy captured (taken from Kay and Schneider, 1992, unit: J/m^2 24h) is plotted vs. the exergy stored (unit J/m^2 or J/m^3), calculated from characteristic compositions of the focal eight ecosystems. The numbers from Table 13.3 are applied to construct this plot.

the use of the available resources does. The inflow of exergy, for instance, will, to a lesser extent, be used for maintenance as the K-strategists are bigger in size which means a smaller specific surface and therefore less respiration relative to the biomass.

Figures from satellite measurements support the description of ecosystem development. A forest captures much more exergy than a desert or a grassland, but a 50-year-old forest or a 200-year-old forest or even an old rain forest all capture approximately about 80% of the incoming solar radiation (Kay and Schneider, 1992). It can, however, be shown (Jørgensen, 1997), that the stored exergy

TABLE 13.3
Exergy Utilisation and Exergy Storage

Ecosystem	% Exergy Utilisation	Exergy Storage kJ/m²
Quarry	6	0
Desert	2	73
Clear cut	49	594
Grassland	59	940
Fir plantation	70	12700
Natural forest	71	26000
Old deciduous forest	72	38000
Tropical rain forest	70	64000

increases when a forest gets older, and a rain forest has more exergy stored than a temperate forest due, for instance, to higher biodiversity (more information).

Table 13.3 shows the exergy utilisation for different types of systems (Kay and Schneider, 1992). In the same table the exergy storage is shown for some typical "average" systems. The results of Table 13.3 are plotted in Figure 13.8. As seen, there is a steep linear relationship between exergy storage and exergy capture when the system is under development from the early to the mature stage. A mature system may still develop its exergy storage, although the exergy captured has attained the practical maximum of about 70 to 80% of the total solar energy received by radiation. This points toward Ex_{bio} rather than ΔEx_{cap} or degradation of exergy as a general optimiser, although the optimisation is parallel when the ecosystem is under development from the early stage to the mature stage.

A parallel to economic systems may be used to illustrate the difference between Ex_{bio} and ΔEx_{cap}. When an enterprise or a country is under development, it is important to increase the turnover of the unit, which is a parallel to ΔEx_{cap} and to maximum power. The turnover is of course dependent on the investment already made. In the long run it is more important for the firm (or country) to increase the *active* investment in infrastructure, production facilities, sales network, and so on. The enterprise or country making the most useful investments will be in the best position for competition. At a particular point in this development the investment in education and information becomes crucial — a clear parallel to the development of ecosystems, where investment in information seems particularly beneficial for the mature ecosystem.

It has furthermore been attempted to describe the development shown in Figure 13.8 by application of a model. The model accounts for the exergy in KJ per m². It has the following state variables: nutrients, plants (*r*-strategists), plants (*K*-strategists), herbivores, carnivores, and detritus. The photosynthesis (uptake of carbon dioxide from the air) is regulated by the exergy flow of the solar radiation in the sense that maximum 80 to 90% of the incoming solar to radiation can be captured by the plants to cover the maintenance (respiration) and the growth and replace the grazing of the herbivores. The model results, see Figure 13.9, show that the *r*-strategists have a fast growth in the first phase, but are, in the long run, replaced by the *K*-strategists. The model also shows a rapid growth in the first phase and a very slow growth of the stored exergy at a later stage, when the incoming flow of exergy in form of solar radiation becomes limiting. These results are consistent with Johnson (1990). He has found that when ecosystems are relatively isolated, competitive exclusion results in a relatively homogeneous system configuration that exhibits low dissipation.

In contrast, physical dissipation is much higher in ecosystems that are less fully developed. Johnson (1990, 1995) concludes that "ecosystem structure is a function of two antagonistic trends: one toward a symmetrical state resulting in least dissipation and the other toward a state of maximum attainable dissipation." The latter is obtained by a rapid growth of biomass and structure, implying a more effective capture of the exergy contained in the solar radiation. The first trend is obtained by a

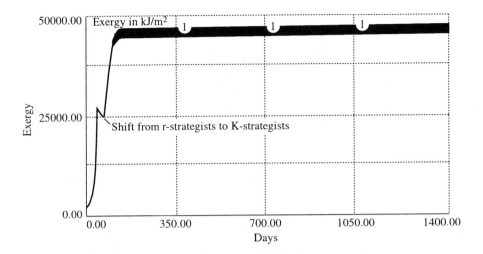

FIGURE 13.9 The graph gives the exergy as f(time) for an ecosystem model limited first by nutrients and later by the amount of exergy captured for the photosynthesis. The model has two types of plants, r-strategists with a fast growth and respiration, and K-strategists with a slower growth but also a lower mortality and respiration. After the system has reached almost 500,000 kJ/m², the exergy of the system grows very slowly and with some minor oscillations. The graph is consistent with the description of an ecosystem where the first phase is in accordance with Kay and Schneider and the later phase with Prigogine and Mauersberger.

reallocation of the biochemical elements, which, due to mass conservation, limit the amount of biomass and structure. A development toward more effective organisms (less dissipation relative to the biomass which, for instance, is the case for bigger organisms; see also Straskraba et al., 1998), requiring less exergy for maintenance, will imply that although (almost) the same amount of exergy is captured, the stored exergy can still increase. Inevitably, the configurations of the interactions become more mutualistic, self-reinforcing, and self-entailing.

This is consistent with Salthe (1985) where three phenological rules of thermodynamically open systems are proposed. As the system develops from immaturity through maturity to senescence,

1. There is an average monotonic decrease in the intensity of energy flow (flow per unit mass) through the system. The gross energy flow increases monotonically against a limit.
2. There is a continual, hyperbolic increase in "complicatedness" (= size + number of types of components + number of organisational constraints), or, generally, an ever-diminishing rate of increase in stored information.
3. There is an increase in its internal stability (its rare of development slows down). originally stated as Minot's law in developmental physiology.

Figure 13.10 attempts, according to Salthe (1985), to summarise these considerations. Clearly the stored exergy measuring the information and biomass (Salthe also uses the expression "the complicatedness") increases over time, while the weight-specific energy flow, after an initial increase, decreases monotonically due to an increasingly better utilisation of the available energy resources.

Ulanowicz (1997) has proposed to approach the two hypotheses, the Fourth Law of Thermo-dynamics and the extended version of the Second Law of Thermodynamics, both described above, by the following formulation: the gradients are steadily broken down in the environment (exergy decreases on this level (the level of cells or of biochemical processes) to be able to build (store) more exergy in the organisms and on the ecosystem level. It is consistent with the criteria that life is a struggle for increase of the gradients, corresponding to growth. It implies increased order and organisation. A few examples illustrate this idea in more detail.

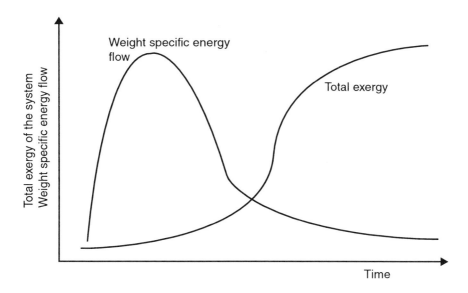

FIGURE 13.10 The development of ecosystems is shown. The total exergy and "complicatedness" increases monotonically. The exergy flow per unit of mass decreases monotonically after an initial increase (Salthe, 1985).

1. The mentioned example of the sequence of microbial oxidation processes according to the amount of stored exergy in form of ATPs, see Table 13.1, also means that the detritus is broken down further to carbon dioxide and water by the oxidation processes first selected, while anaerobic organisms produce methane, which still contains exergy compared with carbon dioxide and water.
2. The major use of exergy for maintenance is applied by plants and trees for respiration and evapotranspiration, which takes place on the cellular level. The coverage of the exergy for maintenance of the plant biomass is crucial however for the development of organisation on the ecosystem level.
3. Caves and the abyss without any light have life, for instance, blind fish. Detritus is broken down to carbon dioxide and water to ensure the highest possible organisation (exergy) under these unique circumstances.
4. The evolution of many species has been toward bigger body size (Raup and Sepkowski, 1982) which means more stored exergy and higher energy demands per organism, but relatively less exergy degradation to the stored exergy. Specific exergy dissipation decreases.

It is also of importance, as already shown, to notice the basic difference between the two descriptors, exergy storage and exergy degradation. The latter is a differential quantity and is measured in a unit which includes time, for instance, $kJ/(m^2 24h)$. As the available exergy per unit of time in the form of solar radiation has an upper limit, exergy degradation per unit of time does not seem an appropriate descriptor of long-term development of ecosystems. Exergy storage is an integrated quantity, measured in a unit not including time, for instance, kJ/m^2. It has, in principle, no upper limit, as the storage of exergy can steadily increase. Each new day brings more exergy, which may be added to the already obtained storage of exergy.

Patten and Fath (2000) have shown that increased cycling implies increased exergy storage at steady-state conditions. It was shown as a general mathematical consequence of steady-state network theory. Figure 13.11 illustrates the results by comparison of two networks at steady state with and without cycling. It is presumed that the process rates are first order donor determined reactions. The cycling implies inevitably that the compartments, the exergy stored, increase in size. The input of exergy (exergy captured) is the same in the two situations that correspond to the

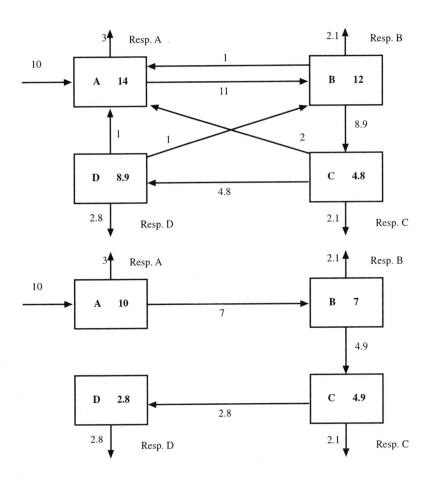

FIGURE 13.11 Two networks at steady state are compared. It is presumed that process rates are first-order, donor-determined reactions. The input of exergy = the total dissipation of exergy is the same in the two situations, but while A has no cycling, B has cycling, B has cycling corresponding to a better utilisation of the first path of exergy. As seen, the flow of exergy through the system and the exergy of the biomass are higher in case B than in case A. The specific exergy dissipation is lower in case B than in case A.

dissipation of exergy as the two networks are in steady state. The specific dissipation of exergy, understood as the exergy dissipated per unit of biomass, will of course decrease as the cycling increases, while the flow of exergy through the system will increase due to the cycling and thereby better utilise the first path of exergy.

Ecosystems are dynamic. It is therefore a simplification to assume steady state. On the other side, steady state may be interpreted as a freezing of the system in a situation assuming that ΔEx_{bio} is 0 (see Equation (25)), which gives a realistic comparison of two situations: with and without cycling. It seems therefore appropriate (according to Patten and Fath, 2000) to set up the following pertinent hypothesis.

Increased cycling implies that the exergy storage, the exergy through flow and the ascendancy increase simultaneously with a decrease in specific exergy dissipation. It is in accordance with the interpretation of the tentative Fourth Law of Thermodynamics.

The role of natural disturbances such as fire or storms should be discussed in this context. When a forest is burned (for details see Botkin and Keller, 1995), complex organic compounds are converted into inorganic compounds. Some of the inorganic compounds from the wood are lost as particles of ash that are blown away or as vapours that escape into the atmosphere and are widely distributed. Other compounds are deposited on the soil surface.

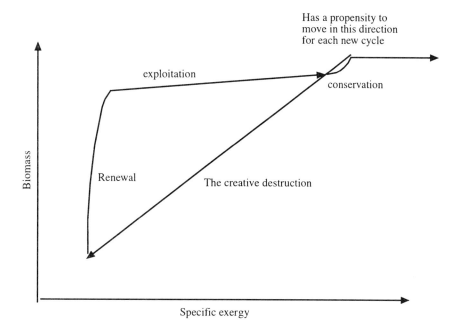

FIGURE 13.12 Holling's four phases are described by use of biomass and specific exergy. This presentation is inspired by Ulanowicz (1997).

These are highly soluble in water and readily available for vegetation uptake. Therefore, immediately after a fire there is an increase in the availability of chemical elements, which are taken up rapidly, especially if there is a moderate amount of rainfall.

The pulse of inorganic nutrients can then lead to a pulse in growth of vegetation. This in turn provides an increase in nutritious food for herbivores. The pulse in chemical inorganic elements can therefore have effects that extend through the food chain. Challenges to find new opportunities to move even further away from thermodynamic equilibrium are therefore created which may explain that the natural disturbances may have a long-term positive effect on the growth of ecosystems in the broadest sense of this concept.

This description is according to Holling's cycle (Holling, 1986). Figure 13.12 is a modified version of this cycle presented by Ulanowicz (1997), where Holling's cycle is modified in almost the same way as Figure 13.12. However, the x-axis is not specific exergy = exergy/total biomass in Ulanowicz's presentation of Holling's cycle, but mutual information of flow structure. The basic idea is the same. The renewal phase corresponds to rapidly increased biomass, exploitation phase to a rapid increase in the level of information, and conservation to a very slow increase in both biomass and information. The destruction phase will, due to an external impact (forcing function), reduce both the amount of biomass and the information stored in this biomass, but new possibilities are thereby created for the utilisation of emergent mutations and sexual recombinations. After each round in Holling's cycle, the biomass can hardly be higher as it is limited by the presence of essential elements, but due to the current test of new mutations and sexual recombination which takes place, new and perhaps better combinations of properties will emerge. Consequently, there is a propensity that the exergy and the specific exergy may increase for every new round in the Holling's cycle.

It is interesting in this context that scientists and visitors to Yellowstone National Park find the rapid recovery of the park remarkable. The park flourishes a decade after fires. Across Yellowstone National Park, tens of millions of trees have sprung from the ashes of what the U.S. assumed was an ecological disaster. The fires of 1988 opened the forest canopy to abundant sunlight and enriched the soil with nutrients from dead tress. Now, 12 years after the fires, the branches of the robust

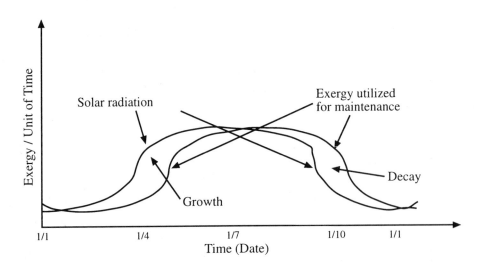

FIGURE 13.13 The seasonal fluctuation of exergy availability as solar radiation and of exergy utilised for maintenance are shown.

saplings, which were not supposed to reproduce for another decade or two, bristle with seed cones. Park officials are even studying whether or not to set controlled burns to clear out more deadwood and other volatile "fuel" in areas missed by the 1988 blazes.

The above-presented development of exergy storage and utilisation assumes that *maximum* storage situations for ecosystems are compared, i.e., for the temperate zone, summer situations are used for the description of development. The seasonal changes, particularly in the temperate and arctic zones, imply that exergy storage and utilisation fluctuate with seasonal changes in exergy availability. The exergy storage is relatively small and may even be negative during winter time, while spring is synonymous with growth in biomass and exergy storage. These seasonal changes in stored exergy and exergy utilisation are shown in Figure 13.13. It is consistent with numerous ecological descriptions of the seasonal variations in vegetation and respiration (see any textbook on ecology).

It has also been examined (Jørgensen, 1995), to find which influence a change in the grazing rate of zooplankton would have on the exergy index of a system consisting of phytoplankton, detritus, nutrients, zooplankton, and fish, usually the applied state variables in eutrophication models. The exergy index is calculated as a function of the grazing rate by the calculation methods presented in the next section. The highest exergy index = (kJ/l) is obtained, see also Figure 13.14, by a realistic grazing rate, gr_{max}. Grazing rates $> gr_{max}$ lead to lower exergy indexes and even chaotic behaviour, while grazing rates $< gr_{max}$ simply give a lower exergy index. At very high grazing rates, the average exergy of the system may be higher than at gr_{max} but the fluctuations of the zooplankton biomass are violent with a high probability to be extinguished. These results are consistent with Kauffman's hypothesis that biological systems have their biggest potential for development (growth) at the edge of the chaos (Kauffman, 1991, 1993). If the grazing rate gets $> gr_{max}$ zooplankton will of course grow faster but it would thereby deplete its own food source and therefore fluctuate between very high and very low biomass concentrations with the probability at the lowest concentrations to be extinguished, for instance, by some random disturbances. So, this example illustrates that zooplankton is forced to take the conditions defined by the other species into account. Phytoplankton as the food source for zooplankton determines the possible growth rate (grazing rate) of zooplankton and the predation by fish is of course also significant for the survival of zooplankton. Similar results are also obtained for other organisms; see, for instance, Jørgensen (1995). Growth rates of organisms adapt to the available resources. This is the highest possible growth rate not causing an over-exploitation of the food source, which may yield chaotic fluctuations. The selection of the growth rate will result in the highest possible (stored) exergy of the system.

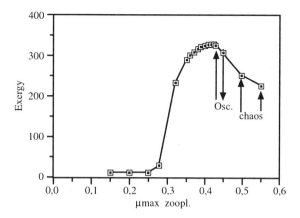

FIGURE 13.14 The exergy is plotted vs. the grazing rate for a realistic model encompassing nutrients, phytoplankton, zooplankton, fish, and detritus.

13.6   ECOSYSTEM PROPERTIES

Table 13.4 summarises the properties and characteristics of developed ecosystems. The hypothetical Fourth Law of Thermodynamics has been used to explain these properties and characteristics. The explanation is included in the last column of the table. This may be considered a support for the hypothetical Fourth Law of Thermodynamics, which does not imply that other system ecological theories are able to explain if not all, then at least most of the observations equally well.

The properties of ecosystems given in Table 13.4 are concerned with the development of the system in its broadest sense by increased organisation and regulation of the system. The better regulation and organisation results in a higher exergy and better exergy utilisation for exergy storage, according to Mauersberger (1995). The trends toward larger organism size explain that the specific entropy, i.e., the entropy production relative to the total biomass, is reduced; see also Mauersberger (1983, 1995), and Jørgensen (1997).

According to Kay (1984) an ecosystem reaches an operating point, which represents a balance between (1) the thermodynamic forces which drive it away from thermodynamic equilibrium and (2) the environmental forces (dissipation of exergy including the exergy produced by metabolism) which tend to disrupt development and drive the system back toward equilibrium. This point of balance between the two forces may be called the optimum operating point. A steady state will, however, never be reached, as the forcing functions are continuously changing. The steady state is a moving target.

If stress is introduced into the system, it will be driven to a new optimum operating point. If the stress is removed, the system will not return exactly to the previous optimum operating point, because the system will have changed its history and the same combination of external and internal factors will, with extremely high probability, never occur again; compare with the steady-space model in Patten et al. (1997). An ecosystem will therefore never return exactly to the same operating point again, because the history and the combination of internal and external factors will never be the same again.

Ecology deals with irreducible systems (Wolfram, 1984a,b; Jørgensen, 1990, 1992a,b, 1997). We cannot design simple experiments that reveal a relationship that can in all detail be transferred from one ecological situation and one ecosystem to another situation in another ecosystem. That is possible, for instance, with Newton's gravity laws because the relationship between forces and acceleration is reducible. The relationship between force and acceleration is linear. Growth of living organisms on the other hand is dependent on many *interacting* factors, which are functions of time. Feedback mechanisms will simultaneously regulate all the factors and rates, and they also interact and are functions of time (Straskraba, 1980).

TABLE 13.4
Characteristics of Developed Ecosystems That Are in Accordance with the Tentative Fourth Law of Thermodynamics

Characteristics	Explanation
1. High biomass	To utilise the available nutrients and water to get the highest possible exergy
2. High level of respiration and evapotranspiration	To maintain the system far from thermodynamic equilibrium
3. Big gradients are developed	The system moves as far away as possible from thermodynamic equilibrium
4. High level of information	To maximally utilise the flow of exergy and the resources
5. High level of specialisation and differentiation	To utilise the heterogeneity in space and time to gain highest possible level of exergy
6. High level of adaptation and buffer capacities	To meet the challenge of the changing forcing functions
7. High complexity of network and high level of organisation	A consequence of the first four characteristics
8. Big size of (some) organisms	To minimise specific entropy production and there by cost of maintenance when the exergy flow becomes limiting
9. Highly developed history	Caused by all the processes of development
10. High indirect/direct effect ratio	A consequence of the complex network
11. Irreversible processes	A consequence of the history of the system
12. Both bottom-up and top-down regulation are applied	To utilise all available avenues to build as much dissipative structure as possible
13. Symbiosis developed	Two or more species move further away from thermodynamic equilibrium
14. Many different strategies are applied	To utilise all available avenues to build as much dissipative structure as possible

An ecosystem consists of so many interacting components that it is impossible to examine all these relationships. Even if we could, it would not be possible to separate one relationship and examine it because the relationship is different when it works in nature with interactions from the many other processes than when we examine it in a laboratory separated from the other ecosystem components.

These observations are indeed expressed in system ecology. A known phrase is "the whole is greater than the sum of the parts" (Allen, 1988), because "everything is linked to everything." It implies that it may be possible to examine the parts by reduction to simple relationships, but when the parts are put together, they will form a whole that behaves differently from the sum of the parts — it will have emergent properties as discussed in Jørgensen et al. (1999).

Allen (1988) claims that this is correct because of the evolutionary potential that is hidden within living systems. The ecosystem has within itself the possibilities of becoming something different, i.e., of adapting and evolving. The evolutionary potential is linked to existence of microscopic freedom, represented by stochasticity and nonaverage behaviour, resulting from the diversity, complexity, and variability of its elements.

Microscopic diversity underlies the taxonomic classification. It adds only to the complexity to such an extent that it will be completely impossible to cover all the possibilities and details of the observed phenomena. We attempt to capture at least a part of the reality by use of models. It is not possible to use one or a few simple relationships, but a model seems to be the only useful tool when we are dealing with irreducible systems.

A little advantage is sufficient to give a far better probability of survival, which is easily demonstrated by simple calculations or by the use of simple models (see, for instance, Jørgensen, 1997).

The selection pressure will rapidly favour even the slightest advantage and an advantage in survival of the young ones is incredibly important for preservation of the selfish gene (Dawkins, 1989).

Biological growth is dependent on a number of factors (see also Chapter 14):

- Some 30 nutrients and micronutrients,
- A number of climatic factors (temperature, wind, etc.) including the amount of energy flowing through the system,
- The transport processes in the systems which are, of course, again dependent on a number of other factors, including the climatic factors,
- All the other biological components in the system.

All these factors are functions of time and space. The conditions on formation of life *are* extremely heterogeneous in time and space. The process of biomass growth never has the same conditions in time or space.

Consequently, there is a need for many different solutions to utilise the exergy flow, and it is therefore not surprising, given the long period available for development, that the ecosystems have high complexity and many different and good solutions. Nor is it surprising that many different mechanisms to find a wide range of solutions have been tested. It may explain why genes also can be modified by the organisms, i.e., that the organisms attempt to change the environment to modify the selection pressure and why a strong coevolution in general has taken place (Patten, personal communication, 1993).

Species in the same ecosystem have lived together for a long time and the influences from the other species have been among the many factors that have determined the selection pressure. The species have "polished each other as an old married couple." As ecosystems are open, the various ecosystems have furthermore influenced each other and have exchanged "knowledge" in the form of genes. Emigration and immigration are common processes in ecosystems.

Symbiosis is favoured, because it gives advantages to two or more species simultaneously and has thereby been considered as a part of the factors mentioned above. A more and more complex interrelationship among the biological components evolves and implies that a complex network develops over time, where the components are more dependent on each other. It explains why the indirect effect becomes so dominant (Patten, 1991), and why the Gaia effect (Lovelock, 1988) has become more pronounced. The presence of a dominant indirect effect can be interpreted as *mutual implication* of *all* or at least most components in a network. Direct negative effects become positive indirect effects and there seems to be a network mutualism. This, again, together with the long period which has been available for evolution, may be able to explain how the Gaia effect emerged (Patten, 1991).

Ecosystems are, however, not entirely deterministic. They are as discussed in Jørgensen et al. (1999) and in Chapter 11 ontic openness. There is a propensity to select pathways determined by the tentative Fourth Law of Thermodynamics. The formation of new pathways contains more random elements, for instance, due to mutations, new sexual recombinations of genes, and dependence on random forcing functions. The development of ecosystems toward higher exergy levels should therefore be seen as a propensity not as a fully determined direction of development. It is consistent with Monod, who claims that development of biological systems is based on a combination of randomness and necessity, the latter being that the system has to follow the tentative Fourth Law of Thermodynamics as a direction of propensity, while new possible pathways for development randomly emerge to a certain extent.

Notice that the change in stored exergy, i.e., the exergy embodied in the organisms, their information, and the structure of the ecosystem, is not necessary ≥ 0. It depends on the changes in the resources of the ecosystem. The proposition claims, however, that the ecosystem has a propensity to reach the highest possible exergy level under the given circumstances and with the available genetic pool ready for this attempt (Jørgensen and Mejer, 1977, 1979). If the exergy needed for maintenance is more than the incoming flux of available exergy, the maintenance will be covered by the stored exergy, which therefore will decrease, which is the situation during the fall in the temperate zone; see Figure 13.13. If the exergy needed for maintenance is lower than the available exergy flux, the surplus exergy will be stored. The tentative Fourth Law of Thermodynamics implies that the selection among the possible processes will attempt to minimise the loss of stored exergy or maximise the gain of stored exergy. Thereby, the maintenance or even growth of the stored exergy will be ensured. That will also guarantee that the highest possible gradient is maintained and the highest possible amount of exergy is captured from the solar radiation, as the latter depends on the size of the ecological structure, which can be measured by the stored exergy.

13.7   STRUCTURAL DYNAMIC MODELS OF ECOSYSTEMS

The idea of this new generation of models is to continuously find a new set of parameters (limited for practical reasons to the most crucial (= most sensitive) parameters), which is better fitted for the prevailing conditions of the ecosystem. "Fitted" is the ability of the species to survive and grow, which may be measured by the use of exergy (see Jørgensen, 1982, 1986, 1988, 1990; Jørgensen and Mejer, 1977, 1979; Mejer and Jørgensen, 1979). Figure 13.15 shows the proposed modelling procedure, which has been applied in the cases presented below.

Exergy has previously been tested as a "goal function" for ecosystem development; see, for instance, Jørgensen (1986). However in all these cases the model applied did not include the "elasticity" of the

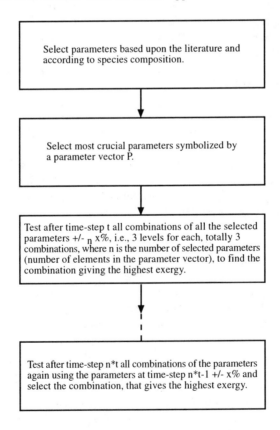

FIGURE 13.15 The procedure used for development of a model with dynamic structure, i.e., a model with currently changing parameters reflecting the adaptation and the changes in the species composition according to the changed conditions.

system, obtained by use of variable parameters and therefore the models did not reflect real ecosystem properties. A realistic test of the exergy principle would require the application of *variable* parameters. If the variation of the parameters resulting from an optimisation is in accordance with the *observed* structural changes (for instance, from one size of phytoplankton to another), it is considered a good support for the tentative Fourth Law of Thermodynamics.

The use of exergy calculations to continuously vary the parameters has now been used in nine cases. All nine cases have been successful in the sense that the variations of the parameters obtained by the exergy optimisation have approximately been in accordance with the observation of structural changes. It has therefore been considered of importance to include some of the structural dynamic case studies in this chapter focusing on the tentative Fourth Law of Thermodynamics.

Four biogeochemical cases of structural dynamic modelling using exergy as goal function are presented in this section. They illustrate clearly the idea behind this type of model and why a good accordance between observations and simulations may be considered a support of the tentative Fourth Law of Thermodynamics. For the five other case studies, see Jørgensen and Padisák (1996), Jørgensen (1994a and 1997), and Jørgensen and de Bernardi (1997).

In the first biogeochemical case the growth of algae was used as the only variable parameter (Jørgensen, 1986). This gave a significantly improved validation of the model, which encouraged further investigation of the possibilities of developing and applying such new modelling approaches. The maximum growth rate μ-max. and the respiration rate, set equal to $0.15 * \mu$-max, were changed in the model relatively to the previously found value by calibration $\mu - c$,

$$\mu\text{-max} = F * \mu - c. \tag{26}$$

The model was run for several F-values and several levels of phosphorus input. The result is plotted in Figure 13.16. As seen, the μ-max, gives maximum exergy decreases when P increases, which is in accordance with ecological observations. When nutrients are scarce, the phytoplankton species compete on the uptake rates of nutrients. Smaller species have a faster uptake due to a greater specific surface and they grow more rapidly. On the other hand, high nutrient concentrations will not favour small species, because the competition focuses on avoidance of grazing, where a greater size is more favourable. The results were used to improve the prognosis published in Jørgensen (1986) by introducing a continuous change of the parameters according to the procedure in Figure 13.15. The validation of the prognosis gave the result that the standard deviation between

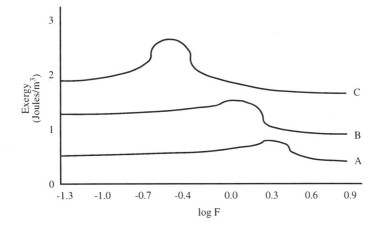

FIGURE 13.16 Exergy is plotted vs. F at different levels of P in a lake. As seen, the maximum exergy occurs at different F-values for different P-loadings. A is an oligotrophic situation that corresponds to a nutrient input of 0.04 mg P/l and 0.3 mg N/l. B corresponds to a nutrient input of mg P/l and 8 mg N/l. C corresponds to a hypereutrophic situation, where the input is 2 mg P/l and 16 mg N/l.

model and measurement was reduced from about 72% to about 33%, although it was also needed to introduce silica as nutrient to account for the appearance of diatoms; see Jøgensen (1986).

The second case where a structural dynamic model was developed by use of exergy as goal function is a lake study, too. The results from Søbygaard Lake (Jeppesen et al., 1990) are particularly fitted to test the applicability of the described approach to structural dynamic models. As an illustration to structural dynamics of ecosystems and the possibilities to capture the flexibility of ecosystems, the case study of Søbygaard Lake will be presented in detail.

Søbygaard Lake is a shallow lake (depth 1 m) with a short retention time (15 to 20 days). The nutrient loading was significantly reduced after 1982, namely, for phosphorus from 30 g P/m²y to 5 g P/m²y. The reduced load did, however, not cause reduced nutrients and chlorophyll concentrations in the period 1982 to 1985 due to an internal loading caused by the storage of nutrients in the sediment (Jeppesen et al., 1990).

However, radical changes were observed in the period 1985 to 1988. The recruitment of plank-tivorous fish was significantly reduced in this period due to a high pH caused by the eutrophication. As a result zooplankton increased and phytoplankton decreased in concentration (the summer average of chlorophyll a was reduced from 700 µg/l in 1985 to 150 µg/l in 1988). The phytoplankton population even collapsed in shorter periods due to extremely high zooplankton concentrations. Simultaneously, the phytoplankton species increased in size. The growth rate decreased and a higher setting rate was observed (Kristensen and Jensen, 1987). The case study shows, in other words, pronounced structural changes. The primary production was, however, not higher in 1985 than in 1988 due to a pronounced self-shading by the smaller algae in 1985. It was therefore very important to include the self-shading effect in the model. This was not the case in the first model version, which gave wrong figures for the *primary production*. Simultaneously, sloppier feeding of the zooplankton was observed, as zooplankton was shifted from *Bosmina* to *Daphnia*.

The aim of the study is to be able to describe by use of a structural dynamic model the continuous changes in the most essential parameters using the procedure shown in Figure 13.15. The data from 1984 to 1985 were used to calibrate the model, and the two parameters that tended to change from 1985 to 1988, got the following values by this calibration.

Maximum growth rate of phytoplankton:	2.2 day-1
Setting rate of phytoplankton:	0.15 day-1

The state variable fish-N was kept constant = 6.0 during the calibration period, but an increased fish mortality was introduced during the period 1985 to 1988 to reflect the increased pH. The fish stock was thereby reduced to 0.6 mg N/l.

A time-step of t = 5 days and $x\%$ = ±10% was applied to examine the change of parameters needed to currently obtain the highest exergy. This means that nine runs were needed for each time-step to select the parameter combination that gives the highest exergy.

The results are shown in Figure 13.17 and the changes in parameters from 1985 to 1988 (summer situation) are summarised in Table 13.5. The proposed procedure (Figure 13.15) is able to approx-imately simulate the observed change in structure. The maximum growth rate of phytoplankton is reduced by 50% from 2.2 to 1.1 day^{-1}, which is approximately according to the increase in size. It was observed that the average size was increased from a new 100 μm³ to 500 to 1000 μm³, which is a factor of about 2 to 3 (Jeppesen et al., 1990). It would correspond to a specific growth reduction by a factor $f = 2^{2/3} - 3^{2/3}$ (see Peters, 1983 or Jørgensen, 1994b).

It means that:

$$\text{growth rate in 1988} = \text{growth rate in 1985} / f \qquad (27)$$

where f is between 1.58 and 2.08, while 2.0 is found by use of the structural dynamic modelling approach.

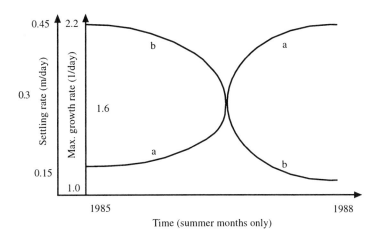

FIGURE 13.17 The continuous changed parameters obtained from the application of a structural dynamic modelling approach on Søbygaard Lake are shown: (a) covers the settling rate of phytoplankton and (b) the maximum growth rate of phytoplankton.

TABLE 13.5
Parameter Combinations Giving the Highest Exergy

	Maximum Growth Rate (day^{-1})	Setting Rate (m*day^1)
1985	2.0	0.15
1988	1.2	0.45

Kristensen and Jensen (1987) observed that the settling was 0.2 m day^{-1} (range 0.02 to 0.4) in 1985, while it was 0.6 m day^{-1} (range 0.1 to 1.0) in 1988. By the structural dynamic modelling approach an increase was found from 0.15 to 0.45 day^{-1}, the factor being the same — three — but with slightly lower values. The phytoplankton concentration as chlorophyll-a was simultaneously reduced from 600 to 200 $\mu g/1$, which is approximately according to the observed reduction. All in all it may be concluded that the structural dynamic modelling approach gave an acceptable result and that the validation of the model and the procedure in relation to structural changes was positive. It may, however, be necessary to expand the model to account for *all* the observed structural changes, including zooplankton, to be able to demonstrate a complete case study. The structural dynamic modelling approach is of course never better than the model applied, and the presented model may be criticised for being too simple and not accounting for the structural dynamic changes of zooplankton.

For further elucidation of the importance to introduce a parameter shift, Researchers have tried to run the 1985 situation with the parameter combination found to fit the 1988 situation, and vice versa. These results are shown in Table 13.6. The results demonstrate that it is of great importance to apply the right parameter set to given conditions. If the parameters from 1985 are used for the 1988 conditions, a lower exergy is obtained and the model to a certain extent behaves chaotically while the 1988 parameters used on the 1985 conditions give a significantly lower exergy. If the parameter set from 1985 was applied for the entire period, 1985 to 1988, it was completely impossible to obtain accordance between observations and simulations.

This is demonstrated in Figure 13.18, where the exergy as a function of time is plotted, when current parameter changes are applied and compared with the 1985 parameters maintained throughout the entire period. For the latter case violent fluctuations of the exergy are observed. Sometimes the

TABLE 13.6
Exergy and Stability by Different Combinations
of Parameters and Conditions

Parameter	Conditions	
	1985	1988
1985	75.0 Stable	39.8 (average) violent fluctuations; chaos
1988	38.7 Stable	61.4 (average) only minor fluctuations

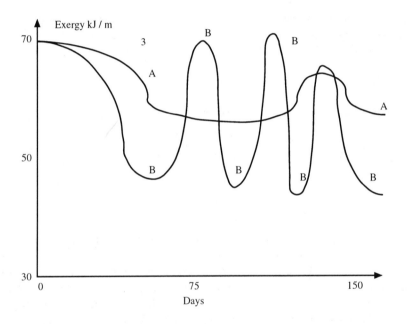

FIGURE 13.18 Exergy changes by two simulations: A is continuously changed parameters and B is the 1985 parameter set maintained. Start of the simulation 1986 summer situation with current changes to the summer situation in 1987.

exergy is above the level of the simulation based on continuously changed parameters, but the final exergy level is significantly lower. As discussed above the change in parameters may play a role in avoidance of chaotic conditions of the system.

The results of the two discussed cases show that it is important for ecological and environmental models to contain the property of flexibility, which we know ecosystems possess. If we account for this property in the models, we obtain models that are better able to produce reliable predictions, particularly when the forcing functions on the ecosystems change and thereby provoke changes in the properties of the important biological components of the ecosystem. In some cases we get completely different results when we apply a continuous change of the parameters compared with the use of fixed parameters. In the first case we get results that are more in accordance with our observations. This is not surprising, as the parameters do actually change in the natural ecosystems.

The property of dynamic structure and adaptable parameters is crucial in our description of ecosystems and should therefore be included in all descriptions of the system properties of ecosystems. The few examples presented here show that it is feasible to account for the adaptability of the properties in models, although a more general experience is needed before clear recommendations on the application can be given.

Lake Annone in Italy offers another interesting case study for structural dynamic modelling. A mass mortality of the most abundant zooplankton-phagous fish due to an infectious gill disease eliminated almost the entire population of this species in the Eastern basin of Lake Annone, in August 1975, while no mortality occurred in the Western basin; see de Bernardi and Giussani (1978). A high density of the planktivorous fish, *Alburnus alburnus alborella*, the bleak, was observed in the Eastern basin, while an almost normal density was observed in the Western basin. Before the fish kill during August 1975 in the Eastern basins a low density of *Daphnia* was sustained by individuals of very small size (this implies a low filtering rate). One month after the fish kill, the *Daphnia* population in the Eastern basin increased rapidly both in density and in individual size reaching the same size values as in the Western basin with low fish density. In 1976 the *Daphnia* population in both basins showed a high density and very important seasonal fluctuations synchronous but opposite the phytoplankton density, suggesting grazing controlled (prey-predator) phytoplankton and zooplankton (mainly *Daphnia*) populations. The individual size of the *Daphnia* population was maintained significantly larger in both basins than the size observed in the Eastern basin before the fish kill; see de Bernardi and Giussani (1978). The described event offers an excellent opportunity to develop and test a structural model and thereby to study the influence of fish predation on planktonic systems (Jørgensen and de Bernardi, 1997). The model should attempt to simulate the shift in growth rate of zooplankton and thereby in the size as a consequence of the mass fish mortality: from about 1000 μm to about 1600 μm; see de Bernardi and Giussani (1978). The questions to be answered by this model exercise may be formulated as follows:

A. Is the model able to simulate the summer situation in the Eastern basin of the lake before and in 1975 and 1976? It has been chosen to limit the test of the model to the summer situation, where the best data are available.
B. Is the model able to simulate the difference in size indicated above which, in accordance to the allometric principles (see, for instance, Peters, 1983), corresponds to a factor $(1.0/1/6)^{-1} = 1.6$ for the growth rate or the grazing rage, when we are using the electivity vs. the prey size for vertebrates found by Zaret (1980)?

The model applied for this case study is a general eutrophication model with nutrients, detritus, phytoplankton, zooplankton, and fish as state variables, as it has been presented several times throughout this volume. The most characteristic features of the model should be mentioned.

1. The sum of inorganic phosphorus and nitrogen are covered by the state variable denoted nutrient. It seems feasible in this case to consider the two main nutrients as one state variable, as the ratio of nitrogen to phosphorus in the lake water is about 6 to 9 as in phytoplankton.
2. The growth of phytoplankton is expressed as ten times the uptake of nutrients, corresponding to an uptake of carbon and other elements, which is ten times the uptake of the two nutrients, P and N. About 91% of the organic matter is converted in accordance with the model to inorganic matter, which is nonnutrient (understood as P and N).
3. The fish predation accounts for the electivity by dividing the predation rate by the grazing rate in the exponent 2. The grazing rate is considered proportional to the size (volume) in 2/3 or to the length in 2; see Peters (1983), which implies that the specific grazing rate, i.e., the grazing rate relatively to the weight, which is the one used in the model, where it is denoted just grazing rate, is therefore proportional to the length in the exponent -1, or

to the volume in $-1/3$. The predation rate becomes therefore proportional to the length 2 in accordance with Zaret (1980).

4. The exergy is currently calculated as detritus $+3.5$ * phytoplankton $+35$ * zooplankton $+325$ * fish.
5. One third of the grazing and the predation goes directly to detritus as it accounts for the nondigested part of the food.

The model has been used to simulate the summer situation in the Eastern basin of Lake Annone in 1975 in accordance with de Bernardi and Giussani (1978). A wide spectrum of the following parameters has been tested: growth rates of fish, zooplankton, and phytoplankton, and mortality rates of fish, zooplankton, and phytoplankton to find the combination that gives the best accordance between observed and modelled values for the state variables during the summer 1975. The difference between the Eastern basin during the summer of 1975 and the Eastern basin in the summer 1976 was simulated by using an abnormally high fish mortality coefficient, 0.4 in the latter case, compared with 0.01 before the observed high fish mortality took place. A significant difference in the fish population between the two cases results. As we know with the growth rate of zooplankton, the grazing rate is the parameter that reflects the reaction to the massive fish mortality. This parameter has been chosen to be changed in accordance with the maximum exergy calculated for the entire model. The structural dynamic modelling approach as represented in Figure 13.15 has now been used for this parameter to attempt to simulate the 1976 summer situation. The model is rather simple compared with many other eutrophication models; see Jørgensen (1994b). The available data are, however, not very detailed, and it seems most appropriate to use the model to simulate only the summer situation and to apply a not too complex model. Figures 13.19 and 13.20 (reproduced from Jørgensen and de Bernardi, 1997) give the results of the simulations with normal and abnormal high fish mortality with different zooplankton grazing rates. The average summer level of exergy and biomass are shown. It is interesting that the highest exergy coincides with the zooplankton growth rate = grazing rate giving a good accordance with the measured data also for the summer of 1976 (the summer average of zooplankton and phytoplankton was found to be about 20 mg/l, respectively, 1,2 mg/l).

These results support the idea to use exergy as a orientor, able to give the combination of properties = the combination of parameters that is best fitted to the conditions. The exergy of the

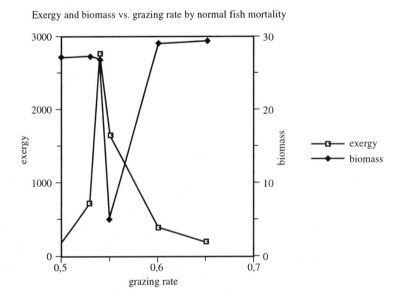

FIGURE 13.19 Exergy (J/l) and biomass (mg/l) are plotted vs. grazing rate for the calibrated model of Lake Annone.

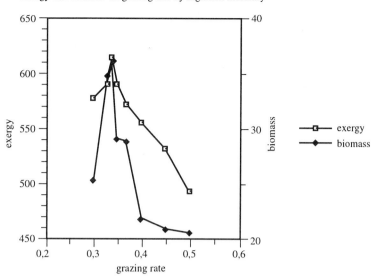

FIGURE 13.20 Exergy (J/1) and biomass (mg/1) are plotted vs. grazing rate for the calibrated model of Lake Annone.

Eastern basin in 1976 is of course significantly lower than in 1975 due to the lack of fish. It is seen that the grazing rate which should be selected in accordance with the highest level of exergy before 1975 is 0.54 (1/24 h), while it is 0.34 (1/24 h) in 1976. The ratio is 1.59. In accordance with the shifts in size — from 1000 μm to about 1600 μm — a 1.6 times lower grazing rate should be expected, which indicates a good accordance between the observed shift in zooplankton size and the simulated shift by the application of exergy as goal function in a structural dynamic modelling approach.

If the total biomass was used, the grazing rate selected would be respectively 0.65 and 0.34 l/day, which would not give a proper description of the observed shift in grazing rate. The exergy seems to be a workable goal function for the development of models with dynamic structure. The results show that it has been possible to develop a model that is able to describe the changes in zooplankton size in Lake Annone from 1975 to 1976 due to an extremely high fish mortality. The use of exergy as goal functions is able to capture the changes in the dynamic structure. It was possible furthermore to assess almost correctly the observed change in size from 1000 to 1600 μm for the zooplankton species, with *Daphnia hyalina* as the dominant species (Jørgensen and de Bernardi, 1997).

The fourth case study of structural dynamic modelling has been applied to predict the space variability of primary producers in a shallow marine water: the Lagoon of Venice. A previous developed model has been modified (Coffaro et al., 1997) to account for the structural dynamics, using exergy as a goal function. Two model components, macroalgae and seagrass, compete for the light and nutrients. An optimisation algorithm was used to find the values of the most crucial parameters that give the highest exergy at different sites of the Lagoon of Venice, characterised by different sets of forcing functions. The state variables are biomass of macroalgae, seagrass leaves, and seagrass rhizomes, the nitrogen content of seagrass and of macroalgae. The competitive differences between macroalgae and seagrass are.

1. Seagrass can take nitrogen up from both water and sediment.
2. Macroalgae can overshadow seagrass in the competition for light.
3. Macroalgae can be washed out of the system by advective transport.

Three forcing functions are used to describe the spatial variability of the Lagoons of Venice: the nitrogen concentration in the water, the hydrodynamics (the kinetic energy during a typical tidal cycle is applied), and the mean depth of the water column.

The surface-volume ratio (S/V was selected as the crucial parameter of macroalgae to represent several physiological properties. The range of S/V was chosen to represent observed macroalgae species in the Lagoon of Venice. The species count is 8000 *Gracilaria confervoides*, 20,000 *Chaetomorfa aerea*, while the upper limit, 40,000, is valid for sheet-like algae as *Ulva rigida* and *Enteromorfa* sp. Opportunistic forms, *r*-strategists, have high S/V ratio, and fast growth rate, but they are sensitive to tissue loss through grazing and abrasion by wave action. Late succession or persistent forms have low S/V ratio, and low growth rate, but they are more resistant to environmental disturbance and have lower nutrient needs. The S/V ratio is related by allometric relationships to the maximum growth rate, maximum rates and half saturation constants for ammonium and nitrate uptake, maximum internal quota of nitrogen, critical internal quota of nitrogen, specific light extinction coefficient, and loss rate and sensitivity to hydrodynamics transport.

The translocation coefficient and the leaf length were selected as the crucial parameters for seagrasses. Three species are present in the Lagoon of Venice: *Zostera marina, Z. noltii*, and *Cymodocea nodosa.*

Observations from 80 stations in the Lagoon of Venice have been applied to validate the structural dynamic modelling approach. It was found that the approach gave the right results with respect to both biomass and community composition in 72 stations. These results are not trivial because they are based on a parameter optimisation of the goal function not on minimisation of the difference between simulated and observed values, and they are therefore *independent* of experimental data. Coffaro et al. (1997) found that the results were not very sensitive to the selection of weighting factors, which implies that it is perhaps not very crucial to the application of the structural dynamic to know the weighting factors with high accuracy.

Coffaro et al. (1997) refer to an attempt to simulate the competition between *Ulva rigida* and *Zostera marina* by application of a traditional competition model without a goal function and with fixed parameters, calibrated to the observations. Approximately the same accordance between model results and observations was obtained as for the structural dynamic model. Coffaro et al. (1997) conclude that the structural dynamic approach is advantageous due to its possibility to give more detailed results by translation of parameters to species (in this case six species were obtained in the structural dynamic model vs. only two species in the traditional competition model) and to its independence of the observations (the use of the goal function replaced the normal calibration).

13.8 A VERY FIRST ATTEMPT TO EXPRESS THE BIOLOGICAL EVOLUTION BY A RELATIVE INFORMATION INDEX

The more species on earth, the more possibilities the ecosphere offers to utilise the available resources and ecological niches and the more possibilities the ecosphere has to adapt to new, expected or unexpected, conditions. The number of species should therefore be an indirect measure of the utilisation of the available resources to move away from thermodynamic equilibrium, or expressed differently, as the width of the ecological information. The same considerations are applied when the survival of species is discussed. Endangered species often have little diversity in the gene pool (Lewin, 1994).

The depth of the ecological information may be expressed by use of the weighting or conversion factors in Table 13.7 for the most developed organism at a given time. We do not know the number of species today — a rough estimation is 10^7 — and we do not have good estimations of the number of species during the evolution. It is therefore proposed to use the number of marine families (Raup and Sepkowski, 1982) as a proper relative measure of the ecological information width. We do

TABLE 13.7
Weighting Factors Applied to Calculate
the Evolution Index Shown in Figure 13.21

−Ma	Applied Weighting Factor
550	30
475	250 (primitive fish)
400	330
375	350 (amphibians)
300	370 (reptiles)
200	400 (mammals)
30	440 (monkeys)
10	500 (apes)
1	716 (human)

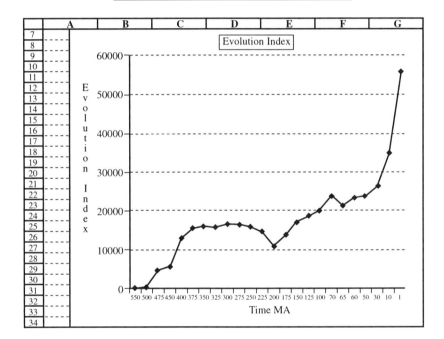

FIGURE 13.21 An introduced evolution index is shown vs. time. The evolution index is a rough estimation of the evolution based on the number of families and the emergence of more and more developed species. Notice the Cambrian explosion, i.e., the rapid increase in the evolution index around 500 Ma ago. It is also possible to see the decrease in the evolution index around 200 Ma and 65 Ma ago due to massive extinctions of many species probably due to a catastrophic event. Compare with Figure 11.21.

know from fossil records approximately when what species emerged. Consequently, we can get a first relative estimation of an evolution index by a multiplication of these possible expressions for the width and the depth of biological information which, in accordance to Section 14.3, is closely related to the relative evolution of the exergy of the biosphere.

Figure 13.21 shows the development over geological time from 550 Ma to today in the above proposed evolution index = (number of marine families) × (weighting factor for the most developed organism at a given time). The applied weighting factors are shown in Table 13.7. Notice in Figure 13.21

the decrease in the evolution index about 220 Ma ago and 65 Ma ago due to extinction of a high number of species (the dinosaurs 65 Ma ago probably by a catastrophic event). The overall trend is, in spite of some fluctuations, probably due to catastrophic events, toward an increase of the evolution index. The plot illustrates of course relatively and very roughly how the ecosphere is moving more and more away from thermodynamic equilibrium by increase of useful knowledge about how to utilise the resources and how to adapt to the currently changed conditions. The amount of biomass has not necessarily changed much over this period of time, except when the terrestrial ecosystems started to be exploited about 400 Ma ago. The amount of inorganic material on earth, available to form complicated and coordinated living matter — this is about 20 elements — has not changed. The change in the exergy of the biosphere is therefore almost solely due to the increased width and depth of information. The plot in Figure 13.21 may therefore also be considered as a plot of the relative change in the exergy of the biosphere. The figure should, however, under all circumstances only be considered a first rough attempt to quantify the evolution.

13.9 CONCLUSIONS AND SUMMARY

A Fourth Law of Thermodynamics has been proposed as a hypothesis. The tentative hypothetical law may be formulated as follows: if a system receives a through-flow of exergy, the system will utilise this exergy flow to move away from thermodynamic equilibrium. If more than one pathway is offered to move away from thermodynamic equilibrium, the one yielding most stored exergy by the prevailing conditions, i.e., with the most ordered structure and the longest distance to thermodynamic equilibrium, will have a propensity to be selected. This hypothesis may also be formulated as an extended version of Le Chatelier's Principle.

The hypothesis, named the tentative Fourth Law of Thermodynamics, has been shown to be in accordance with several observations. Where more pathways are competing, greater support exists for the presented tentative hypothesis, named the Fourth Law of Thermodynamics, as the pathways selected are the ones giving the highest amount of stored exergy.

The hypothesis is also in accordance with the general observations of ecosystem development, described, for instance, by Odum (1969) and including Holling's cycle.

The tentative Fourth Law of Thermodynamics makes it possible to unite Kay and Schneider's maximum dissipation hypothesis, the extended Second Law of Thermodynamics, and Prigogine's and Mauersberger's minimum entropy production hypothesis, because the two hypotheses are valid at two different stages of ecosystem development. The extended Second Law of Thermodynamics can explain that ecosystems in the development stage have rapid development of gradients and structure (increase of exergy storage) and corresponding rapid increase in exergy dissipation to maintain the growing gradients and structure. When the ecosystems on the other side are mature, they capture all the exergy they possibly can from the solar radiation and can therefore not expand the exergy dissipation further. They can, however, by a development toward less exergy dissipation rather than to develop structure, "save" more exergy for storage, mainly in the form of information included by a better allocation of the limited resources. The two stages of development, the early stage and the mature stage, can be explained by the extended Second Law of Thermodynamics and Prigogine's and Mauersberger's minimum dissipation principle, but the observations of *both* the stages are in accordance with the tentative Fourth Law of Thermodynamics.

The hypothesis is furthermore supported by ten examined cases of structural dynamic modelling. Four of the nine cases have been presented here as an illustration of this type of model. In all nine cases, a good accordance between observed and simulated structural changes was obtained. As the simulated structural changes are based on exergy optimisation, these results may be considered a good support of the tentative Fourth Law of Thermodynamics. In addition, it was not possible to get a workable model in most cases if the parameters were not made variable to account for the shift in species composition and adaptation which we actually do know take place in ecosystems.

REFERENCES

Allen, P. M., Evolution: Why the whole is greater than the sum of the parts. In: W. Wolff, C.-J. Soeder, and F. R. Drepper (eds.), *Ecodynamics: Contributions to Theoretical Ecology, Part I: Evolution. Proceedings* of an International Workshop, 19–20 October 1987, Jülich, Germany. Springer-Verlag, Berlin, 2–30, 1988.

Aoki, I., Entropy laws in ecological networks at steady state. *Ecol. Modelling,* 42: 289–303, 1988.

Aoki, I., Ecological study of lakes from an entropy viewpoint Lake Mendota. *Ecol. Modelling,* 49: 81–87, 1989.

Aoki, I., Inclusive Kullback index — a macroscopic measure in ecological systems. *Ecol. Modelling,* 66: 289–300, 1993.

Boltzmann, L., The Second Law of Thermodynamics. (Populare Schriften, Essay no. 3, address to Imperial Academy of Science in 1886). Reprinted in English in: *Theoretical Physics and Philosophical Problems,* Selected Writings of L. Boltzmann. D. Reidel, Dordrecht, 1905.

Botkin and Keller., *Environmental Science, Earth as a Living Planet.* John Wiley & Sons, NY, 630 pp., 1995.

Brown, J. H., Marquet, P. A., and Taper, M. L., Evolution of body size: consequences of an energetic definition of fitness. *Am. Nat.,* 142: 573–584, 1993.

Brown, J. H., *Macroecology.* The University of Chicago Press, Chicago, IL, 1995.

Cavalier-Smith, T., *The Evolution of Genome Size.* Wiley, Chichester. 480 pp., 1985.

Caffaro, G., Bocci, M., and Bendoricchio, G., Structural dynamic application to space variability of primary producers in shallow marine water. *Ecol. Modelling,* in press.

Dawkins, R. D., *The Selfish Gene,* second edition. Oxford University Press, Oxford, 1989.

de Bernardi, R. and Giussani, G., The effect of mass fish mortality on zooplankton structure in a small Italian lake. *Verh. Int. Ver. Limnol.,* 20: 1045–1048, 1978.

Fonseca, J. C. C., Application of nucleus DNA to estimate exergy of organisms. Thesis at the Faculty of Science and Technology, Coimbra University, Portugal, 1999.

Gerson, U. and Chet I., Are allochthonous and autochtonous soil microorganisms r- and K-selected? *Rev. Écol. Biol. Sol.,* 18: 285–289, 1981.

Givnish, T. J. and Vermelj, G. J., Sizes and shapes of liana leaves. *Am. Nat.,* 110: 743–778, 1976.

Heal, O. W. and Dighton J., Nutrient cycling and decomposition in natural terrestrial ecosystems. In: M. J, Mitchell, J. P., Nakas (eds.): *Mocrofloral and Faunal Interactions in Natural and Agro Ecosystems.* Nijhoff & Junk, Dordrecht, 1473 pp., 1986.

Holling, C. S., The resilience of terrestrial ecosystems: Local surprise and global change. In: Clark W. C., and Munn, R. E., (eds.). *Sustainable Development of the Biosphere.* Cambridge University Press, Cambridge, pp. 292–317, 1986.

Jeppesen, E. J., et al. Fish manipulation as a lake restoration tool in shallow, eutrophic temperate lakes. Cross-analysis of three Danish case studies. *Hydrobiologia,* 200/201: 205–218., 1990.

Johnson, L., The thermodynamics of ecosystem. In: O. Hutzinger, (ed.). *The Handbook of Environmental Chemistry,* vol. 1. The Natural Environmental and the Biogeochemical Cycles. Springer-Verlag, Heidelberg, pp. 2–46, 1990.

Johnson, L., The far-from-equilibrium ecological hinterlands. In: B. C. Patten., S. E. Jørgensen, and Auerbach, S. I. (eds.). *Complex Ecology. The Part-Whole Relation in Ecosystems.* Prentice-Hall PTR, Englewood Cliffs, NJ, pp. 51–104, 1995.

Jørgensen, S. E., A holistic approach to ecological modelling by application of thermodynamics. In: W. Mitsch et. al. (eds.). *Systems and Energy.* Ann Arbor, MI, 1982.

Jørgensen, S. E., Structural dynamic model. *Ecol. Modelling,* 31: 1–9, 1986.

Jørgensen, S. E., Use of models as an experimental tool to show that structural changes are accompanied by increased exergy. *Ecol. Modelling,* 41: 117–126, 1988.

Jørgensen, S. E., *Modelling in Ecotoxicology.* Elsevier, Amsterdam, 1990.

Jørgensen, S. E., Development of models able to account for changes in species composition. *Ecol. Modelling,* 62: 195–208, 1992a.

Jørgensen, S. E., Parameters, ecological constraints and exergy. *Ecol. Modelling,* 62: 163–170, 1992b.

Jørgensen, S. E., Review and comparison of goal functions in system ecology. *Vie Milieu,* 44: 11–20, 1994a.

Jørgensen, S. E., *Fundamentals of Ecological Modelling* (second edition) (Developments in Environmental Modelling, 19). Elsevier, Amsterdam, 628 pp., 1994b.

Jørgensen, S. E., The growth rate of zooplankton at the edge of chaos: ecological models. *J. Theor. Biol.,* 175: 13–21, 1995.

Jørgensen, S. E., *Integration of Ecosystem Theories: A Pattern.* 2nd edition. Kluwer Academic Publ., Dordrecht. 400 pp. (1st edition, 1992), 1997.

Jørgensen, S. E. and Mejer, H. F., Ecological buffer capacity. *Ecol. Modelling,* 3: 39–61, 1977.

Jørgensen, S. E. and Mejer H. F., A holistic approach to ecological modelling. *Ecol. Modelling,* 7: 169–189, 1979.

Jørgensen, S. E. and Padisák, J., Does the intermediate disturbance hypothesis comply with thermodynamics? *Hydrobiologia,* 323: 9–21, 1996.

Jørgensen, S. E., Nielsen, S. N., and Mejer, H., Emergy, environ, exergy and ecological modelling. *Ecol. Modelling,* 77: 99–109, 1995.

Jørgensen, S. E. and Bernardi, de R., The application of a model with dynamic structure to simulate the effect of mass fish mortality on zooplankton structure in Lago di Annone. *Hydrobiologia,* 356: 87–96, 1997.

Jørgensen, S. E., Patten, B. C., and Straskraba, M., Ecosystem emerging: 3. Openness. *Ecol. Modelling,* 117: 41–64, 1999.

Jørgensen, S. E., Patten, B. C., and Straskraba, M., Ecosystem emerging: 4. Growth. *Ecol. Modelling,* in press, 2000.

Kauffman, S. A., Antichaos and adaptation. *Sci. Am.,* 265 (2): 64–70, 1991.

Kauffman, S. A., *Origins of Order. Self Organization and Selection in Evolution.* Oxford University Press, Oxford, 1993.

Kay, J., Self Organization in Living Systems, Thesis, Systems Design Engineering, University of Waterloo, Ontario, Canada, 1984.

Kay, J. and Schneider, E. D., On the applicability of non-equilibrium thermodynamics to living systems [internal paper]. Waterloo University, Ontario, Canada, 1990.

Kay, J. and Schneider, E. D., Thermodynamics and measures of ecological integrity. In: *Proc. Ecological Indicators,* Elsevier, Amsterdam, pp. 159–182, 1992.

Kristensen, P. and Jensen, P., Sedimentation and resuspension in Søbygård Sø (lake). Special report in Danish published by Århus University, 1987.

Lewin, B., *Genes V.,* Oxford University Press, Oxford, 620 pp., 1994.

Li, W.-H. and Grauer, D., *Fundamentals of Molecular Evolution.* Sinauer, Sunderland, MA, 430 pp., 1991.

Lotka., A. J., Contribution to the energetics of evolution. *Proc. Natl. Acad. Sci. U.S.A.,* 8: 147–150, 1922.

Lovelock, J. E., *The Ages of Gaia.* Oxford University Press, Oxford, 1988.

Margalef, R., *Perspectives in Ecological Theory.* Chigaco University Press, Chicago, IL, 1968.

Marquet, P. A. and Taper, M. L., On size and area: Patterns of mammalian body size extremes across landmasses. *Evol. Ecol.,* 12: 127–139, 1998.

Mauersberger, P., General principles in deterministic water quality modeling. In: G.T. Orlob (ed.). *Mathematical Modeling of Water Quality: Streams, Lakes and Reservoirs* (Intenational Series on Applied Systems Analysis, 12). Wiley, New York, 42–115, 1983.

Mauersberger, P., Entropy control of complex ecological processes. In: B. C. Patten and S. E. Jørgensen (eds.) *Complex Ecology: The Part-Whole Relation in Ecosystems.* Prentice-Hall, Englewood Clifs, NJ, 130–165, 1995.

Mejer, H. F. and Jørgensen, S. E., Energy and ecological buffer capacity. In: S. E. Jørgensen (ed.). State-of-the-Art of Ecological Modelling. (Environmental Sciences and Applications, 7.). *Proceedings of a Conference on Ecological Modelling,* 28 August–2 September 1978, Copenhagen. International Society for Ecological Modelling, Copenhagen, 829–846, 1979.

Morowitz, H. J., *Energy Flow in Biology.* Academic Press, New York, 1968.

Morowitz, H. J., *Beginnings of Cellular Life.* Yale University Press, New Haven, CT, 1992.

Odum, E. P., The strategy of ecosystem development. *Science,* 164: 262–270, 1969.

Odum, H. T., *Environment, Power, and Society.* Wiley Interscience, New York, 1971.

Odum, H. T., *System Ecology.* Wiley Interscience, New York. 510 pp., 1983.

Pahl-Wostl, C., *The Dynamic Nature of Ecosystems: Chaos and Order Entwined.* Wiley, Chichester, 400 pp., 1995.

Patten, B. C., Network ecology: indirect determination of the life-environment relatinship in ecosystems. In: M. Higashi and T. P. Burns (eds.). *Theoretical Studies of Ecosystems: The Network Perspective.* Cambridge University Press, Cambridge, 288–351, 1991.

Patten, B. C., Synthesis of chaos and sustainability in a nonstationary linear dynamic model of the American black bear (*Ursus americanus* Pallas) in the Adirondack Mountains of New York. *Ecol. Modelling,* 100: 11–42, 1997.

Patten, B. C., Straskraba, M., and Jørgensen, S. E., Ecosystem Emerging: 1. Conservation. *Ecol. Modelling,* 96: 221–284, 1997.

Patten, B. C. and Fath, B. C., Environ theory and analysis. Submitted to *Ecol. Modelling,* 2000.

Peters, R. H., *The Ecological Implications of Body Size.* Cambridge University Press, Cambridge, 1983.

Prigogine, I., *From Being to Becoming: Time and Complexity in the Physical Sciences.* Freeman, San Fransisco, CA, 260 pp., 1980.

Raup, D. M. and Sepkowski, J. J., Mass extinctions in the marine fossil record. *Science,* 215: 1501–1503, 1982.

Recknagel, F., Petzhold, T., Haeke, O., and Krusche, F., Hybrid expert system DELAQUA — toolkit for water quality control of lakes and reservoirs. 71:17, 1994.

Reeves, H., *The Hour of Our Delight. Cosmic, Evolution, Order and Complexity.* Freeman, New York, 246 pp., 1991.

Reynolds, C. S., The plant life of the pelagic. *Verh. Int. Verein. Limnol.,* 26: 97–113, 1996.

Reynolds, C. S., The plant of the pelagic, pp. 97–114, in International Association of Theoretical and Applied Limnology. Volume 26 Part 1. Presented at the 26, Congress in Sao Paulo, 1995. Edited by W. D. Williams and A. Sladeckova, Stuttgart, 1996.

Russell, L. D. and Adebiyi, G. A., *Classical Thermodynamics.* Saunders College Publishing. Harcourt Brace Jovanovich College Publishers, Fort Worth, 620 pp., 1993.

Salomonsen, J., Properties of exergy. Power and ascendency along a eutrophication gradient. *Ecol. Modelling,* 62: 171–182, 1992.

Salthe, S. N., *Evolving Hiearchical Systems.* Columbia University Press, New York, 344 pp., 1985.

Schlesinger, W. H., *Biogeochemistry. An Analysis of Global Change,* 2nd edition. Academic Press, San Diego, 680 pp., 1997.

Shieh, J. H. and Fan, L. T., Estimation of energy (enthalpy) and energy (availability) contentis in structurally complicated materials. *Energy Resources,* 6: 1–46, 1982.

Shugart, H. H., *Terrestrial Ecosystems in Changing Environments.* Cambridge University Press, Cambridge, 534 pp., 1998.

Straskraba, M., Natural control mechanisms in models of aquatic ecosystems. *Ecol. Modelling,* 6: 305–322, 1979.

Straskraba, M., The effects of physical variables on freshwater production: analyses based on models. In: E. D. Le Cren and R. H. McConnell (eds.), *The Functioning of Freshwater Ecosystems* (International Biological Programme 22). Cambridge University Press, Cambridge, 13–31, 1980.

Straskraba, M., Jørgensen, S. E., and Patten, B. C., Ecosystem Emerging: 2. Dissipation. *Ecol. Modelling,* in press.

Svirezhev, Y., Exergy as a measure of the energy needed to decompose an ecosystem. Presented at ISEM's (International Society of Ecol. Modelling) International Conference on the State-of-the-Art of Ecological Modelling, 28 September–2 October, Kiel, 1992.

Svirezhev, Y., Thermodynamic orientors: how to use thermodynamic concepts in ecology. In: F. Müller and M. Leupelt (eds.). *Eco Targets, Goal Functions, and Orientors.* Springer, New York, pp. 102–122, 1998.

Szargut, J., Morris, D. R., and Steward, F. R., Exergy Analysis of Thermal, Chemical and Metallurgical Processes. Hemisphere Publishing Corporation, New York; Springer-Verlag, Berlin, 312 pp., 1988.

Szargut, J., 1998. Exergy Analysis of Thermal Processes; Ecological Cost. Presented at a workshop in Porto Venere, May 1998.

Ulanowicz, R. E., *Growth and Development. Ecosystems Phenomenology.* Springer-Verlag, New York, 204 pp., 1986.

Ulanowicz, R. E., Formal agency in ecosystem development. In: M. Higashi and T. P. Burns (eds.). *Theoretical Studies of Ecosystems: The Network Perspective.* Cambridge University Press, Cambridge, 340 pp., 1991.

Ulanowicz, R. E., *Ecology, the Ascendent Perspective.* Columbia University Press, New York, 201 pp., 1997.

Wolfram, S., Cellular automata as models of complexity. *Nature,* 311: 419–424, 1984a.

Wolfram, S., Computer software in science and mathematics. *Sci. Am.,* 251: 140–151, 1984b.

Zaret, T. M., *Predation and Freshwater Communities.* Yale University Press, New Haven, CT, 187 pp., 1980.

*Through analysis
to the specific statement —
through synthesis
to the general statement*

CHAPTER 14

The core of this chapter is an exergy calculation of the biosphere, based on information from *Chemical Evolution of Earth* by Vinogradov (1959). The chemical differences between the biosphere and the crust determine how much more exergy the biosphere has. The specific average exergy of the total biomass is found by this method to be about 309 kcal/g, while the specific exergy found from the calculations presented in Chapter 13 is about 308 kcal/g. If we anticipate that most of the biomass is trees, bushes, and similar vegetation, which has a β value (see Chapter 13) of 70, it means 70 times more exergy than detritus. As detritus contains in average about 4.4 kcal/g, the average exergy, including the embodied information, will as indicated above correspond to 308 kcal/g. The author calls this coincidence very interesting!

In addition, the chapter focuses on quantification of entropy fluxes using the Costanza and Neil (1982) input–output model of the biosphere.

The entropy fluxes are divided into three main groups: entropy fluxes caused by the water cycle and sunlight, entropy fluxes of natural carbon dioxide, oxygen, and biomass, and exergy fluxes of other mineral materials. The calculation leads to an interesting conclusion: decreasing entropy will mean increasing rarity of commodities. The author expresses what takes place in the biosphere from an energy point of view as follows: "the biosphere selects commodities in a solar supermarket." Fortunately, the cost of the commodities in the form of exergy dissipation can be paid by the solar radiation and we can even add to the storage of exergy in the biosphere as discussed in Chapter 13. This can be seen directly from the large storage of exergy in the biosphere, determined by two different methods with approximately the same result.

14 Thermodynamics of the Biosphere

Yuri M. Svirezhev

CONTENTS

14.1   Diversity of the Biosphere 351
14.2   Exergy of the Biosphere 353
14.3   Is the Biosphere in a Steady State? Some Thermodynamic Calculations.........................355
14.4   Constraints on Information Entropy Measures of Embodied Solar Energy
        Fluxes in the Biosphere 357
References 363

14.1   DIVERSITY OF THE BIOSPHERE

Life on the Earth is presented by various forms, and it is necessary to support all this diversity. This is a main thermodynamic role of solar energy. Otherwise, in accordance to the Second Law, life would be fully uniform, and, what is more, it would not exist. How can we estimate what kind of work and how much of it are performed by sunlight? For this the following simple thermodynamic model is used (see also Svirezhev and Svirejeva-Hopkins, 1997).

Let the biota contain n different classes. These classes could be represented by different taxonomic units: biomes, ecosystems, communities, species. Each class has its own mass N_i, so that the total mass of biota is equal to

$$N = \sum_{i=1}^{n} N_i.$$

(For instance, if these classes are biological species then $n \sim 3 * 10^6$). We assume that at some initial moment of time, they were mixed up as some "prebiosphere" substance and this "prebiosphere" system had no structure. What is the manner in which such an ordered structure as the biosphere appeared? We think that it is a result of work of one super-being, the "*Biosphere Demon*" (let us remember Maxwell's Demon). He feeds on negative entropy (*negentropy*) of sunlight, and distributes n different types of biosphere "particles" into n boxes, removing them from one "pre-biosphere" pool.

As a result, each ith box contains N_i particles of the ith type with unit mass, and the new structure which appears may be called the "*biota*." Certainly, in the general case the individual masses of different particles must differ from each other, but, as a first approximation, we assume their equality. The transition from a fully mixed system ("pre-biosphere chaos") to the "structured" biosphere is accompanied to the entropy reduction by the value (Landau and Lifshitz, 1964)

$$S = -\kappa N \sum_{i=1}^{n} p_i \ln p_i \qquad (1)$$

where $p_i = N_i/N$ and κ is a specific entropy of the mass unit of some "pre-biosphere" substance. In our case this substance is a mixture of chemical elements, from which the living matter can be constructed: 106 *molecules of CO_2 + 90 molecules of H_2O + a few molecules of some other substances.* Note when we speak about a "mixture," we have in view a real physical mixture, most likely a sum of elements that is already existing but, not still reacting, and the "living matter" is still not formed. If we remember that the specific molar entropies for CO_2 and H_2O (vapour) are equal to 50 cal/mol*K and 16 cal/mol*K, correspondingly (at 15°C), then $k \approx 1.07$ cal/g*K.

Let us now consider some energetic properties of this process. In order to produce N' units of a new biomass, the same amount of old biomass must be eliminated; i.e., corresponding amount of dead organic matter is decomposed (since the biosphere is in a steady state). Therefore, the value of N' must be equal to the net annual production of the stationary biosphere, Pr. As a result of decomposing processes, a quantity of heat is equal to the caloric equivalent of carbon contained in N'. If to assume that the entropy production, S', accompanying these processes is approximately equal to their thermal effect, then $S' = Pr_c/T$ where Pr_c (in calories) is the carbon equivalent of the net annual production and $T = 287$ K is the annual average planetary temperature.

In accordance to the principle of dynamic equilibrium, the entropy production S' must be balanced by the entropy production caused by exchange processes between the biosphere and its environment. We assume (and this is our basic hypothesis) that this entropy production is equal to S, i.e., the decrease of entropy which is described by Formula (1) and corresponds to the process of new biomass formation. In other words, the value Pr_c is a work performed by the "Biosphere Demon" during one year.

If we remember that the value $H = -\Sigma_{i=1}^n p_i \ln p_i$ is the information entropy or diversity of the system then

$$S = \kappa N H = S' = Pr_c/T \qquad \text{and} \qquad H = (Pr_c/N) \cdot (1/\kappa T). \qquad (2)$$

Since N is the annual biosphere production, Pr, expressed in grams of dry biomass then the ratio $Pr_c/N = Pr_c/Pr$ is the energy (carbon) contents of 1 g of dry biomass. (Speaking about a dry biomass, we take into account not only carbon, nitrogen, and phosphorus, but also a fixed water.) It is known that 1 g of biomass contains 0.5 g of carbon that corresponds to 4500 cal. Then $Pr_c/N = 4500$ cal/g and

$$H = (4500 \text{ cal/g})/((1.07 \text{ cal/g*K}) * 287 \text{ K}) = 14.65 \approx 15. \qquad (3)$$

If we multiply this value by 1.44, we get the result in bits: $H_b = 21$.

Let us estimate the probability of spontaneous creation of the biosphere. It is equal to

$$W = e^{-H} = e^{-15} = 3 * 10^{-7}, \qquad (4)$$

i.e., is very small. But if the contemporary biosphere is a result of a sufficiently large number of attempts L, in accordance to the simple probabilistic model (Chernavsky and Chernavskaya, 1984), the probability of its creation will be equal to,

$$W_L = LW/(1 + LW). \qquad (5)$$

How to evaluate the number of attempts? The photosynthetic biosphere with vegetation exists approximately during the last 10^9 years (Rutten, 1971). The average time of the biosphere renovation is equal to $\tau = B/Pr_c$ where $Pr = 5.4 * 10^{20}$ cal/year is the annual net production in caloric units, N is the total biomass of the biosphere in the same units, equal to $8.3 * 10^{21}$ cal (Svirezhev et al., 1985). Then $\tau = 15$ years, and, if we assume that one attempt is nothing else than one cycle of the biosphere renovation, then $L = 10^9/15 = 0.66 * 10^8$ and

$$W_L = (0.66 * 10^8 * 3 * 10^{-7})/(1 + 0.66 * 10^8 * 3 * 10^{-7}) = 0.95. \qquad (6)$$

Thus, the probability of the biosphere creation is close to one. In other words, *there is nothing surprising in the existence of the contemporary biosphere.*

It is also very interesting that the probability W depends neither on the mass of the biosphere nor on its productivity (you can see it from the Formula (2) for H). The probability W depends on only two factors.

1. The work of climatic machine which determines the Earth temperature.
2. The type of main energetic reaction which determines the κ-value.

Photosynthesis reaction uses two gases as the basic substance for formation of living matter: carbon dioxide and water vapour. Certainly, we can imagine other hypothetical reactions which would use other elements and substances (for instance, silicon instead of carbon), but this would give other values of k and, as a consequence, other values of diversity. As a result, the probabilities of existence of such the virtual biospheres would differ from the similar probability for the really existing one.

Note, however, that nevertheless the probability W_L depends in an implicit way on the total mass of the biosphere and its productivity, because the number of attempts (L) is defined by these values.

Knowing the value of H, the number of components (elements, elementary units, etc.) of the biosphere could be estimated. If these elements are relatively independent and they occur with almost the same frequencies, then $H \approx \ln n$ and $n \approx e^H = e^{15} = 3.3 * 10^6$.

It is interesting that this number coincides with the number of biological species on our planet. An impression appears that our "Demon" uses species as boxes. On the other hand, if elementary units inside the biosphere are organised in hard structures (like trophic chains and trophic levels) with the exponential distributions of frequencies p_i, then $n \approx H$, i.e., a number of these elementary units would be relatively small.

14.2   EXERGY OF THE BIOSPHERE

Reading the book *Chemical Evolution of Earth* by Vinogradov (1959), I turned my attention to the table containing the comparison of the chemical compositions for living and nonliving matter (biota and Earth crust, correspondingly). You can see a part of this table (original table contains the information for almost all chemical elements) in Table 14.1.

If we assume that the Earth crust (nonliving matter) is a system in thermodynamic equilibrium, then we can calculate the exergy of living matter, i.e., the exergy of the biosphere, where the nonliving matter of the Earth crust is considered as some reference state. In other words, we consider the biosphere as some chemical system (for instance, an "active membrane"), which either concentrates or disperses chemical elements in comparison with their basic concentrations in the Earth's crust.

TABLE 14.1
Chemical Composition of Living (Biota) and Nonliving (Crust) Matter (in % to weight)

Element		C	Si	Al	O	N	P
Biota	$1.05*10^1$	$1.8*10^1$	$2.0*10^{-1}$	$5.0*10^{-3}$	$7.0*10^1$	$3.0*10^{-1}$	$7.0*10^{-2}$
Crust	1.00	$3.56*10^{-1}$	$2.6*10^1$	7.45	$4.9*10^1$	$4*10^{-2}$	$1.2*10^{-1}$

Let s_i be the concentration of ith element in the biota and be s_i the same in the crust. In accordance to S. E. Jørgensen (1992), the exergy of the biosphere will be

$$Ex = B \cdot RT \sum_{i=1}^{n} \left[\frac{s_i}{\mu_i} \ln\left(\frac{s_i}{s_i^*} \cdot \frac{B}{B^*} \right) - \frac{s_i - s_i^*(B^*/B)}{\mu_i} \right] \tag{7}$$

where T is the annual average temperature of the biosphere, 288 K, R is the gas constant, 1.987 cal/K*mol, μ_i is the atomic weight of ith element, B and B^* are the total amount of matter in the biosphere and Earth's crust, correspondingly. It is obvious that we can introduce the new value $e_x = Ex/B$. This value can be called a specific exergy and it is equal to the amount of exergy contained in 1 g of living matter.

It is very important that a specific exergy depends on not only the relative elements composition of both the living matter and the crust one (i.e., the percentage content of different chemical elements in these systems, in the biosphere and in the crust), but, in the significant degree, its value depends on the ratio $r = B/B^*$, i.e., on the relative value of crust matter involved into the global bio-geochemical cycles. It will be clearer to rewrite (7) in the form

$$e_x = RT \left\{ \sum_{i=1}^{n} \frac{s_i}{\mu_i} \ln\left(\frac{s_i}{s_i^*} \right) + \left(\ln \frac{B}{B^*} - 1 \right) \cdot \sum_{i=1}^{n} \frac{s_i}{\mu_i} + \frac{B^*}{B} \sum_{i=1}^{n} \frac{s_i^*}{\mu_i} \right\}$$

or

$$e_x = K + \sigma(\ln r - 1) + \frac{\sigma^*}{r} \tag{8}$$

where

$$K = RT \sum_{i=1}^{n} \frac{s_i}{\mu_i} \ln\left(\frac{s_i}{s_i^*} \right), \qquad \sigma = RT \sum_{i=1}^{n} \frac{s_i}{\mu_i}, \qquad \sigma^* = RT \sum_{i=1}^{n} \frac{s_i^*}{\mu_i}.$$

Taking into account that all concentrations in Vinogradov's table are evaluated in the relative weight units (%) and using Formula (8), we can calculate the values of K, σ, and σ^*: $K = 185$, $\sigma = 94$, $\sigma^* = 31$. All the values are measured in cal/gram.

Let us calculate the value of specific exergy for different values of r. The first simplest hypothesis is if we assume that the total amount of matter has not been changed in the course of transition from nonliving state to living one, i.e., the peculiar conservation law of matter is realised and $B = B^*$, i.e., $r = 1$. In other words, there is dynamic equilibrium between the biota and crust, between the living and nonliving matter of the biosphere (that is understood in Vernadsky's wide sense), where all the matter of crust has been involved into the "Big Living Cycle." Let us remember that according to Vernadsky (1926), the Earth crust is a result of the biosphere activity, the trace of the past biospheres. By setting $r = 1$ in (8) we immediately get the value of specific exergy for 1 g of living matter. It is equal to 122 cal/g. Since $e_x = \sum_{i=1}^{n} e_x^i$, where e_x^i is the contribution of partial exergy, corresponding to ith element, into the total exergy, it is interesting to compare these contributions (see Table 14.2).

TABLE 14.2
Partial Specific Exergy for Different Chemical Elements
(in cal/g of living matter)

Element	H	C	Si	Al	O	Na	Mg	Fe
e_x^i	86.9	25.3	5.07	1.57	1.4	0.57	0.49	0.42

In fact, the main contribution belongs to hydrogen (~71%). These are water, carbonhydrates, etc. The second place is occupied by carbon (~21%). Then silicon, aluminium, and oxygen are disposed (4.15%, 1.3%, and 1.12%). The contribution of others is negligible.

Since the summary contribution of hydrogen, carbon, and silicon is equal to 96.2%, we can say our Biosphere is on the hydrogenous–carbonate–silicon biosphere. If we compare the "carbon exergy" of 1 g of living matter (25.3 cal) and so-called "carbon equivalent" (~2 kcal per 1 g of raw biomass), we can conclude that the part of the "structural," "creative" exergy is equal to 1.25%, i.e., this is very small in comparison with "heat" exergy). The latter is equal to the number of calories obtained in the process of a biomass burning.

However, S. E. Jørgensen (1995) has recently suggested a new measure of exergy based on the genetic complexity of different organisms in respect to detritus. In accordance to this, if the "exergy cost" of detritus is equal to 1, then the "exergy cost" of most plants and trees will be situated in the vicinity of 30 to 70. Note that the global vegetation is a leading actor of our biosphere. And if the free energy of 1 g of detritus is equal to 4.4 kcal/g, then the specific exergy of living matter of the biosphere must be equal to 140 to 300 kcal/g. Is there a contradiction? How can we resolve this?

Let us remember Vinogradov's estimations (Vinogradov, 1959) for the total amount of crust matter, 10^{25} g, and for the total biomass of the biosphere, 10^{21} g, so that $r = 10^{-4}$. Then the specific exergy will be equal to ≈ 309 kcal/g. Comparing this value with the above-mentioned Jørgensen's specific exergy, we see these values are close. It seems to me this coincidence is very interesting. Note that the main contribution into the specific exergy gives the term σ^*/r, i.e., the term corresponding to processes that are working against the entropy and that separate a thin film of living matter from the immense mass of the crust. The latter, in turn, is the entropy storage of the past biospheres. It is clear, if the film is thinner, the ability of one unit of living matter to perform such a work must be higher. In other words, if the value of r is less, the specific exergy must be bigger. But this situation is typical for the old biosphere when it was in an equilibrium long ago.

Let us imagine the young biosphere that is developed on the thin and young crust. This young biosphere is very aggressive and all the crust matter is involved into processes of chemical interaction and exchange with the biosphere. Then we get the case with $r = 1$ mentioned above. We see that the main role in the formation of comparatively low exergy is played by the term K (see (8)) determined by the chemical composition of living matter. In other words, at the first stages of the biosphere formation, the exergy of living matter is determined, mainly, by its chemical composition and, as a consequence, the sort of chemical processes used by life for the formation of its own matter.

14.3   IS THE BIOSPHERE IN A STEADY STATE? SOME THERMODYNAMIC CALCULATIONS

You can see that the hypothesis about the quasi-stationary state of the contemporary biosphere plays one of the most important roles in globalistics (Svirezhev, 1997). Should we test this statement in some way? Since various estimations show that the energy balance of the biosphere is fulfilled with sufficient accuracy, then this speaks to an advantage of stationary hypothesis, but this is not sufficient. From the viewpoint of the thermodynamics of open systems, the balance between the internal entropy production and its export into environment must also be fulfilled. In order to estimate the balance we suggest using one thermodynamic criterion based on the "zero-dimension thermodynamic model of the biosphere."

The model contains the following compartment: atmosphere (A), biota (B), pedosphere (P), and hydrosphere (H). The model characteristic time is equal to 10^3 years.

Let dS_{iE} be the annual production of entropy by the open system, "Earth," and dS_{eE} be the annual export of entropy into "Cosmos." If the system "Earth + Cosmos" is in a dynamic equilibrium then $dS_{iE} = - dS_{eE}$, where

$$dS_{eE} = 4/3 * dE * (1/T_s - 1/T_E) \sim - 1.8 * 10^{22} \text{ J/K} * \text{year},$$

$$T_s \cong 5800 \text{ K}; \qquad T_E \cong 260 \text{ K}.$$

Let dS_j be the entropy production by jth compartment and dS_{kj} be the export of entropy from jth compartment to kth one. Calculating compartments entropy and entropy fluxes between compartments, we use ideas and methods from Morowitz's brilliant book (1968). These calculations are often cumbersome, so we shall omit their details. These can be found in the work of Venevsky (1991) and Svirezhev (1992).

For the model compartments we get:

Atmosphere. We use the polytropic model of the "static" atmosphere (Khrgian, 1983) which is the mixture of ideal gases N_2, O_2, CO_2, argon, and H_2O vapours. The calculation of entropy for the mixture of gases, if their molar concentrations and specific entropy for each component are known, is a standard operation in thermodynamics (see, for instance, Landau and Lifshitz, 1964). Then the total entropy contents in the atmosphere is equal to $S_A = 3.49 * 10^{22}$ J/K. Since carbon dioxide is one of the "life gases," we separately calculate the corresponding entropy $S_A^{CO2} = 1026 * 10^{19}$ J/K.

Biota (Biosphere). We assume that "Biota" is submitted into a thermostat with the temperature $T = 14°C$, i.e., with the temperature equal to the annual average temperature of our planet. If we consider the following standard (averaging) composition of biomass: liquid H_2O − 44%, fixed H_2O −6%, cellulose −37.5%, proteins −8.4%, carbonhydrates, lipids, etc. −4.1%, then $S_B = 9.05 * 10^{18}$ J/K. Excluding the water entropy, we get immediately the entropy of dry biomass equal to $S_B = 2.9 * 10^{18}$ J/K.

The main uncertainty is how to estimate the fraction of H_2O (from 40% up to 90% for different plant species). We developed the special method in order to minimise an influence of the uncertainty, so that the corrected value $S_B = 1 * 10^{19}$ J/K (if it takes into account the marine biota).

Calculation of entropy balance for biota gives (data were taken from Costanza and Neil, 1982).

- Solar radiation utilised by vegetation (directly): $dS_{eB} = 2.4 * 10^{20}$ J/K* year.
- Water balance for vegetation: $dS_{HB} = 2.6 * 10^{20}$ J/K* year.
- Atmosphere–Biota interaction. The value, dS_{AB}, is added of the entropy fluxes accompanying the processes of carbon dioxide and oxygen diffusion through stomata and evapotranspiration

$$dS_{AB} = dS_{AB}^{CO2} + dS_{AB}^{O2} + dS_{AB}^{H2O}$$

$$= 1.1 * 10^{18} + (-1.05 * 10^{18}) + (-4.96 * 10^{20}), \text{ in J/K* year.}$$

- Soil–Biota interaction: $dS_{PB} = -8.25 * 10^{17}$ J/K* year.

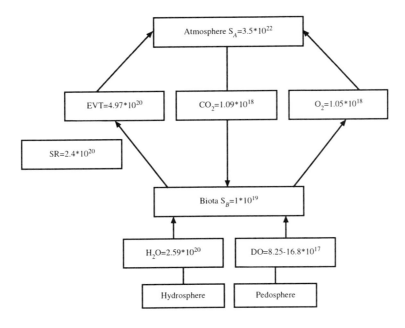

FIGURE 14.1 Annual entropy balance for the biosphere (biota) in the 1970s. SR = solar radiation; CO_2 = carbon dioxide (net); H_2O = liquid phase of water; DO = dead organic matter. All storage is in J/K, all fluxes are in J/K*year.

Since all these values possess different exponents, it is senseless to sum them with the accuracy up to characteristics. We shall sum only their exponents. As a result we get:

Exponent 20: $dS_1 = dS_{eB} + dS_{HB} + dS_{AB}^{H2O} = 3.85 * 10^{18}$ J/K*year.

Exponent 18: $dS_1 = dS_{AB}^{CO2} + dS_{AB}^{O2} = 4.4 * 10^{18}$ J/K* year.

The value of dS_1 would be less if we take into account an "entropy jump" caused by the two phase transitions: fixed water $\Rightarrow$ liquid water $\Rightarrow$ vapour. In this case $dS_1 =$

$$1 \div 2 * 10^{18} \text{ J/K* year.}$$

Thus, with the accuracy up to two exponents, the entropy balance for the biosphere (or, more correct, for biota) in year 1970 was equal to zero; i.e., the biosphere was in a dynamic equilibrium. You can see this in detail in Figure 14.1.

14.4 CONSTRAINTS ON INFORMATION ENTROPY MEASURES OF EMBODIED SOLAR ENERGY FLUXES IN THE BIOSPHERE

As mentioned above, all the Earth's natural systems operate because they receive the external flux of solar energy from outer space. So, if we imagine that the overall solar flux absorbed by our planet consists of small "coins" (for example, light quanta), then we can compare the biosphere with a big market at which different customers (biosphere systems) can buy different commodities (energy and material fluxes) or exchange these commodities one for another. Each of the commodities circulating at such a "market" (for example, phosphorus, nitrogen, carbon, etc.) has its own cost in energy units.

TABLE 14.3
Definition of Processes and Commodities

Processes

Economic processes (urban ecology)	Human populations, industries, households, and other man-made structures and institutions
Agriculture	Soils, crops, agricultural infrastructure
Natural plants	Natural primary producers, including marine producers and managed forestry
Animals	Wild and domestic animals
Soil and surface waters	Soil on uncultivated land, surface water, and accessible groundwater
Deep ocean	Water below the euphotic zone
Surface ocean	Water in the euphotic zone
Atmosphere	
Deep geology	Deep geologic formations and storage of mineral resources

Commodities

Goods and services	Manufactured G&S in 1970 U.S. dollars
Agricultural products	Agricultural crop biomass
Natural products	Uncultivated crop biomass (including forestry products) and animal biomass
Nitrogen	Nitrogen in simple inorganic nitrogenous compounds
Carbon dioxide	Carbon in carbon dioxide
Phosphorus	Phosphorus in simple inorganic phosphorus-containing compounds
Water vapour	
Fresh water	Fresh liquid water
Fossil fuel	Carbon in fossil fuel

Source: Modified from Costanza, R. and Neil, C., 1982.

Costanza and Neil (1982) suggested an input–output (I–O) model of the biosphere to estimate the direct and indirect solar energy cost of different natural and artificial commodities. There are ten commodities and ten processes (which we can interpret as ten compartments) in this I–O model (see Table 14.3). The solution of the corresponding linear equations is shown in Table 14.4.

One can use these results to recalculate all the fluxes of type number i (commodity number i in Table 14.3) in energy units x_{ijk}, where j is the number of input compartments, are k is the number of output compartments.

We try to show that the distribution of aggregated fluxes of different types among compartments can reflect physical and functional ordering in the biosphere. The functional ordering can be found in the values of embodied energy. We can divide the set of commodities into three groups according to the order of magnitude of energy intensity of the commodity (or energy cost of the material unit), presented in Table 14.5.

The first group includes the commodities with energy intensities lying in the interval from zero to 10^4 kcal per unit of commodity, such as sunlight, water, and water vapour. The second group consists of natural and agricultural products, and carbon dioxide. It occupies the energy cost interval from 10^4 kcal to approximately 10^5 kcal per material unit. The commodities included in the third group are nitrogen, phosphorus, fossil fuels, manufactured goods, and services.

The studies of thermodynamic fluxes in the biosphere (Svirezhev, 1992) showed that physical entropy fluxes could also be divided into three main groups ordered by quantitative values.

1. Entropy fluxes caused by the water cycle and sunlight;
2. Entropy fluxes of natural carbon dioxide, oxygen, and biomass;
3. Entropy fluxes of other mineral materials.

TABLE 14.4
Process by Commodity Input–Output Table for the Biosphere

Commodities	Urban Ecology	Agriculture	Natural Plants	Animals	Soil	Deep Ocean	Surface Ocean	Atmosphere	Deep Geology	Net Outputs	Totals
					Processes						
Inputs											
Goods and services (10^{12} \$)	2.71	0.08								1.19	3.98
Agriculture prod. (10^{15} g of dry weight, gdw)	1.28	4.55		3.27							
Natural products (10^{15} gdw)	1.18		163.4	27.9	103.4	34.6			0.16		167.3
Nitrogen (10^{12} g)	55.0	62.4	208.0		493.6		168.0	389.3			1,376.5
Carbon dioxide (10^{15} g C)		8.2	147.0			15.6	37.2	110.3			318.3
Phosporus (10^{12} g)	12.6	28.5	1,345.7		8.4		21.0	9.5	13.0		1,438.7
Water vapor (km^3)					111,419		424,700	496,100			496,100
Fresh water (km^3)	1,008	15,490	51,226							2,000	605,843
Fossil fuel (10^{15} g C)	5										5
Sunlight (10^{18} kcal)		23	227					606			856
Outputs											
Goods and services (10^{12} \$)	3.98	9.1									3.98
Agriculture prod. (10^{15} gdw)				3.9							9.1
Natural products (10^{15} gdw)			163.4						6		167.3
Nitrogen (10^{12} g)	80.0	31.0		295.0	340.5		182.0	448.0			1,376.5
Carbon dioxide (10^{15} g C)	5.0	6.1	73.6	14.0	46.5	15.6	49.5	108.0			318.3
Phosporous (10^{12} g)	14.2				241.3	1,161.1		9.5	12.6		1,438.7
Water vapor (km^2)	79	5,931	50,740		14,650		424,700				496,100
Fresh water (km^3)	929	9,829			98,985		496,100				605,843
Fossil fuel (10^{15} g C)									0.07	4.93	5

Source: Modified from Costanza, R. and Neil, C., 1982.

TABLE 14.5
Embodied Energy Intensities

Commodities (Inputs)	Embodied Energy Intensity (kcal solar/g)
Goods and services (10^{12} \$)	191.2 E6 kcal solar/\$
Agriculture products (10^{15} gdw)	13.9 E3
Natural products (10^{15} gdw)	39.2 E3
Nitrogen (10^{12} g)	0.63 E6
Carbon dioxide (10^{15} g C)	57.1 E3
Phosphorus (10^{12} g)	1.17 E6
Water vapour (km^3)	0.55 E18 kcal solar/km^3 = 0.55 E3 kcal solar/g
Fresh water (km^3)	0.55 E18 kcal solar/km^3 = 0.55 E3 kcal solar/g
Fossil fuel (10^{15} g C)	96.4 E3

Source: Modified from Costanza, R. and Neil, C., 1982.

One can easily see that both sets, i.e., partitioned either by physical attributes or by functional attributes, are similar. How can we introduce information entropy measures for distributions of fluxes among the compartments?

We assume that the total flow through in the system (expressed in kcal of solar energy) $T = \sum_i^{10}\sum_j^{10}\sum_k^{10} x_{ijk}$ forms some probabilistic space. Let us consider different partitions of the space ξ_m, where

$$\xi_0 = \left\{ \bigcup_{\forall j, k}\left[\sum_i^{10} x_{ijk}\right]\right\}, \qquad \xi_m = \left\{\left[\bigcup_{\forall j, k}\left(\sum_{i_1,\ldots i_{l_m}}^{10} x_{ijk}\right)\right]\cup\left[\bigcup_{\forall j, k}\left(\sum_{i \neq i_1,\ldots, i_{l_m}}^{10-l_m} x_{ijk}\right)\right]\right\}, \qquad (9)$$

m is the number of group, $i_1,\ldots, i_{l_m}$ is the number of commodities l belonging to the m group, l_m is the number of commodities in the m group (see an example in Figure 14.2).

Let us introduce the notations

$$T_{jk}^0 = \sum_i^{10} x_{ijk}, \qquad T_{jk}^m = \sum_{i_1,\ldots, i_{l_m}}^{10} x_{ijk}, \qquad \overline{T}_{jk}^m = \sum_{i \neq i_1,\ldots, i_{l_m}}^{10-l_m} x_{ijk},$$

where T_{jk}^m is the part of the total flux entering compartment j from comaprtment k, including commodities of the group m (for example: sunlight, water vapour, and fresh water for the group 1), $\overline{T}_{jk}^m$ is the balance of the total flux.

We can now define the normalised entropy of partition ξ_m:

$$R(\xi_m) = \frac{H(\xi_m)}{H_{max}^m}, \qquad (10)$$

where

$$H(\xi_0) = -\sum_{jk}\frac{T_{jk}^0}{T} * \ln\sum_{jk}\frac{T_{jk}^0}{T},$$

$$H_{max}^0 = \ln(100), \quad H_{max}^m = \ln(200),$$

$$H(\xi_m) = -\left[\sum_{jk}\frac{T_{jk}^m}{T} * \ln\sum_{jk}\frac{T_{jk}^m}{T}\right] - \left[\sum_{jk}\frac{\overline{T}_{jk}^m}{T} * \ln\sum_{jk}\frac{\overline{T}_{jk}^m}{T}\right].$$

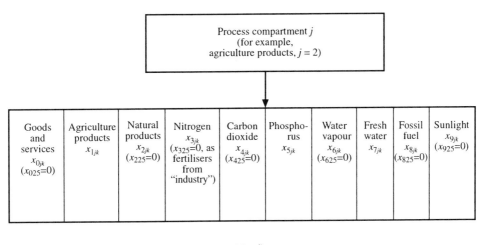

FIGURE 14.2 Optional partitions of the defined probability space for the flux entering compartment *j* from compartment *k*.

Another useful informational measure is the normalised conditional entropy of partition ξ_m with respect to partition ξ_{m-1}.

$$R(\xi_m/\xi_{m-1}) = \frac{H(\xi_m/\xi_{m-1})}{H_{\max}^{m/m-1}}, \tag{11}$$

Conditional entropy of events X with respect to Y that one can calculate from conditional probabilities (Pierce, 1980):

$$H(X/Y) = -\sum_{i=1}^{n}\sum_{k=1}^{m} p(y_k)p(x_i/y_k)\ln\ p(x_i/y_k). \tag{12}$$

Conditional probabilities of events from partition ξ_m with respect to events from partition ξ_{m-1} for the flux coming from the compartment k to compartment j is

$$p(x_m/y_{m-1}) = \frac{T_{jk}^{m}}{T_{jk}^{m-1}}. \tag{13}$$

Therefore, we can calculate conditional entropy of partition ξ_m with respect to partition ξ_{m-1} as

$$H(\xi_m/\xi_{m-1}) = -\sum_{jk}\frac{T_{jk}^{m-1}}{T}*\left[\frac{T_{jk}^{m}}{T_{jk}^{m-1}}*\ln\frac{T_{jk}^{m}}{T_{jk}^{m-1}} + \frac{\overline{T}_{jk}^{m}}{T_{jk}^{m-1}}\ln\frac{\overline{T}_{jk}^{m}}{T_{jk}^{m-1}}\right]$$

$$-\sum_{jk}\frac{\overline{T}_{jk}^{m-1}}{T}*\left[\frac{T_{jk}^{m}}{\overline{T}_{jk}^{m-1}}*\ln\frac{T_{jk}^{m}}{\overline{T}_{jk}^{m-1}} + \frac{\overline{T}_{jk}^{m}}{\overline{T}_{jk}^{m-1}}\ln\frac{\overline{T}_{jk}^{m}}{\overline{T}_{jk}^{m-1}}\right], \tag{14}$$

$$H_{\max}^{m/m-1} = \ln(200),$$

$$H(\xi_1/\xi_0) = -\sum_{jk}\frac{T_{jk}^{0}}{T}*\left[\frac{T_{jk}^{m}}{T_{jk}^{0}}*\ln\frac{T_{jk}^{m}}{T_{jk}^{0}} + \frac{\overline{T}_{jk}^{m}}{T_{jk}^{0}}\ln\frac{\overline{T}_{jk}^{m}}{T_{jk}^{0}}\right] - \sum_{jk}\frac{\overline{T}_{jk}^{0}}{T}*\left[\frac{T_{jk}^{m}}{\overline{T}_{jk}^{0}}*\ln\frac{T_{jk}^{m}}{\overline{T}_{jk}^{0}} + \frac{\overline{T}_{jk}^{m}}{\overline{T}_{jk}^{0}}\ln\frac{\overline{T}_{jk}^{m}}{\overline{T}_{jk}^{0}}\right],$$

$$H_{\max}^{1/0} = \ln(200). \tag{15}$$

The hypothesis is that if we are moving from the first partition to the next in the sequence discussed above, then we have two sequences of inequalities:

$$1. \quad R(\xi_0) > R(\xi_1) > R(\xi_2) > R(\xi_3), \tag{16}$$

$$2. \quad R(\xi_1/\xi_0) > R(\xi_2/\xi_1) > R(\xi_3/\xi_2). \tag{17}$$

The first one means that each distribution of aggregated fluxes among the compartments in chosen groups is increasingly distant from the equi-distribution. The second one means that information of one distribution about another is increasing.

The results of calculations for the chosen aggregation of fluxes among the compartments into three groups are the following:

$R(\xi_0)$	$R(\xi_1)$	$R(\xi_2)$	$R(\xi_3)$
0.3956	0.3673	0.1612	0.1378

$R(\xi_1/\xi_0)$	$R(\xi_2/\xi_1)$	$R(\xi_3/\xi_2)$
$0.6123*10^{-2}$	$0.3399*10^{-02}$	$0.6985*10^{-03}$

So our proposition that the distribution of aggregated, different-type fluxes among the compartments in groups can reflect the physical and functional ordering of the biosphere is true.

The following combinatorial experiment was made to investigate how many variants of unification of ten commodities in three groups have such informational constraints. We fixed the commodity "sunlight" in the first group, and the commodity "manufactured goods and services" in the final third group and then we calculated entropy measures for all the variants of unification of different types of commodities in the three groups. It is easily seen that there are $3^8 = 6561$ different variants of unification. Only 28 of them have the same informational constraints for distributions of aggregated fluxes for our chosen unification in three groups. As the informational ordering is meaningful it will be no wonder if one can find more rigorous informational constraints on unification of commodities in groups, which will leave only one unique variant.

It is interesting that the transfer of commodity "fossil fuel" from the third group to the second in the process of entropy calculation leads to

$$R(\xi_2) \ll R(\xi_0), R(\xi_1), R(\xi_3). \tag{18}$$

The "fossil fuel" flux is used by man in the biosphere as a mineral resource to supply industry and agricultural ecosystems. It has the big embodied energy value. But if the flux of fossil fuel to the biosphere from environment increases rapidly, the embodied energy value of the commodity will diminish. This can change the ordering scheme of fluxes as, in that case, the commodity "fossil fuel" is removing in the second group, which have functional and physical status "groups of commodities connected with a biosphere natural productivity" due to its embodied energy value. Hence, the last inequality will take place depicting human stress on the natural productivity, which leads to disorder in the whole system from the viewpoint of the information theory and thermodynamics.

It is seen that this informational ordering takes place in accordance with the physical structure of the biosphere, expressed in orders of thermodynamic characteristics of the energy and matter fluxes (either embodied solar energy or physical entropy). Consequent decreasing of the normalised entropy and conditional entropy by partitions of the fluxes can reflect the evolutionary path of the Earth as a self-organizing system.

Indeed, let our "Biosphere Demon" try to select commodities in a "solar supermarket" for constructing of a "*biosphera machina*." First of all he will take "less expensive" commodities to run biogeochemical cycles of the "*machina*" and then will involve new "more expensive" elements. Hence, decreasing entropy will mean increasing rarity of commodities and, therefore, evolutionary sequence of involvement in the biosphere for different sets of elements. Of course, this is only the hypothesis to be discussed and proved by other investigations.

REFERENCES

Chernavsky, D. S. and Chernavskaya, N. S., Problem of the new information in evolution, in *Thermodynamics and Control of Biological Processes*. A. Zotin, Ed.: Nauka, Moscow, 247–254, 1984.

Costanza, R. and Neil, C., The energy embodied in the products of the biosphere, in *Energy and Ecological Modelling*, W. Mitsch, R. Bosserman, and A. Klopathek, Eds.: Elsevier, Amsterdam, 743–755, 1982.

Jørgensen, S. E., *Integration of Ecosystem Theories: A Pattern,* Kluwer Academic Publishers, Dordrecht, 1992.

Jørgensen, S. E., Exergy and ecological buffer capacities as measures of ecosystem health. *Ecosyst. Health* 1: 150–160, 1995.

Khrgian, A. G., *Physics of the Atmosphere* (in Russian), v.I: Gidrometeoizdat, Leningrad, 1983.

Landau, L. D. and Lifshitz, E. M., *Statistical Physics* (in Russian), Nauka, Moscow, 1964.

Morovitz, H. J., *Energy Flow in Biology,* Academic Press, New York, 1968.

Pierce, J. R., *An Introduction to Information Theory,* Dover Publications, New York, 1980.

Rutten, M. G., *The Origin of Life by Natural Causes,* Elsevier. Amsterdam, 1971.

Svirezhev, Yu. M., Thermodynamic Ideas in Ecology, Lecture on the School of Mathematical Ecology, ICTP, Trieste, 1992.

Svirezhev, Yu. M. and Svirejeva-Hopkins, A., Diversity of the biosphere, *Ecol. Modelling,* 97: 145–146, 1997.

Svirezhev, Yu. M., Krapivin, V. F., and Tarko, A. M., Modelling of the main biosphere cycles, in *Global Change.* T. F. Malone and J. G. Roederer, Eds., Cambridge University Press, Cambridge, 298–313, 1985.

Venevsky, S. V., Entropy as a function of the biosphere. *J. Gen. Biol. Moscow,* 52 (6), 900–915, 1991 (in Russian with English resume), 1991.

Vernadsky, V. I., *The Biosphere* (in Russian), Gostekhizdat, Leningrad, 1926.

Vinogradov, A. P., *Chemical Evolution of the Earth,* The USSR Academy Scientific Publisher, Moscow, 1959.

Index

A

Action
    hypothesis, 249, 277
    principles, 251
Adaptability, 196, 203
Agricultural machinery, 92
Agricultural production
    emergy evaluation of, 77
    phase, 87
    total, 85
Agriculture
    arable, 276
    disaster, 150
    ecological, 149
    Italian, 77, 86
    no-till, 149
    products, 360
    sustainable, 149
Agroecosystem, 135
    change of information entropy in, 142
    energy efficiency of, 137
    evolution, 141
    flow diversity of, 138
    industrial, 143, 147
    information entropy indices for, 146
    low structure of, 140
    production, annual total, 148
    structure of energy flows of, 144
Agrosystem(s)
    industrialised, 152
    infrastructure, 136
    partially closed, 150
Alburnus alburnus alborella, 339
Algae
    assimilation, 30
    competitors of, 20
    green, 37
    process of growth in, 45
    reserves
        maximum specific synthesis rate of, 23
        specific maintenance costs for, 24
    role of in *Daphnia* nutrition, 27
    steady-state density of, 53
Almonds, 85, 86
Amino acids, 295, 311
Anabolic power, 35
Anabolism, 293
Animal
    diversity, 266
    -dominated ecosystem, 247
Animat
    experiments, 206
    orientor satisfaction, 201
    Wilsons', 195
Anthropogenic ecosystems, 217
Aphotic zone, 179
Arable agriculture, 276
Arctic charr, 268
Arctic systems, uniqueness of, 240
Artificial energy, 135, 142, 147, 148
Ascendancy, 138
Assimilation power, 35
Atmosphere–Biota interaction, 356

B

Bacteria
    assimilation, 30
    Gram-negative, 48
    maximum assimilation from, 42
    maximum specific ingestion rate of, 23
    reserves, specific maintenance costs for, 24
    -rich environments, durable wood found in, 13
    structural biomass, growth of, 46
Baffin Bay–Davis Strait, bowhead whales of, 244
Balaena mysticetus, 240, 244
Balanus glandula, 259
Big Living Cycle, 354
Binding probability, 37
Biocalorimetry, 167
Biodiesel, from soybeans, 102
Bioenergetics, 167
Bioethanol, 102
Biofuels production, 100, 101
Biological events, interpretation of, 253
Biological systems, 251, 306
Biomass
    Accumulation Ratio, 260
    formation of, 306
    growth of bacterial structural, 46
    increase, 320
    proper, 8
Biosphere
    Demon, 351
    exergy of, 353
    function of in homeorhetic system, 267
    long-term evolutionary development of, 257
    machina, construction of, 363
    organization of, 7
    Space-Time Activity (BIOSTA), 77
    structured, 351
    study of, 5
    total biomass of, 352

Biosphere, thermodynamics in, 351–364
    diversity of biosphere, 351–353
    embodied solar energy fluxes in biosphere, 357–363
    exergy in biosphere, 353–355
    thermodynamic calculations, 355–357
BIOSTA, see Biosphere Space-Time Activity
Biota, 351, 356
Boltzmann constant, 156, 161
Boltzmann entropy, 118
Bowhead whales, of Baffin Bay–Davis Strait, 244
Broken symmetry, 254
Buffer capacities, 115

C

Calluna vulgaris, 258
CAM, see Crassulacean acid metabolism
Canadian Arctic, 232
Canadian Northwest Territories, lakes of, 233, 234
Carbon
    equivalent, 355
    life, formation of, 295
Carbon dioxide production, 57
Catabolism, 293
Chaetomorfa aerea, 342
Change
    asymmetry of, 9
    -discharge cycle, of molecular cascade, 294
Chetaev instability theorem, 119, 124
Chlorophyll, 12, 337
City(ies)
    hierarchical organization of sectors within, 72
    land near, 77
Clementsian climax, 258
Climate energy, 11
Climax concept, 256
Closed system, 246
Coexistence, 196, 199
Collision, 199
Commodities, definition of, 358
Concept emergy, definition of, 317
Concept exergy, 310
Conservation law for elements, 48
Conservation Law of Thermodynamics, 69
Cope's Rule, 272
Corals, 3, 13
Coregonus clupeaformis, 234, 235, 237, 238
Coriolis force, 269
Corn, 84, 86
Crassulacean acid metabolism (CAM), 307
Creative exergy, 355
Crop(s)production systems, 135
    herbaceous, 99
    production, 84, 139
        assets for, 83
        electricity used for, 81
        goods and assets for, 79
        labor for, 81

    lubricants, 81
    sustainable, 151, 152
Crystallization, 252
Cyclops scutifer, 267
Cymodocea nodosa, 342

D

DAB, see Daphnia, Algae, and Bacteria
DAB community
    chemical compounds of, 28
    steady-state control of consumer assimilation in, 54
    steady-state distribution of carbon and nitrogen in, 49,
        50
Daphnia
    Algae, and Bacteria (DAB), 17, 22
    corpses, 25
    embryonic, 38
    feeding rates for, 39
    half-saturation constant of, 23
    hyalina, 341
    population structure, 44
    reproduction, 43
    structural mass density of, 23
Darwin, 229, 270, 277
Data-gathering efforts, 67
DEB, see Dynamic Energy Budget
DEB model, 26, 34
    assumptions leading to, 38
    for daphnids, 33
    energy fluxes specified by, 34
Decomposition products, inorganic, 158
Deer, entropy outflow from, 170
Deep geology, 358
Developmental physiology, Minot's law in, 326
Diatoms, 37
Diesel, 91, 94
    fuel, for transport of beets, 96
    for livestock, 81
    for process heat, 97, 98
Dimethyl sulfide (DMS), 56
Direct energy, 136
Diseases, impact of, 262
Disorganising system, 142
Dissipating power, 35, 40
Dissipation heat, 51
Dissipative function, 219
Dissipative structure, 161
Diversity, 263, 264
    animal, 266
    index of, 160
DMS, see Dimethyl sulfide
DNA, 43, 312
Dominant species, characteristics of, 247
Droop model, 22
Duality principle, 128
Dutch farms, 143
Dynamic Energy Budget (DEB), 21, 22, 56

E

Earth
cycle, 66, 80, 90
energy flow in, 176
history of solid, 6
Eco-energetics analysis, 135
Ecological agriculture, 149
Ecological Law of Thermodynamics (ELT), 194
Ecological succession, 8
Ecology
energy and, 1
founding father of British, 257
Ecology, thermodynamics and, 213–227
average rate of entropy production and equilibrium in
average, 225–226
entropy production and chemical load, 224–225
exergy and entropy, 221–222
exergy and information, 222–224
physical approach, 214–219
systems far from thermodynamic equilibrium, 219–221
Economy machine, 105
Ecosystem(s)
animal-dominated, 247
anthropogenic, 217
biological components of, 338
changes in, 6
closed, 249
development, 316, 320, 324, 327
descriptor of, 153
goal function for, 334
by use of thermodynamic functions, 2
dynamics, 1, 328
emigration in, 333
energy-flow analysis in, 167
evaluating of for economic questions, 73
exergy flows of, 191
hierarchies in, 71
immigration in, 333
integrity, 107
lake, 173, 174, 232
mature, 321
models, internal reserves ignored in, 26
natural, 224
nondeterministic, 299
properties, 331
state changes, 215
structural dynamic models of, 334
successively closed, 217
Ecosystems, application of thermodynamic concepts to
real, 135–152
change of information entropy in agroecosystems,
142–143
change of information entropy and energy load,
145–146
description of agroecosystems using energy flows, eco-
energetics analysis, 135–137
farms, 143–145
Hungarian agriculture, 148–150
information measures defined according to typical
structure of agroecosystem, 139–141
limits of agriculture intensification and entropy cost,
147–148
partially closed agrosystem, 150–152
Ulanowicz's ascendancy and flow diversity of
agroecosystem, 138–139
Ectropy, 278
ED, see Empower Density
Effectiveness, 196
EIR, see Emergy Investment Ratio
Electricity, 92, 224
crop production, 81
livestock, 81
water removal from ethanol, 97
ELR, see Environmental Loading Ratio
ELT, see Ecological Law of Thermodynamics
Emergy, 11, 316
accounting, see Large-scale ecosystems, emergy
accounting of human-dominated
algebra, 69
analysis, example of useful findings from, 106
as basis for policies, 105
conservation of, 71
density, 89
diagram, of Italian agricultural and livestock system, 78
evaluation
of agricultural production, 77
of national economic system, 101
exported, 103
imported, 103
indices and flows based on, 106
Investment Ratio (EIR), 104
solar, 64, 65
Sustainability Index, 67, 99, 103
Yield Ratio (EYR), 61, 67, 89
Emigration, in ecosystems, 333
Empower Density (ED), 61, 67
Endosomatic energy, 3, 8
Energy
acquisition, 273
applied, 79
artificial, 135, 142, 147, 148
climate, 11
conservation laws, 34, 139
-conserving systems, 252
demand, in agricultural phase, 99
direct, 136
dissipated, 220
ecology and, 1
efficiency, coefficients of, 147
endosomatic, 3, 8
entropy-free, 155
exosomatic, 3, 8, 10, 11
flow(s), 135
in earth, 176
global, 229, 269
streamlining of, 271
systems described by, 142
weight-specific, 326
fluxes, specified by DEB model, 34
Gibbs free, 156
grey, 216

historical change, 6
indirect, 136
inflow, into human body, 171
load, 211
load, change of, 145
memory, 69
nonphotosynthesis, 12
quality of, 71
ratio, without residues, 99
sink, stability development within, 268
solar, 66, 80, 136, 351
Sustainability Index, 61
turnover, 51
ultimate source of, 316
zooplankton, 317
Engraulis ringens, 266
Entropy
    degradative part of, 224
    excess, 226
    fee, 150
    flows, calculations of, 165
    -free energy, 155
    homeostasis, 171
    outflow, from deer, 170
    overproduction of, 150
    principle, in living systems, 181
    production, 119, 165, 168, 294
        ecological systems, 180
        human body life span, 172
        Lake Biwa, 174, 177
        lake ecosystems, 177
        Lake Mendota, 177, 178
        leaf, 170
        lizard, 170
        multiphases in, 181
        nature, 169
        white-tailed deer, 171
    pump, 214, 216, 225
    recovery of, 7
Entropy and exergy principles, in living systems, 167–190
    comparative study of entropy production in lake
            ecosystems, 177–181
        entropy production in ecological systems, 180–181
        Lake Biwa and Lake Mendota, 177–180
    entropy and entropy production, 168–169
    entropy flow and entropy production in nature, 169–177
        animal, 170–171
        calculations, 175–177
        earth, 175
        human body, 171–173
        lake ecosystem, 173–175
        plant, 169–170
    entropy principle in living systems, 181–182
    exergy principle in living systems, 182
Environment(s)
    adapting cognitive system to changing, 203
    basic properties of normal, 195
    indicators, 199
    inputs, 87
    prebiological, 221
    properties of, 197

reliability, 200, 204
as scarce resource, 75
stability, 200, 204
variety, 200, 204
work, valuing, 74
Environmental Loading Ratio (ELR), 61, 67, 86
Enzymes, 295
Equilibrium coordinates, equations for, 127
Ergodynamics, 229, 252
Erosion
    nitrogen loss with, 93, 98
    phosphate loss with, 93
    potash loss with, 93
    soil, 149
Ethanol
    produced, 90, 97
    water removal from, 97
Euphotic zone, 56, 179
European virgin forests, 245
Eusocial insects, 3, 13
Evapotranspiration, energy of in terrestrial plants, 10
Evolution
    concept explaining, 191
    Darwinian, 229
    directionality in, 277
    index, calculation of, 343
    laws of, 8
    models, phenomenological thermodynamics of, 129
Exergy, 65, 155–163, see also Entropy and exergy
            principles, in living systems;
            Multidimensional system orientation, exergy
            and
    balances, 158–160, 203
    changes, 338
    concept, 131, 310
    consumption, 159, 293
    contribution of from prokaryotic cell, 312
    cost, 355
    creative, 355
    definition of, 2, 155–158, 309
    degradation of, 318
    destruction, 182
    dissipation, optimised, 318
    dissipative structure, 161–162
    exosomatic, use of and capacity to organize space, 14
    expenditure, 195
    gradients, 198
    heat, 355
    index, 304, 314
        calculation of, 308
        concept exergy used to compute, 310
    information and, 160–161, 222
    as limiting factor, 300
    loss, 199
    as Lyapunov function, 123
    maximum principle, 211, 221, 222, 223
    nonflow chemical, 157
    principles, in living systems, 182
    storage, 325, 330
    total specific, 126
    use of as orientor, 340

utilisation, 325
Existence, 196
Exosomatic artifacts, 8
Exosomatic energy, 3, 8, 10, 11, 14
Exosomatic structures and captive energies, relevant in
        succession and evolution, 5–15
    asymmetry of change, 9
    constraints of physical world, 6–7
    ecological succession, 8
    everyday work, 10–12
    need for plain physics, 9–10
    present ecological theory, 5–6
    properties of space required for obtaining work from
        exosomatic energy, 12–14
    use of exosomatic energy and capacity to organize
        space, 14
    what life is and how biosphere becomes organized, 7–8
Exploitation functions, 76
EYR, see Emergy Yield Ratio

F

Farms, 143
    Dutch, 143
    Hungarian, 143
    Lithuanian, 143
Fat exosomatic energy, 11
Fertilizers, 133, 148
    nitrogen, 82, 87
    organic, 151
    phosphate, 82, 91
    potash, 82, 91
Fir plantation, 325
First Law of Thermodynamics, 155, 289, 290
Fish, 314
    contribution of to exergy of ecosystem, 313
    fauna, world's most diverse, 264
    growth rates of, 340
    species, top-carnivorous, 318
Fisher fundamental theorem, of natural selection, 130
Fisher–Haldane–Wright equations, 130
Flows matrix, 139
Fodder
    industrial, 83
    self-produced, 83
Food
    density, 44
    production, transformities of global, 78
    source, over-exploitation of, 330
    uptake, 196
    webs, 247
Forage, 83, 84, 86
Forest(s)
    European virgin, 245
    natural, 325
Forrester's world model, 195
Fossil fuel, 78, 98, 133, 211, 363
Fourth Law of Thermodynamics, tentative, 305–347
    calculation of exergy index, 308–316
    development of ecosystems, 320–330

ecosystem properties, 331–334
first attempt to express biological evolution by relative
        information index, 342–344
observations applied to test hypothesis, 307–308
orientors describing growth of organism or ecosystem
        development, 316–320
structural dynamic models of ecosystems, 334–342
Freedom of action, 196, 202
Fruits, 84, 86

G

Galileo, 9
Gasoline, 81, 87, 92
Gavia Lake Arctic charr, 241
Genes, nonrepetitive, 313
Genetic combination, emergence of new, 321
Geochelone gigantaea, 240, 243
Geology, deep, 358
Gestalt stability, 263
Giant land tortoises, 240
Gibbs free energy, 156
Glaciation, 269
Global cycles, problems concerning, 5
Global energy flow, 229, 269
GNP, see Gross National Product
Gödel's theorem, 298, 299
Gouy–Stodola theorem, 168, 182
Gracilaria confervoides, 342
Grapes, 86
Grassland, 325
Green algae, 37
Grey energy flow, 216
Gross National Product (GNP), 104

H

Heat
    death, 161
    dissipation, 51
    exergy, 355
Herbaceous crops, 99
Herbicides, 87, 147
Hierarchy(ies)
    ecosystem, 71
    structuring, 145
    thermodynamics of, 71
Homeokinesis, 248
Homeorhetic system, 249, 250
Homo sapiens, 13
Hoplostethus atlanticus, 240, 244, 245
Human
    body
        energy inflow into, 171
        life span, entropy production for, 172
    -dominated processes, 76
    labor, 92, 98, 136
Humankind, role of cultural transmission in, 3
Hungarian agriculture, 148

Hungarian farms, 143
Hungarian puszta, 148
Hydroporus polaris, 267

I

Ideal gases, 119, 160
Immigration, in ecosystems, 333
Imperfect symmetry, 231–285
    action hypothesis, 249–255
        action principles, 251
        biological systems, 251–253
        broken symmetry, 254–255
        homeorhetic system, 249–250
        move away from thermodynamic equilibrium,
            253–254
        Onsager's reciprocal relations, 250
        origins, 253
    Darwin's legacy, 277–278
    evolution and natural selection, 270–275
        evolution, 270–275
        natural selection, 275
    global energy flow, 269–270
    premises, 232–246
        probative system, 234–240
        stepping stones of Rosetta, 232–234
        uniqueness of Arctic systems, 240–246
    strong inference, 246–249
        autonomous lake, 246–247
        categories of ecosystems, 247
        characteristics of dominant species, 247–248
        interspecific interactions, 248–249
    testing of hypothesis, 255–269
        action as inherent stabilizing property of living
            things, 262
        diversity, 263–268
        production/biomass ratio, 259–260
        r- and *K*-selection, 260–262
        stability, 262–263
        stability and diversity, 268–269
        succession, 255–259
Index of diversity, 160
Indirect calorimetry, 51
Indirect energy, 136
Industrial agroecosystems, 143, 147
Industrial fodder, 83
Industrial production phase, 88
Inflation, 104
Information
    diversity index, 118
    entropy, 133, 145
    exergy and, 160, 222
    theory, 141
Inorganic decomposition products, 158
Inorganic soup, 131
Insecticides, 87, 91
Irreducible systems, 298
Irrigation
    fuel for, 88
    surface water for, 93, 94

Isolated system, 246
Italian agriculture, 77
    emergy yields ratio for crops in, 86
    livestock system and, emergy diagram of, 78
Italian economic system, monitoring performance of, 104

J

Jørgensen's reference state, 224

K

Kopotkin, Prince, 277
K-selection, 260
K-strategy, 122

L

Labor
    for crop production, 81
    human, 92, 98
    for livestock, 82
Lake(s)
    Arctic, 268
    Canadian Northwest Territories, 233
    density of fish in Arctic, 236
    ecosystem, 173, 174, 177
    trout, 237
    whitefish, 237
Lake Biwa, entropy production in, 174, 177, 178
Lake Mendota, entropy production in, 177
Land use(s)
    characteristics of, 73
    metabolism, 73
Large-scale ecosystems, emergy accounting of human-
        dominated, 63–113
    emergy concepts and definitions, 64–69
        emergy algebra, 69
        energy memory, 69
        procedure for emergy accounting, 67–69
    environmental and human-dominated processes,
        76–107
        emergy evaluation of agricultural production,
            77–101
        emergy evaluation of national economic system,
            101–107
    evaluating ecosystems for economic questions, 73–76
        assessing sustainability, 76
        evaluating renewability and replacement of
            resources, 75–76
        valuing environmental work, 74–75
    hierarchies, 70–73
        hierarchies in ecosystems including humans, 71–73
        thermodynamics of hierarchies, 71
    maximum empower, 70
    thermodynamics of oscillating systems, 107–108
Laws of Ergodynamics, 229, 252
Laws of evolution, 8
Le Chatelier principle, 123, 306, 344

Lemons, 85, 86
Life
    conditions for creation of, 294
    formation of carbon, 295
    processes, temperature range needed for, 292
Life in closed bottle, how light and nutrients affect, 19–60
    canonical to natural communities, 55–56
    DAB system, 27–29
    DEB model, 34–46
        alga, 45–46
        bacterium, 46
        Daphnia population structure, 44–45
        dissipating power, 40–41
        feeding and time budget, 35–40
        growth and reserve kinetics, 41–44
        mineral fluxes and mass balance, 46
    DEB representation of DAB community, 22–27
    energy turnover and dissipation heat, 51–53
    mass turnover, 47–50
    SU-extended Monod model, 29–34
    top-down vs. bottom-up control, 53–54
Light, see also Life in closed bottle, how light and nutrients affect
    absorption, 180
    availability, 40
    measures, differences in, 37
    quanta, 357
Linum usitatissimum, 242
Lithuanian farms, 143
Little Nauyuk Lake, 265
Livestock
    assets for, 83
    diesel for, 81
    electricity used for, 81
    goods and assets for, 79
    labor for, 82
    production, 85
Living systems, see Entropy and exergy principles, in living systems
Lizard, entropy production in, 170
Lotka-Volterra model, 12
Lubricants, 81, 92
Lyapunov function, 115, 119, 123, 131

M

Machinery manufacture, 96
Macrophytes, 3, 13
Mankind, 13
Man-made capital, 106
Man-made systems, 321
Marr-Pitt model, 22
Mass transport processes, 294
Matter, 6
Maturity growth, 34
Maximum Empower Principle, 108
Maximum Power Principle, 70
MCA, see Metabolic Control Analysis
Mean Malthusian parameter, 129
Mechanical equipment, 82

Memory algebra, 69
Metabolic Control Analysis (MCA), 54
Michaelis Menten kinetics, 37
Microbial loop, 20
Microeconomic theory, 74
Microzooplankton, 55
Migration patterns, movement of biological material during, 269
Mineral fluxes, 46
Miniaturization, 9
Minimum entropy principle, 316
Minot's law in developmental physiology, 326
Model
    DEB, 26, 34
        assumptions leading to, 38
        for daphnids, 33
        energy fluxes specified by, 34
    Droop, 22
    Forrester's world, 195
    ideal gas, 119
    Lotka-Volterra, 12
    Monod, 22, 29, 54
    Parr-Pitt, 22
    resource building, 74
    thermodynamic, of vegetation, 218
Monod model, 22, 29, 54
Multidimensional system orientation, exergy and, 193–209
    emergent goal functions, 196–198
        basic orientors, 197
        orientor as implicit goal functions, 198
        orientor satisfaction and exergy, 198
        properties of environment, 197
        properties of orientors, 198
    evolution and intelligent use of exergy, 194–196
    quantification of environmental indicators and system orientors for minimal ecosystem, 199–204
        adaptability, 203–204
        effectiveness, 202
        environmental reliability, 200
        environmental stability, 200–201
        environmental variety, 200
        existence, 201–202
        freedom of action, 202–203
        measures of animat orientor satisfaction, 201
        resource availability, 199–200
        security, 203
    simulations with minimal ecosystem, 204–207
        emergence of basic value orientations, 206
        emergence of individual differences in value orientation, 206–207
        sample results, 204–206
Mutatis mutandi, 8
Mytilus californianus, 259

N

National economic system, emergy evaluation of, 101
Natural disturbances, 328
Natural energy, of sun, 135
Natural forest, 325

Natural selection, 130, 270, 275
Necromass, 8
Needs, definition of, 195
Negative entropy, 165
Negative exergy balance, 203
Negentropy, 278, 293, 351
Nitrogen, 91
    fertilizer, 82, 87
    harvested in residues, 94
    loss, with erosion, 93, 98
    minerals, in oligotrophic systems, 50
Nonphotosynthesis energy, 12
Nonrepetitive genes, 313
Normative intelligence, 195
No-till agriculture, 149
Nutrients, see Life in closed bottle, how light and nutrients
        affect

O

Ohm's law, 253
Oligotrophic systems, nitrogen minerals in, 50
Olives, 85, 86
Onsager relations, 127, 130, 253
Ontic openness, 296
Openness, 287
    ontic, 296
    physical, 293
Open system, 246
Orange roughy, 245
Oranges, 85, 86
Ordered systems, 298
Organic fertilisers, 151
Organic matter, use of light to produce, 19
Orientor(s)
    adaptability, 203
    basic, 197
    coexistence, 199
    exergy used as, 340
    properties of, 198
    satisfaction, 198, 200
Oscillating systems, thermodynamics of, 107
Ostwald's Principle, 175
Oxygen
    consumption, 42, 52, 57
    flux, 43

P

PAR, see Photosynthetically active radiation
Pelecanus occidentalis thagus, 266
Peptidoglycan, 48
Persistence, 117
Pest(s)
    control, 149
    impact of, 262
Pesticides, 82, 87, 91, 147, 148
Phalocorax bougainvilli, 266

Phosphate
    fertilizer, 82, 91
    loss, 93
Phosphorus, harvested in residues, 94
Photo-inhibition, 24
Photosynthesis, 12, 53
Photosynthetically active radiation (PAR), 10
Physical openness, 293
Phytoplankton, 314
    concentration, 337
    contribution of to exergy of ecosystem, 313
    growth rates of, 340
    maximum growth rate of, 336
Plankton
    effects of UV radiation on, 56
    seawater, 20
Postbreeding survival, 261
Potash, 87
    fertilizer, 82, 91
    harvested in residues, 94
    loss, with erosion, 93
Prebiological environment, 221
Pre-biosphere chaos, 351
Prevital state, 125, 223
Prey
    -handling time, 38
    -predator system, 214
Prigogine theorem, 117, 119, 213
Primary production, 336
Primeval atmosphere, reactions of, 291
Principle of Least Action, 251, 252, 274, 275, 278
Principle of Least Time, 254
Principle of Most Action, 252, 253, 272
Probative system, 234
Process heat
    demand, 97
    diesel for, 98
Processes, definition of, 358
Production/biomass ratio, 229, 259
Proteins, 295
Protoorganisms, 254
Protozoa, 272
Pulsing paradigm, 107

Q

Quantum mechanics, uncertainty relations in, 296
Quasi-organism, 256

R

Radiation
    photosynthetically active, 10
    solar, 170, 179, 194, 229
        exergy received from, 305
        utilised by vegetation, 356
        work potential inherent in, 290
Rain chemical potential, 80, 90
Rain forests, 50, 325

Random systems, 298
Rapeseed, 86
Rate of living, 50
Reciprocal relations, 127, 130, 250
Redundancy index, 141, 143
Reference state, 125, 223
Relative exergy index, 315
Relative information index, 342
Renewable resources, 79, 90
Renewable substitutes, rate of creation of, 107
Reserve kinetics, 41
Resource(s)
    availability, 199
    building, model of, 74
    evaluating renewability and replacement of, 75
    flow, supporting biological system, 126
    inorganic, 131
    inputs, to process, 65
    renewable, 79, 90
    scarce, 197
    used up, 74
Respiration
    /biomass, 319
    coefficient, mean, 215
    index, 273
    intensity, 272
    Quotient (RQ), 47, 52
    rates, 270
    ratio, 52
Rice, 84, 86
River Amazon, 264
r–K selection spectrum, 261
RNA, 312
RQ, see Respiration Quotient
r-selection, 260
r-strategy, 122

S

Sala variegata, 266
Salvelinus
    alpinus, 234, 238, 239, 265, 268
    namaycush, 234, 235, 237
SDA, see Specific Dynamic Action
Seawater plankton, 20
Second class energy, 156
Second Law of Thermodynamics, 115, 156, 159, 162, 213, 289, 344
Security, 196
Selection, rules for, 10
Self-organization, 63
    examples of, 291
    optimum pattern of, 109
    system, 9, 70, 207
Sense vector, 199, 204
Social Darwinists, 277
Soil
    acidification, 149, 150
    amount of solar energy reaching, 68
    erosion, 149

Solar emergy, 64, 65
Solar energy, 66, 80, 90, 136
    amount of reaching soil, 68
    thermodynamic role of, 351
Solar entropy pump, 216
Solar radiation, 170, 229
    absorption of, 179
    energy, 179
    exergy received from, 305
    input, 194
    utilised by vegetation, 356
    work potential inherent in, 290
Solar supermarket, 363
Solar transformity, 65, 66
Somatic growth, 34, 35
Somatic maintenance, 34
Soybeans, 86, 102
Species diversity, 55, 142
Specific Dynamic Action (SDA), 47
Spontaneous processes, 220
SS, see Suspended solid
Stability
    concept
        in ecology, 117
        in thermodynamics, 122
    gestalt, 263
    mathematical theory of, 115
Stability, thermodynamics and theory of, 117–132
    exergy and Lyapunov function, 123–124
    extreme properties of Volterra equations for competing
        species, 121–122
    generalisation of exergy concept, 131–132
    Lyapunov function, 119–121
    phenomenological thermodynamics of evolutionary
        models, 129–130
    problem of additivity and thermodynamic constraints,
        124–126
    stability concept in ecology, 117–119
    stability concept in thermodynamics, 122–123
    thermodynamic basis of Volterra's equations for
        competing species, 126–129
Stabilization, 257
Steppe community, 148
Stromalites, 3
Structured biosphere, 351
Substrate rejection, 36
Succession, 255
Successionally closed ecosystems, 217
Sugarbeet, 84, 86
    diagram of biofuel production from, 101
    ethanol production from, 90
    production, 87, 95
    residue, 89
Sugar residue, 89
Summation theorems, 54
Sun, natural energy of, 135
Sunflowers, 85, 86
Surface wind, 66
Suspended solid (SS), 179
Sustainability, 76, 151
Sustainable agriculture, 149